Tree Biology and Biotechnology

Rajwant K. Kalia • Rakesh Pathak
Editors

Tree Biology and Biotechnology

Springer

Editors
Rajwant K. Kalia
ICAR-Central Arid Zone Research Institute
Jodhpur, Rajasthan, India

Rakesh Pathak
ICAR-Central Arid Zone Research Institute
Jodhpur, Rajasthan, India

ISBN 978-981-96-0001-4 ISBN 978-981-96-0002-1 (eBook)
https://doi.org/10.1007/978-981-96-0002-1

This Springer imprint is published by the registered company Springer Nature Singapore Pte Ltd.
The registered company address is: 152 Beach Road, #21-01/04 Gateway East, Singapore 189721, Singapore

Preface

Trees are dominant species in several ecosystems on the land and their economic relevance is well known in terms of production of food, feed, fodder, fuel, timber, and other products. They also deliver various ecosystem services, including temperature regulation, mitigation of soil erosion, and managing and filtering rainwater along with providing habitats for different organisms. They require some specific strategies for breeding and propagation because of their long life and reproductive cycle. Biotechnology has played an important role in the tree breeding and propagation in the recent years. Several techniques including somatic embryogenesis, propagation in bioreactors, genetic transformation, proteomics, genomics, synthetic seeds, etc. have paved remarkable roles in the tree breeding and its improvement. Systematic studies in various tree species are getting attention of researchers and other stakeholders. Presence of genetic diversity within the forest trees and trees outside forest (ToF) is a matter of debate among the scientific fraternity and policy makers. Several institutes and organization across the world is carrying out research and development works in the improvement of tree species. Literature search and websites revealed scattered information on various aspects of tree breeding, biology, and genetics. Compilation of reviews/reports on diverse aspects of tree biology and biotechnology written by experts in the field is the need of the hour. Keeping in view, the editors have thought to provide comprehensive information on the latest developments in the field of tree biology and biotechnology under a single cover.

The book offers compilation of critical reviews and reports on various aspects of tree biology, breeding, genetics, biotechnology, etc. written by the experts. The book is organized into 18 chapters covering molecular biology, genomics, physiology, and innovative approaches for tree improvement. Chapter 1 reviews the successful applications of slow-growth and cryopreservation techniques for the conservation of tree species in light of recent understandings. Chapter 2 explores the achievements of somatic embryogenesis in the *Abies* genus focusing on the use of the embryogenic tissue for cryopreservation, genetic advances, and other research studies. Chapter 3 reviews the status of natural dyes utilization, production, and prospective of major dye yielding agroforestry tree species in India, techniques for dye extraction, importance of tissue culture, and standardization of tissue culture approaches in dye yielding species. Chapter 4 discusses the utilization of tissue culture techniques in the multiplication of African tropical hardwood tree species including several methods for improved productivity at a time of increasing demand for tree products and services. Chapter 5 aims to explore the vital contributions of soil microorganisms to nutrient recycling in forest ecosystems, highlighting their role in transforming organic matter, cycling essential nutrients, and sustaining the intricate web of life within these verdant landscapes in forest management and conservation. Chapter 6 reports the biosynthesis of guggulsterone, discovery, prospects, and associated challenges with respect to plant tissue culture technology for sustainable production of guggulsterone. Chapter 7 explores the paradigm shift occurring in tree improvement by harnessing the biotechnological tools including molecular markers, genetic transformation, tissue culture, genomic sequencing, and marker-assisted selection. Chapter 8 reviews the genomic selection in detail and discusses different methods and its perspectives in forest trees. Chapter 9 highlights the unique species compositions, endemic and rare plants, and the ecological services provided by various sacred groves across different regions of India. The biology, origin, distribution, ecology, morphology, growth management of *A. tortilis* and its mutual associations with other organisms such as arbuscular mycorrhizal fungi and nitrogen-fixing bacteria along with biotechnological studies carried out in the species have been outlined in Chap. 10. Chapter 11 summarizes the paradigms of clonal forestry in the different areas of interest in agroforestry. Chapter 12 reviews various types of land cover along with comparison between agroforestry and Trees outside Forest (ToF) and concept of ToF including community forestry, social forestry, and urban forestry. Chapter 13 underscores the importance of community engagement and policy interventions in promoting sustainable tree health management practices against detrimental factors, ensuring their resilience and longevity in the face of evolving environmental challenges. Chapter 14 reviews the insect nuisances of timberland and their monitoring with the aim of forest pest management using strategies that are ecologically sound, economically viable, and socially and environmentally acceptable. Chapter 15 delves into the realm of eco-tech strategies, focusing on the intersection of biological and biotechnological innovations in forest insect pest control. Chapter 16 gives a comprehensive exploration of genetic resistance and marker-assisted selection, genome editing, RNA interference, plant immune system activation, and omics approaches in forest disease management. Chapter 17 highlights the major abiotic stresses in trees, its impacts, response mechanisms, and strategies to mitigate these issues in trees while Chap. 18 addresses different abiotic stresses along with nitric oxide mediated response to various abiotic stresses.

All the included chapters present reviews and reports concerning in the advances made in the field of tree biology research. The editors of the book believe that the wide range of topics on tree biology having a great relevance in biotechnology covered in a single volume would be highly useful for the agricultural students, researchers, teachers, and professionals interested in the field of agro-forestry, horticulture, silviculture, and tree improvement. This book may be an important contribution to agricultural colleges and university libraries and research centers across the globe, where tree biology and biotechnology are being instructed. The editors thank every author wholeheartedly for contributing their valuable chapters for the excellent outcome of this book. We are extremely thankful to Springer Nature, for assisting us to publish the work globally.

Jodhpur, Rajasthan, India Rajwant K. Kalia
Jodhpur, Rajasthan, India Rakesh Pathak

Acknowledgments

Dr. Rakesh Pathak is acknowledged for his contribution as Corresponding Editor of the book entitled *Tree Biology and Biotechnology*.

Contents

Editors and Contributors

About the Editors

Rajwant K. Kalia is Principal Scientist (Agroforestry) at ICAR-Central Arid Zone Research Institute (CAZRI), Jodhpur, India. She is a graduate of University of Delhi and has obtained Ph.D. from ICFRE-Forest Research Institute, Dehradun. She has more than 20 years of post-doctoral research experience in the field of tree/plant biology and biotechnology including post-doctoral research at ICAR-NIPB, Delhi; University of Delhi; and ICAR-NBPGR, Delhi. She has been a recipient of BOYSCAST Fellowship to work at University of California, Riverside, and Purdue University, West Lafayette, USA. She has handled more than 15 research projects on several aspects including micropropagation, conservation, cryopreservation, molecular characterization, and improvement of important tree and crop species. She has also supervised Ph.D. students on these research aspects. Dr. Kalia has published more than 70 research papers and book chapters and is serving as peer reviewer of high-impact national and international research journals.

Rakesh Pathak obtained Ph.D. from the Jai Narain Vyas University, Jodhpur. He is Assistant Chief Technical Officer in the ICAR-Central Arid Zone Research Institute, Jodhpur, Rajasthan, and is a Fellow of Institution of Chemists (India). Dr. Pathak has published more than 100 research papers and book chapters on molecular characterization and genetic diversity studies of several crops and trees in journals and books of national and international repute. He has molecularly characterized and submitted more than 300 novel gene sequences belonging to plant varieties/germplasm, macro-fungi, fungal pathogens, and PGPRs with NCBI, USA, to address Indian biodiversity. Dr. Pathak has delivered several oral and poster presentation in national and international conferences. He has been regularly reviewing manuscripts as a referee for high-impact journals.

Contributors

Jayashree Anandakumar ICAR-National Institute of Veterinary Epidemiology and Disease Informatics, Bengaluru, Karnataka, India

A. Arunachalam ICAR-Central Agroforestry Research Institute, Jhansi, Uttar Pradesh, India

M. N. Ashwath College of Forestry, Thrissur, Kerala Agricultural University, Thrissur, Kerala, India

K. Aswini Division of Microbiology, ICAR-IARI-Pusa Campus, New Delhi, India

Ankita Bhardwaj Department of Silviculture & Agroforestry, College of Forestry, Kerala Agricultural University, Thrissur, Kerala, India

Shiwani Bhatnagar Forest Protection Division, ICFRE-Arid Forest Research Institute, Jodhpur, Rajasthan, India

K. Bindyalaxmi College of Forestry, Thrissur, Kerala Agricultural University, Thrissur, Kerala, India

Jorge Canhoto Centre for Functional Ecology, Associate Laboratory Terra, Department of Life Sciences, University of Coimbra, Coimbra, Portugal

Tiziano Caruso Department of Agricultural, Food and Forestry Sciences, University of Palermo, Palermo, Italy

Luciana R. Chappa Department of Economics, Swedish University of Agricultural Sciences, Uppsala, Sweden

Gladys Cheruto Environment and Natural Resource Management Programme, Kenya Agricultural and Livestock Research Organization, Nairobi, Kenya

Amoghavarsha Chittaragi Department of Plant Pathology, CCSHAU, Hisar, Haryana, India

Manisha Choudhary Biotechnology Unit, Department of Botany, UGC – Centre of Advanced Study (CAS), Jai Narain Vyas University, Jodhpur, Rajasthan, India

Mariana Correia Centre for Functional Ecology, Associate Laboratory Terra, Department of Life Sciences, University of Coimbra, Coimbra, Portugal

K. Darshan Forest Protection Division, ICFRE-Tropical Forest Research Institute, Jabalpur, Madhya Pradesh, India

Umakanta Dash Indian Agricultural Research Institute, Dirpai, Assam, India

Jerry C. Dlamini Unit for Environmental Sciences and Management, Faculty of Natural and Agricultural Sciences, North-West University, Potchefstroom, South Africa

Constantin Dushimimana Independent Consultant, Kigali, Rwanda

Maria Antonietta Germanà Department of Agricultural, Food and Forestry Sciences, University of Palermo, Palermo, Italy

Harun I. Gitari Department of Agricultural Science and Technology, School of Agriculture and Environmental Studies, Kenyatta University, Nairobi, Kenya

Shruti Godara Genetics and Tree Improvement Division, Forest Research Institute, Haldwani, Uttarakhand, India

A. K. Handa ICAR-Central Agroforestry Research Institute, Jhansi, Uttar Pradesh, India

Nourhene Jouini Department of Agricultural, Food and Forestry Sciences, University of Palermo, Palermo, Italy

Ruth Kahuthia-Gathu Department of Agricultural Science and Technology, School of Agriculture and Environmental Studies, Kenyatta University, Nairobi, Kenya

S. Kala ICAR-IISWC-Research Centre, Kota, Rajasthan, India

Tarun Kant Genetics and Tree Improvement Division, Arid Forest Research Institute, Jodhpur, Rajasthan, India

Drishti Kataria Genetics and Tree Improvement Division, ICFRE-Tropical Forest Research Institute, Jabalpur, Madhya Pradesh, India

Abhishek D. Kattappanavar Forest Protection Division, Institute of Wood Science and Technology, Bengaluru, Karnataka, India

Kavya Department of Agriculture and Environmental Sciences, National Institute of Food Technology Entrepreneurship and Management, Sonipat, Haryana, India

Priscilla Kimani Department of Biosciences, Kenya Forestry Research Institute, Nairobi, Kenya

Ruby Susana Kujur ICFRE – Institute of Forest Productivity, Ranchi, Jharkhand, India

ICFRE – Institute of Wood Science and Technology, Bengaluru, Karnataka, India

Alok Kumar ICFRE – Institute of Forest Productivity, Ranchi, Jharkhand, India

K. Kumaran Department of Agronomy, ADAC&RI, TNAU, Trichy, Tamil Nadu, India

Ravinder R. Lall ICFRE – Institute of Forest Productivity, Ranchi, Jharkhand, India

Diksha Loona Department of Plant Pathology, Punjab Agricultural University, Ludhiana, Punjab, India

Joseph Machua Department of Biosciences, Kenya Forestry Research Institute, Nairobi, Kenya

Akshay F. Madiwalar College of Forestry, Sirsi, University of Agricultural Sciences, Dharwad, Karnataka, India

Sagar Maitra Department of Agronomy and Agroforestry, Centurion University of Technology and Management, R.Sitapur, Odisha, India

Manojkumar S. Department of Silviculture & Agroforestry, College of Forestry, Kerala Agricultural University, Thrissur, Kerala, India

Desha Meena Genetics and Tree Improvement Division, ICFRE-Arid Forest Research Institute, Jodhpur, Rajasthan, India

Rekha Meena Department of Agriculture and Environmental Sciences, National Institute of Food Technology Entrepreneurship and Management, Sonipat, Haryana, India

Suresh Kumar Meena Genetics and Tree Improvement Division, ICFRE-Arid Forest Research Institute, Jodhpur, Rajasthan, India

Shobha Mehra Genetics and Tree Improvement Division, Arid Forest Research Institute, Jodhpur, Rajasthan, India

G. P. Mishra Division of Seed Science and Technology, ICAR-Indian Agricultural Research Institute, New Delhi, India

Sourav Ranjan Mohapatra College of Forestry, Thrissur, Kerala Agricultural University, Thrissur, Kerala, India

Somnath Singh Munda ICFRE – Institute of Forest Productivity, Ranchi, Jharkhand, India

Riziki Mwadalu Central Highlands Eco-Region Research Programme, Kenya Forestry Research Institute, Nairobi, Kenya

J. Nandini Forest Protection Division, Institute of Wood Science and Technology, Bengaluru, Karnataka, India

Jamal Nasar College of Agriculture, Institute of Rice Industry Technology Research, Guizhou University, Guiyang, P. R. China

Sonali Nigam Department of Botany and Microbiology, St. Aloysius College, Jabalpur, Madhya Pradesh, India

Emmanuely Z. Nungula Department of Environment and Sustainable Development, Mzumbe University, Morogoro, Tanzania

Lukelysia Nyawira World Agroforestry, Nairobi, Kenya

Suman Parihar Plant Biotechnology and Molecular Biology Laboratory, Department of Botany (UGC-CAS), Jai Narain Vyas University, Jodhpur, Rajasthan, India

Ashok Kumar Parmar Genetics and Tree Improvement Division, Arid Forest Research Institute, Jodhpur, Rajasthan, India

Ashok K. Patel Biotechnology Unit, Department of Botany, UGC – Centre of Advanced Study (CAS), Jai Narain Vyas University, Jodhpur, Rajasthan, India

Swathi S. Patil University of Agricultural Science, GKVK, Bengaluru, Karnataka, India

Siddu Lakshmi Prasanna Agricultural College, Bapatla Acharya N G Ranga Agricultural University, Guntur, Andhra Pradesh, India

Priyanka Department of Agricultural Entomology, University of Agricultural Sciences, Bengaluru, Karnataka, India

Kishore Pujar Department of Agricultural Entomology, University of Agricultural Sciences, GKVK, Bengaluru, Karnataka, India

Manoj K. Rai Department of Environmental Science, Indira Gandhi National Tribal University, Amarkantak, Madhya Pradesh, India

Asha K. Raj AICRP on Agroforestry, Department of Silviculture & Agroforestry, College of Forestry, Kerala Agricultural University, Thrissur, Kerala, India

Ambika Rajendran Division of Genetics, ICAR-IARI-Pusa Campus, New Delhi, India

Athulya Ramakrishnan Forest Protection Division, Institute of Wood Science and Technology, Bengaluru, Karnataka, India

Komal Rani Genetics and Tree Improvement Division, ICFRE-Tropical Forest Research Institute, Jabalpur, Madhya Pradesh, India

Shivani Ranjan Department of Agronomy, Dr. Rajendra Prasad Central Agricultural University, Samastipur, Bihar, India

V. Rathna ICFRE – Institute of Forest Productivity, Ranchi, Jharkhand, India

A. A. Safeena Majeed Indian Institute of Science, Bengaluru, Karnataka, India

E. Santhoshinii University of Agricultural Sciences, Dharwad, Karnataka, India

Chandra Shekhar Seth Departments of Botany, University of Delhi, New Delhi, India

Ishaka Sharma Biotechnology Unit, Department of Botany, UGC – Centre of Advanced Study (CAS), Jai Narain Vyas University, Jodhpur, Rajasthan, India

Lata Sharma Plant Biotechnology and Molecular Biology Laboratory, Department of Botany (UGC-CAS), Jai Narain Vyas University, Jodhpur, Rajasthan, India

G. S. Shekhawat Plant Biotechnology and Molecular Biology Laboratory, Department of Botany (UGC-CAS), Jai Narain Vyas University, Jodhpur, Rajasthan, India

Narpat S. Shekhawat Biotechnology Unit, Department of Botany, UGC – Centre of Advanced Study (CAS), Jai Narain Vyas University, Jodhpur, Rajasthan, India

R. Shifin College of Forestry, Thrissur, Kerala Agricultural University, Thrissur, Kerala, India

Susmita Shil Department of Silviculture & Agroforestry, College of Forestry, Kerala Agricultural University, Thrissur, Kerala, India

K. Shiran College of Forestry, Thrissur, Kerala Agricultural University, Thrissur, Kerala, India

ICAR-CAZRI-Regional Research Station, Jaisalmer, Rajasthan, India

S. Shreedevasena ICAR-Directorate of Medicinal and Aromatic Plants Research, Anand, Gujarat, India

Anusree Sobhanan Department of Agriculture and Environmental Sciences, National Institute of Food Technology Entrepreneurship and Management, Sonipat, Haryana, India

Sumit Sow Department of Agronomy, Dr. Rajendra Prasad Central Agricultural University, Samastipur, Bihar, India

V. Subbulakshmi ICAR-CAZRI-Regional Research Station, Bikaner, Rajasthan, India

Suresh Ramanan S. ICAR-Central Agroforestry Research Institute, Jhansi, Uttar Pradesh, India

B. Swathi ICAR-Krishi Vignana Kendra, Chikkaballapur, Karnataka, India

Aditi Tailor Genetics and Tree Improvement Division, ICFRE-Arid Forest Research Institute, Jodhpur, Rajasthan, India

Gadigavarahalli Subbareddy Uma Forest Protection Division, Institute of Wood Science and Technology, Bengaluru, Karnataka, India

S. Vennila ACRI- TNAU, Thiruvannamalai, Tamil Nadu, India

Ankit Verdiya ICAR-Central Agroforestry Research Institute, Jhansi, Uttar Pradesh, India

Neelam Verma Forest Protection Division, ICFRE-Arid Forest Research Institute, Jodhpur, Rajasthan, India

Abbreviations

ABA	Abscisic acid
AC	Active charcoal
AFLP	Amplified fragment length polymorphism
AFRI	Arid Forest Research Institute
AGF	Agroforestry system
AL medium	*Abies lasiocarpa* medium (Kvaleen et al. 2005)
Al	Aluminum
AM	Arbuscular mycorrhizal
AMF	Arbuscular mycorrhizal fungi
AMT	Agrobacterium mediated transformation
APX	Ascorbate peroxidase
ASALs	Arid and semi-arid lands
ATREE	Ashoka Trust for Research in Ecology and the Environment
BA	Benzyladenine
BAP	6-Benzylaminopurine
BLG	Culture medium by Verhagen & Wann (1989)
Bt	*Bacillus thuringiensis*
Bti	*Bacillus thuringiensis* subsp. *israelensis*
Btk	*Bacillus thuringiensis* subsp. *kurstaki*
Ca	Calcium
CAD	Cynnamyl alcohol deshydrogenase
cADPR	Cyclic ADP-ribose
CAld5H	Conifer aldehyde 5-hydroxylase
CAT	Catalase
CAZRI	*Central Arid Zone Research Institute*
CBD	Convention on Biological Diversity
CDA	Chitin deacetylase
CDPKs	Ca^{2+}-dependent protein kinases
cGMP	Cyclic guanosine monophosphate synthesis
CIS	Continuous immersion system
CKs	Cytokinins
CO	Carbon monoxide
CO_2	Carbon dioxide
COMT	Caffeate/5-hydroxyferulate O-methyl transferase

CPREEC	C.P.R. Environmental Education Centre, Chennai
CPTs	Candidate plus trees
CPVs	Cytoplasmic polyhedroviruses
CRISPR	Clustered regularly interspaced short palindromic repeats
Cas9	CRISPR associated protein 9
CSO	Clonal seed orchards
CSVPA	Cultural and Spiritual Values of Protected Areas
Cu	Copper
DBH	Diameter at breast height
DCR	Douglas Fir Revised Medium by Gupta and Durzan (1985)
DHAR	Dehydroascorbate reductase
DKW	Driver Kuniyuki Walnut medium by Driver and Kuniyuki (1984)
DMAPR	Directorate of Medicinal and Aromatic Plants Research
DMSO	Dimethyl sulfoxide
DNA	Deoxyribonucleic acid
dsRNA	Double stranded RNA
EAB	Emerald ash borer
EC	Embryogenic callus
EcM	Ectomycorrhiza
EcMF	Ectomycorrhizal fungi
ECPGR	European Cooperative Program for Plant Genetic Resources
ENVIS	Environmental Information System
EPB	Entomopathogenic bacteria
EPF	Entomopathogenic fungi
EPNs	Entomopathogenic nematodes
EPVs	Entomopathogenic viruses
ERM	Ericoid mycorrhizal
ET	Embryogenic tissue
FAO	Food and Agriculture Organization
Fe	Iron
FPP	Farnesyl diphosphate
FRS	Forestry system
FSI	Forest Survey of India
FYM	Farmyard manure
G × E	Genotype × environment
GAS	Gene assisted selection
G-BLUP	Genomic Best Linear Unbiased Prediction
GBS	Genotyping-by-sequencing
GDH	Glutamate dehydrogenase
GEBV	Genomic estimated breeding value
GFRA	Global Forest Resource Assessment
GM	Genetically modified
GMOs	Genetically modified organisms
gRNA	Guide RNA
GS	Genomic selection

GSH	Glutathione
GSNO	S-Nitrosoglutathione
GST	Glutathione-S-transferase
GVs	Granular viruses
GWAS	Genome-wide association studies
H_2O_2	Hydrogen peroxide
H_2S	Hydrogen sulfide
$HgCl_2$	Mercuric chloride
HMGR	3-Hydroxy-3-methyl glutaryl coenzyme A reductase
HO	Heme oxygenase
HPLC	High-performance liquid chromatography
hsp / HSPs	Heat shock protein
IAA	Indole-3-acetic acid
iap	Inhibitor of apoptosis protein genes
IBA	Indole-3-butyric acid
ICAR	Indian Council of Agricultural Research
ICCAs	Indigenous and Community Conserved Areas
ICFRE	Indian Council of Forestry Research & Education
ICRAF	International Centre for Research in Agroforestry
IDRC	International Development Research Centre
IGRs	Insect growth regulators
IPCC	Intergovernmental Panel on Climate Change
IPDM	Integrated pest and disease management
IPM	Integrated pest management
ISSR	Inter Simple Sequence Repeat
IUCN	International Union for the Conservation of Nature and Natural Resources
JA	Jasmonic acid
K	Potassium
KIN	Kinetin
LD	Linkage disequilibrium
LdM-NPV	Lymantria dispar multicapsid nuclear polyhedrosis virus
LGs	Linkage groups
LiDAR	Light detection and ranging
MAB	Man and Biosphere Programme
MALDI	Matrix-assisted laser desorption/ionization
MAPKs	Mitogen-activated protein kinases
MARS	Marker-assisted recurrent selection
MAS	Marker-assisted selection
MDA	Malondialdehyde
MDHAR	Monodehydroascorbate reductase
Mg	Magnesium
mLV	Medium LV by Litvay et al. (1985)
Mn	Manganese
Mo	Molybdenum

MS	Murashige and Skoog medium by Murashige and Skoog (1962)
NAA	1-Naphthalene acetic acid
NaCl	Sodium chloride
$NAOCl_2$	Sodium hypochlorite
NDVI	Normalized difference vegetation index
NEC	Non-embryogenic callus
NGS	Next-generation sequencing
NH_3	Ammonia
NH_4^+	Ammonium
NiR	Nitrite reductase
NO	Nitric oxide
NO_2^-	Nitrite
NO_3^-	Nitrate
NOS	Nitric oxide synthase
NPV	Nuclear polyhedrosis virus
NR	Nitrate reductase
NSKE	Neem seed kernel extract
O_3	Ozone
OA1	Ocular albinism type 1 gene
OBs	Occlusion bodies
P	Phosphorus
PAHs	Polycyclic aromatic hydrocarbons
PCD	Programmed cell death
PEG	Polyethylene glycol
PGPRs	Plant growth-promoting rhizobacteria
PGRs	Plant growth regulators
PI	Protease inhibitors
PIAs	Plant immune activators
PM	Proliferation medium
PO_4^{3-}	Phosphate ions
PPI	Plant-pathogen interactions
ppm	Parts per million
PPV	Plum pox virus
PR proteins	Pathogenesis-related proteins
PTC	Plant tissue culture
PTGS	Post-transcriptional gene silencing
PTMs	Post-translational modifications
PVS	Plant vitrification solution
QTLs	Quantitative trait loci
RAD-seq	Restriction-site-associated DNA sequencing
RAPD	Random amplified polymorphic DNA
RFLP	Restriction fragment length polymorphism
RNA	Ribonucleic acid
RNAi	RNA interference
RNS	Reactive nitrogen species

ROS	Reactive oxygen species
RR-BLUP	Ridge regression best linear unbiased prediction
SA	Salicylic acid
SAMPL	Selective amplification of microsatellite polymorphic loci
SCAR	Sequence characterized amplified region
SE	Somatic embryogenesis
SG	Soluble granules
SH	Schenk and Hilderbrandt medium by Schenk and Hilderbrandt (1972)
siRNAs	Small interfering RNAs
SIT	Sterile insect technique
SMT	Sterol-C-methyltransferase
SNPs	Single nucleotide polymorphisms
SO_2	Sulfur dioxide
SO_4^{2-}	Sulfate
SOD	Superoxide dismutase
SOFO	State of the Worlds' Forests
SPA	Seed Production Area
SPGS	Sawlog Production Grant Scheme
SPS	Sanitary and phytosanitary measures
SSRs	Simple sequence repeats
TBP	Tree Biotechnology Programme
TDZ	Thidiazuron
TIS	Temporary Immersion System
TOF	Tree outside forest
TRX H	Thioredoxin h-type
UAE	United Arab Emirates
UNESCO	United Nations Educational, Scientific and Cultural Organization
UNFCCC	United Nations Framework Convention on Climate Change
USPGR	US Pyrus Genetic Resources
VIR, Russia	N.I. Vavilov All-Russian Institute of Plant Genetic Resources
VOCs	Volatile organic compounds
WAC	World Agroforestry Centre
WHO	World Health Organization
WP	Wettable powder
WPM	Woody plant medium by Lloyd and McCown (1980)
WSC	Water soluble concentrate
WTO	World Trade Organization
WW	Weight by weight
XOR	Xanthine oxidoreductase
ZEA	Zeatin
Zn	Zinc
°C	Degree Celsius
2,4-D	2,4-Dichlorophenoxyacetic acid
2iP	6-(γ,γ-Dimethylallylamino) purine
4CL	4-coumarate-CoA ligase

Exploring Plant Tissue Culture for In Vitro Conservation of Trees: Recent Progress

1

Ashok K. Patel, Ishaka Sharma, Manisha Choudhary, Narpat S. Shekhawat, and Manoj K. Rai

Abstract

Anthropogenic activities and overexploitation of plant resources have led to the deterioration/depletion of natural ecosystems, putting many tree species at risk. In this frightening situation, conservation strategies, including both in situ and ex situ, became necessary to successfully recover, restore, or reintroduce the declining phytodiversity. In vitro strategies have emerged as the most promising and efficient tools of biotechnology for germplasm conservation of various plant species, including commercially important, and rare and endangered tree species. These techniques enable the cultivation/multiplication of trees despite constraints such as limited starting material, seed dormancy, self-incompatibility, poor seed setting, and recalcitrant or non-viable seeds. Different in vitro approaches, such as slow growth for short to medium-term and cryopreservation for long-term conservation, can assist significantly in the conservation as well as reintroduction of many tree species, especially threatened species. The successful applications of slow growth and cryopreservation techniques in the conservation of tree species, recent insights, and their comprehensive compilation are presented in this chapter.

A. K. Patel · I. Sharma · M. Choudhary · N. S. Shekhawat
Biotechnology Unit, Department of Botany, UGC – Centre of Advanced Study (CAS), Jai Narain Vyas University, Jodhpur, Rajasthan, India

M. K. Rai (✉)
Department of Environmental Science, Indira Gandhi National Tribal University, Amarkantak, Madhya Pradesh, India
e-mail: manoj.rai@igntu.ac.in

R. K. Kalia, R. Pathak (eds.), *Tree Biology and Biotechnology*,
https://doi.org/10.1007/978-981-96-0002-1_1

Keywords

Cryopreservation · In vitro conservation · Growth retardants · Minimal growth · Tree species

1.1 Introduction

With the advancement in biotechnology, conservation science achieved a phenomenal height, putting its impact on protecting and safeguarding the biodiversity of nature and maintaining a stable ecosystem on this planet. In terms of biotechnology, the word conservation emphasizes germplasm conservation, which ensures the preservation of genetic diversity. Many anthropogenic activities and environmental stresses result in habitat loss, fragmentation, and degradation of biodiversity (Rajasekharan and Wani 2020; Rai et al. 2021). It indicates the alarming extinction rate of endangered plant species due to genetic erosion. Hence, to seize the loss of genetic diversity, the conservation of plant resources is necessary. Many tree species, especially forest and fruit trees, have enormous economic and medicinal importance, and many of them are being cultivated for their multipurpose uses (Patel et al. 2020; Arora et al. 2022). Despite the global importance, many tree species are declining quickly, especially in the wild. Therefore, the conservation and management of such tree species need more attention (Shekhawat et al. 2014).

The simplest approach to preserve plant germplasm is to store desiccated seeds at low temperatures, but this method cannot be used to preserve cultivars of plant species that are vegetatively propagated in order to preserve their distinct genomic makeup or crops that do not produce seed or have recalcitrant seed. In addition, the seed propagation system also hinders the conservation and management of many tree species due to poor germination and recalcitrant nature (Sharma et al. 2019). These disadvantages lead to an advanced form of in vitro germplasm conservation through slow growth and cryopreservation strategies. The primary germplasm collection of tree species is held in field gene banks that are vulnerable to diseases and environmental stresses. In addition, the conventional in vitro culture requires continuous subculture, medium preparation, and clean-up of the growth room, which increases the chances of genetic instability. Hence, a secure in vitro conservation process is needed for the safeguarding of these genetic resources.

Depending on the storage duration, different in vitro conservation methods are employed. For short-term and medium-term storage, the main aim is growth reduction, which is achieved by subculture modifications and environmental changes like photoperiod, temperature and culture vessels, etc., and this is done by slow-growth culture method (Chauhan et al. 2019; Rai 2022). The in vitro storage under a slow-growth process using low temperatures of more than zero (around 4 °C) and without transfer to a new media can prevent the genetic instability of target plant species, whereas for long-term storage, the germplasm (usually the meristematic tissue) is preserved in liquid nitrogen at an optimal temperature of −196 °C, which is termed as cryopreservation (Engelmann 2000; Sakai and Engelmann 2007; Yamamoto

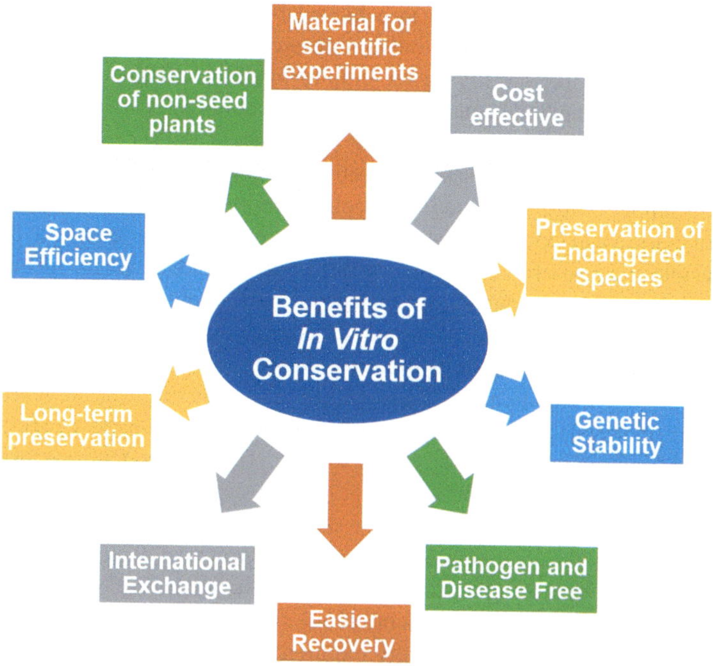

Fig. 1.1 Potential applications of in vitro conservation

et al. 2011). Plant cryopreservation techniques have advanced in recent years. It is an in vitro conservation technique that allows long-term storage of plant genetic resources, which can be essential for their micropropagation and plant breeding programs. Various cryopreservation techniques have been used in recent years for the preservation of many forest trees and fruit trees. In this chapter, we focus on applications of slow growth and cryopreservation methods of plant genetic resources, emphasizing recent progress in tree species. A schematic illustration of applications of in vitro conservation is presented in Fig. 1.1.

1.2 Strategies for In Vitro Conservation of Trees

1.2.1 Short- to Medium-Term Conservation

This is the most straightforward method of limiting the growth and development of in vitro materials and is usually applied to differentiated plantlets or shoot cultures (Rajasekharan and Sahijram 2015; Rai 2022). The main objective of this technique is to store plant genetic resources by increasing the subculture period without causing any damage to the tissue and survival after storage. This in vitro conservation approach allows us to store plant material for several months to 1–3 years without subculture. Short- to medium-term conservation can be stored under growth-limiting conditions, which is achieved by reducing the growth of plant material and increasing the time interval between subcultures. Various growth reduction strategies like modifying the culture medium, using growth retardants, and changing environmental conditions such as low temperature or decrease in light intensity have been used to conserve plant resources for the short to medium term (Engelmann 1997, 2011; Rajasekharan and Sahijram 2015; Rai 2022). The slow-growth technique causes reduced metabolic activity in cells, resulting in slow growth of culture and extending the subculture period. Thus, the technique becomes more feasible and cost-effective as it reduces the labor cost of frequent subculturing. Slow growth is a routine method of short-term conservation and can be achieved either using any one method or a combination of more than one methods (Table 1.1).

The low-temperature storage is the most common way to store cultures under slow-growth conditions. In most cases, storage at 4 °C is ideal for the regrowth of stored plant materials (Table 1.1). However, the optimum storage temperature is determined by the cold sensitivity of the species. The storage temperature ranges from 5 to 9 °C for temperate species, but it is often slightly higher for tropical species (Chauhan et al. 2019; Rai 2022). The storage of germplasm at lower temperatures without freezing is recommended because the growth of plant tissues is slowed down at lower temperatures by decreasing the metabolic activity of cells, therefore limiting the growth of the plants. Owing to the slow growth of culture, the subculture of stored explant is not frequently required. In some cases, this method may also accompany the reduction in light intensity (Koc et al. 2014). In tree species, the temperatures applied ranged from 4 °C, i.e., *Crataegus monogyna* (Arbeloa et al. 2017), *Pistacia lentiscus* (Koc et al. 2014), *Sequoia sempervirens* (Ozudogru et al. 2011a), to 10–22 °C, i.e., *Tetraclinis articulata* (Juan-Vicedo et al. 2022), *Ceratonia siliqua* (Mascarello et al. 2015), *Medusagyne oppositifolia* (Marriott and Sarasan 2010). Using low-temperature (4 °C) strategies, Arbeloa et al. (2017) successfully stored the in vitro generated shoots of 18 fruit tree species of 8 genera (*Crataegus*, *Cydonia*, *Eriobotrya*, *Ficus*, *Malus*, *Prunus*, *Pyrus*, and *Punica*) for 7 to 12 months.

In tropical countries, low-temperature storage is a costly option. Loss of stored germplasm can be threatened by power/equipment breakdowns, particularly in summer. In such cases, the growth of the cultured plant material can be reduced by altering the medium compositions, i.e., decreasing the concentrations of nutrient medium or increasing/decreasing sucrose concentration (Rai 2022). At normal temperatures, the growth of the cultured shoots can be slowed down by adding osmotic active substances such as mannitol and growth retardants. For instance, Gomes et al. (2021) reported medium-term in vitro conservation of *Castanea* spp. in a medium containing different concentrations of mannitol and sucrose for 12 months. In another study, El-Bahr et al. (2016) used three osmotic substances, i.e., mannitol, sorbitol, and sucrose, for slow-growth conservation of embryogenic cultures of date palms. For medium-term storage of germplasm, partial desiccation of somatic embryos has also been recommended. The desiccation tolerance of cells is enhanced by ABA treatment (Rai et al. 2011). A combination of more than one growth treatment can also be used for the storage of many plant species. The combination of slow growth treatment can vary from species to species. Growth limitation is

Table 1.1 Progress made in slow growth conservation of germplasm of tree species

Plant species	Storage conditions	Explant used	Storage period	Reference[a]
Azadirachta indica	12 °C	Encapsulated nodal segments	4 weeks	Padilla et al. (2021)
Balanites aegyptiaca	24 °C ± 2 °C	Nodal segments	12 months	Hwida and Abd El-Kader (2012)
Balanites aegyptiaca	4 °C	Shoots (synthetic seeds)	4 months	Varshney and Anis (2014)
Castanea sativa	8 °C	Shoots	48 months	Capuana and Di Lonardo (2013)
Castanea spp.	Use of sucrose and mannitol in minimal-growth medium	Nodal segments	12 months	Gomes et al. (2021)
Ceratonia siliqua	18 °C and low concentration of PGRs	Shoots	6 months	Mascarello et al. (2015)
Citrus jambhiri	22 °C and different concentrations of sucrose	Micro-cuttings of in vitro-grown plants	12 months	Carvalho et al. (2022)
Crataegus monogyna	4 °C	Shoots	12 months	Arbeloa et al. (2017)
Cydonia oblonga	4 °C	Shoots	7–12 months	Arbeloa et al. (2017)
Eriobotrya japonica	4 °C	Shoots	12 months	Arbeloa et al. (2017)
Ficus carica	4 °C	Shoots	12 months	Arbeloa et al. (2017)
Malus domestica	4 °C	Shoots	12 months	Arbeloa et al. (2017)
Medusagyne oppositifolia	22 ± 2 °C	Shoots	6 months	Marriott and Sarasan (2010)
Pistacia lentiscus	4 °C in dark	Shoot tips	6 months	Koc et al. (2014)
Phoenix dactylifera	Use of mannitol, sorbitol, and sucrose	Embryogenic cultures	12 months	El-Bahr et al. (2016)
Pyrus communis	4 °C	Shoots	12 months	Arbeloa et al. (2017)
Sequoia sempervirens	4 °C	Encapsulated buds & shoots	12 months	Ozudogru et al. (2011a)
Tetraclinis articulata	4 °C & 10 °C	Apical shoots	28 weeks	Juan-Vicedo et al. (2022)
Ziziphus jujuba	4 °C	Nodal segments	10 months	Sota and Kongjika (2014)

[a] References include only some selected studies from 2010 onwards

usually achieved by decreasing the temperature and commonly combined with osmotic regulators, minimal growth medium, mainly increasing or decreasing concentrations of sucrose or use of growth retardants (Rai 2022).

Another biotechnological method that is currently widely utilized to conserve and propagate the germplasm of various economically significant species is the synthetic seed or encapsulation technology (Ara et al. 2000; Rai et al. 2009). The synthetic seeds are prepared by artificial encapsulation of vegetative explants such as apical or axillary shoot buds or embryogenic explants like somatic embryos. The key approach in the synthetic seed storage system is to incubate the seeds at various temperatures and use of minimal growth medium by adding growth retardants in the medium or minimizing nutrient components or sugar in the medium (Sharma et al. 2019; Faisal and Alatar 2019). These encapsulated synthetic seeds can then be stored for a few weeks to several months at different temperatures, varying from species to species, including tree species (Rai et al. 2008; Ozudogru et al. 2011a; Hung and Trueman 2011; Varshney and Anis 2014; Gomes et al. 2021; Padilla et al. 2021).

1.2.2 Long-Term Conservation: Cryopreservation

Cryopreservation is a viable method for long-term conservation of plant genetic resources at an ultra-low temperature (−196 °C) in liquid nitrogen. At this ultra-low temperature, all metabolic activities, including cell division, are halted, therefore ensuring germplasm preservation for longer periods of time (Bhojwani and Dantu 2013; Rai 2022). Cryopreservation involves eliminating freezable water from cryopreserved tissues through dehydration and then rapid freezing. Cryopreservation is the most revolutionary biotechnological strategy for the long-term preservation of phytodiversity without any risk of genetic alterations (Benelli 2021). Also, it is a cost-effective alternative to conventional ex situ conservation approaches due to its ability to store germplasm in a limited space with minimum labor and maintenance (Yamamoto et al. 2011; Sharma et al. 2019). Maintaining vegetatively propagated crops in a field or greenhouse renders them vulnerable to various biotic and abiotic hazards. Therefore, plant gene banks are increasingly using cryopreservation methods to conserve the vegetatively propagated species (Wang et al. 2021).

Plant cryopreservation techniques have advanced rapidly in recent years, allowing for the potential long-term storage of important genetic resources of various crops, including fruit and forest trees. Tree biodiversity serves as a source of food, medicine, wood, fuel, etc. Therefore, preserving tree biodiversity is essential for modern and classical plant breeding programs. Cryopreservation is a potential option for deciduous trees and shrubs because it makes use of the natural dormancy of cold-hardy crops (Li et al. 2018; Tanner et al. 2021). Different types of cells, tissues, and organs including seeds, shoot-tips, dormant-buds, pollens, cell suspensions, embryogenic cultures, somatic and zygotic embryos can be cryopreserved (Engelmann 2004; Reed et al. 2011; Benelli et al. 2013). Cells from suspension cultures and with high cytoplasmic content are the most resistant to freezing;

whereas the cells having larger vacuoles are highly susceptible to cryoprotectants and, therefore, should be avoided (Wang et al. 2021). For vegetatively propagated crops, the most frequently used organs for cryopreservation are shoot tips, obtained from apical/axillary buds of in vitro raised shoots (Lambardi and De Carlo 2003). For woody plant species, the preservation of progenies generated from meristems via vegetative propagation exhibits genetic stability, and therefore, meristematic tissues are considered ideal for cryopreservation. The physiological state of the plants is equally critical in any conservation method. For instance, dormant buds must be deeply dormant, and in vitro shoots must be healthy and cold-acclimated for successful cryopreservation (Zhao et al. 2008). For maintaining the genetic stability with the mother/donor plant, shoot tips are the most commonly used explants in cryopreservation studies (Table 1.2), especially in fruit and forest trees (Engelmann 2000; Zhao et al. 2008; Benelli et al. 2013).

Plants with recalcitrant seeds are frequently cryopreserved using embryos and embryonic axes. Many tropical and subtropical tree species lack inherent resistance to the cold since they are not adapted to the cold season. Additionally, larger seeds from trees like mango, avocado, and durian are sensitive to desiccation, chilling, and freezing stress. Therefore, they are unsuitable for seed storage or cryopreservation. In such plant species, shoot-tip cryopreservation is a common technique that ensures the clonal integrity of germplasm (O'Brien et al. 2021). Successful dormant bud cryopreservation was also reported for several fruit tree genera including *Diospyros* (Matsumoto et al. 2001), *Malus* (Jenderek et al. 2011), *Juglans* sp. (Jenderek et al. 2013), *Pyrus* (Kovalchuk et al. 2014). Cryopreservation of embryogenic tissues has been successful in tree species like pedunculate oak (*Quercus robur*) (Martínez et al. 2003), black pine (*Pinus nigra*) (Salaj et al. 2007), and holm oak (*Quercus ilex*) (Barra-Jiménez et al. 2015) or hybrid firs (*Abies alba* × *A. cephalonica*, *Abies alba* × *A. numidica*) (Salaj et al. 2010). Cryopreservation by two different methods (direct dormant bud cryopreservation and cryopreservation of shoot tips from in vitro culture) was used for wild species of apple (*Malus*) and pear (*Pyrus*) (Höfer and Flachowsky 2023). In apples, a total of 180 and 116 samples (belonging to 32 species) showed an average survival rate of 39% and 40%, respectively, after cryopreservation for 10 years using dormant buds. However, in the case of wild pear species, cryopreservation was tested using both techniques in a total of 35 samples (belonging to 21 species), and they reported that shoot tips showed better results as compared to the dormant buds.

Cryopreservation focuses on protecting cells by minimizing ice crystal formation during cooling. There are two methods for cryopreserving plant tissues: (i) slow or step-wise freezing and (ii) Ultra-rapid freezing. Both approaches dehydrate cells prior to freezing, resulting in concentrated protoplasm that prevents ice crystal formation. Therefore, a pre-culture method usually improves cells' desiccation tolerance (Bhojwani and Dantu 2013). Successful cryopreservation relies on inducing dehydration and cold tolerance in cells exposed to freezing conditions. For this, tissues are cultured for 2–4 days on a medium having additives, e.g., sucrose, sorbitol, mannitol, and proline. Also, it is helpful in enhancing the desiccation tolerance of the cells. Low-temperature therapy also helps cell adaptation to high osmotic

Table 1.2 Progress made in long-term conservation or cryopreservation of tree species

Plant species	Explant used	Techniques	References[a]
Abies hybrids	Embryogenic callus	Two-step cooling	Salaj et al. (2010)
Betula lenta	Seeds and in vitro shoot tips	Droplet-vitrification	Rathwell et al. (2016)
Castanea sativa	Shoot tips	Vitrification	Vidal et al. (2010)
Diospyros kaki	Shoot tips	Encapsulation-droplet vitrification	Yanli et al. (2012)
	Dormant shoots (1 year old)	D cryo-plate	Matsumoto et al. (2015)
	Shoot tips	Modified droplet vitrification	Niu et al. (2010)
Elaeis guineensis	Polyembryoids	Droplet vitrification	Gantait et al. (2015)
Eucalyptus spp.	In vitro raised shoot culture	Encapsulation dehydration, droplet-vitrification	Kaya et al. (2013)
Fraxinus excelsior	Embryogenic callus	Two-step freezing	Ozudogru et al. (2010)
Malus domestica	In vitro axillary buds	Droplet-vitrification	Condello et al. (2011)
Persea americana	Embryogenic calli and somatic embryos	Droplet-vitrification and slow freezing	Guzmán-García et al. (2013)
Phoenix dactylifera	In vitro raised shoot tips	Vitrification, encapsulation-vitrification and droplet-vitrification	Fki et al. (2013)
Picea omorika	Embryogenic callus	Pre-growth-desiccation	Hazubska-Przybył et al. (2010)
Picea sitchensis	Somatic embryos	Encapsulation dehydration	Gale et al. (2013)
Pinus nigra	Embryogenic callus	Two-step cooling	Salaj et al. (2011)
Pinus pinaster	Embryogenic callus	Two-step cooling	Álvarez et al. (2012)
Pinus sylvestris	Embryogenic callus	Two-step cooling	Latutrie and Aronen (2013)
Populus trichocarpa	Dormant buds	Two-step cooling	Bonnart et al. (2014)
Prunus cerasifera	Shoot tips	Droplet vitrification	Vujović et al. (2011)
Prunus cerasifera and *P. domestica*	Shoot tips	V and D cryo-plates	Vujović et al. (2015)
Pyrus pyraster	Shoot tips	Encapsulation-dehydration	Condello et al. (2009)

(continued)

Table 1.2 (continued)

Plant species	Explant used	Techniques	References[a]
Quercus ilex	Embryogenic cultures	Vitrification	Barra-Jiménez et al. (2015)
Quercus robur	Somatic embryos	Pre-growth desiccation	Fernandes et al. (2008)
Quercus robur	Somatic embryos	Vitrification	Vidal et al. (2010)
Salix spp.	Dormant buds	Two-step cooling	Bonnart et al. (2014)
Sequoia sempervirens	Apical and basal buds	Vitrification, encapsulation-vitrification and droplet-vitrification	Ozudogru et al. (2011b)
Thuja koraiensis	Somatic embryos	Employing sorbitol and DMSO	Ahn et al. (2019)
Ulmus americana	Shoot tips	Vitrification	Uchendu et al. (2013)

[a] References include only some selected studies from 2010 onwards

stress by inducing the accumulation of heat shock proteins (HSPs), cold regulatory proteins, and dehydrins. Likewise, high sucrose concentration affects the cellular protein pattern and protects freeze-labile enzymes (Bhojwani and Dantu 2013). In preculture therapy, Dimethyl sulfoxide (DMSO) is utilized as a cryoprotectant. According to Zhao et al. (2008), preculture on a medium containing 5% DMSO for 48 hours is routinely utilized in all controlled freezing and vitrification techniques.

Different types of cryopreservation methods have been developed, namely, slow cooling, dehydration, encapsulation-dehydration, vitrification, encapsulation vitrification, droplet vitrification, D-cryo-plate, V-cryo-plate, and cryo-mesh (Fig. 1.2) (Engelmann 2000; Sakai and Engelmann 2007; Yamamoto et al. 2011; Bhojwani and Dantu 2013). Vitrification, encapsulation-vitrification, and encapsulation-dehydration are the most common methods of cryopreservation applied in many tree species (Table 1.2). In the vitrification method, the sample is first pretreated with a vitrifying solution of cryoprotectants such as PVS2 or PVS3 for 2–4 days and then treated with a loading solution normally high in sucrose molarity and placed in liquid nitrogen for freezing (Sakai and Engelmann 2007; Bhojwani and Dantu 2013; O'Brien et al. 2021). In the encapsulation-dehydration method, explants are initially encapsulated in alginate gel and then pretreated with high sucrose before dehydration. The alginate beads are dehydrated in a laminar air flow hood or silica gel before immersion in liquid nitrogen (Zhao et al. 2008; Bhojwani and Dantu 2013; O'Brien et al. 2021). In the encapsulation-vitrification method, shoot tips are encapsulated in alginate beads followed by being treated with vitrification solutions such as PVS2 or PVS3 prior to immersion in liquid solution (Matsumoto 2017; O'Brien et al. 2021). Droplet-vitrification and cryo-plate methods have recently become more popular among cryobiologists. These two methods are modifications of classical vitrification methods in which explants are placed either on an aluminium foil containing a droplet of 5–10 μl of cryoprotective solution or are attached to

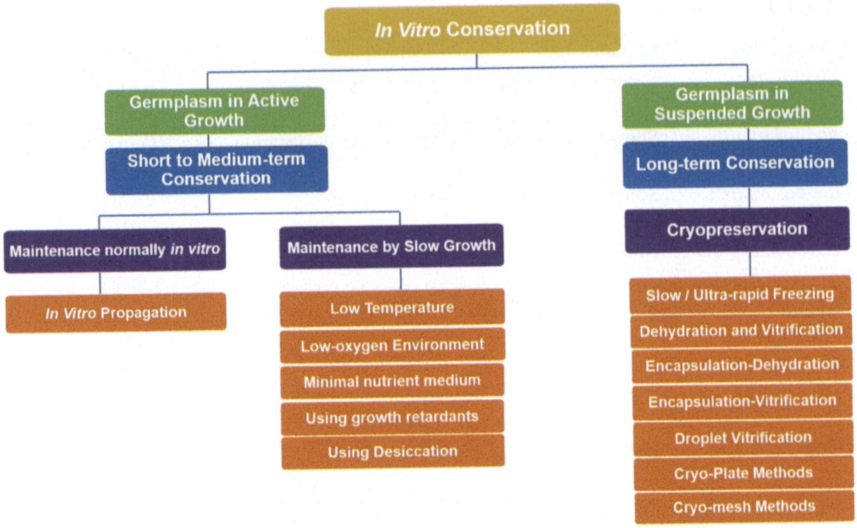

Fig. 1.2 Schematic illustration of different strategies of in vitro conservation

aluminium cryoplates with calcium alginate beads in uniform wells (Rai 2022). These two methods are nowadays efficiently applied in the cryopreservation of many tree species (Table 1.2). Both techniques are designed in such a way that intracellular ice crystal formation can be minimized and increase the regrowth ability of explants after storage (Rai 2022).

The high regrowth potential of cryopreserved explants is crucial for the success of cryopreservation of any plant species. The regrowth rate is highly dependent on the explants, cryoprotectants, and techniques used. For instance, cryopreservation of winter buds of Japanese birch by slow freezing to conserve the genetic resources of economically and ecologically important forestry trees of boreal areas of Japan such as *Betula platyphylla*, *B. maximowicziana*, and *B. Armani* (Endoh et al. 2023). The study compared the regrowth rates of shoot primordia subjected to slow freezing from −2 to −40 °C and then cryopreserved in liquid nitrogen (−196 °C) with the regrowth rates of shoot primordia that was not subjected to slow freezing and cryopreservation. The observed regrowth rate of isolated shoot primordia of *B. platyphylla* was approximately 80%, while shoot primordia of *B. maximowicziana* and *B. ermanii* showed 30% regrowth rate (Endoh et al. 2023). Five bird cherry (*Padus* Mill.) varieties were examined for their response to cryopreservation in liquid nitrogen vapor (−183 to −185 °C) from the bird cherry gene bank at the N.I. Vavilov All-Russian Institute of Plant Genetic Resources (VIR, Russia) (Verzhuk et al. 2023). The viability was compared to those seen during cuttings' standard mid-term preservation procedure at the VIR gene bank, which involves storage of cuttings at a low temperature (5 °C). The initial viability of winter cuttings ranged from 86.7% to 93.3%. The viability decreased to 53.3–86.7% and 43.3–60%, respectively, after 6 months of cold storage and cryopreservation, which was above the 40% viability threshold in all varieties tested. Cuttings recovered in the lab following cold storage had a higher mean viability (80%) than those retrieved in the field (58%). Cryopreservation for two Finland native elm species (*Ulmus glabra* and *U. laevis*) was tested through slow controlled cooling and regeneration of thawed buds by micropropagation (Välimäki et al. 2021). Navya et al. (2023) performed the cryopreservation of pollen grains of *Artocarpus heterophyllus* and evaluated the effectiveness of cryopreservation using post-storage germination testing and in vivo pollination using cryopreserved pollen. Embryogenic tissues of *Abies alba* were cryopreserved using the slow-freezing approach (Salaj et al. 2022). All lines that were evaluated recovered, but there were differences in regrowth frequencies among cell lines (91.67–100%). The recovered tissues exhibited characteristics that were similar to those of the control group (non-pretreated and non-cryopreserved tissues). The cryopreserved tissues generated viable cotyledonary somatic embryos.

It is also important to assess the stability of the cryopreserved germplasm. Modern molecular marker techniques can be used for the assessment of the genetic stability of cryopreserved or tissue culture-raised plants (Phulwaria et al. 2014; Arora et al. 2022; Rai 2023). For instance, Ahn et al. (2019) developed a cryopreservation system for the long-term storage of cultures of Korean arbor vitae (*Thuja koraiensis*), a critically endangered coniferous tree in Korea, and assessed the genetic stability of regenerated plantlets from cryo-stored embryogenic cell lines and found no genetic variations in the plantlets generated from non- and cryopreserved embryogenic cell lines.

1.3 Conclusion

Significant progress has been made in conserving many fruit and forest tree species in the last two to three decades using tissue culture-mediated slow growth and cryopreservation techniques. Despite the success in conserving the germplasm of many nonorthodox and vegetatively propagated tree species, in vitro conservation has some limitations. One of the major limitations of this technique is to evaluate the storage conditions for each species because each species requires a specific conservation strategy and sometimes the same strategy may be detrimental for other species also. In a few species, the low regrowth potential of explants after storage is another drawback of this technique. Therefore, the evaluation of a proper slow-growth or cryopreservation conservation approach for any species is necessary. Moreover, choosing an appropriate conservation method for any species should focus on high regrowth potential after storage. Nowadays, newly developed cryo-plate and cryo-mesh cryopreservation techniques have revolutionized plant conservation biology. Utilization of these most modern techniques will help in the conservation of tree species, especially commercially important and rare and endangered tree species.

References

Ahn CH, Heo K, Park HS, Choi YE (2019) *In-vitro* propagation and cryopreservation of *Thuja koraiensis* Nakai via somatic embryogenesis. In Vitro Cell Dev Biol-Plant 55:605–614

Álvarez JM, Cortizo M, Ordás RJ (2012) Cryopreservation of somatic embryogenic cultures of *Pinus pinaster*: effects on regrowth and embryo maturation. CryoLetters 33:476–484

Ara H, Jaiswal U, Jaiswal VS (2000) Synthetic seed: prospects and limitations. Curr Sci 78:1438–1444

Arbeloa A, Marín JA, Andreu P, García E, Lorente P (2017) *In-vitro* conservation of fruit trees by slow growth storage. Acta Hortic 1155:101–106

Arora K, Rai MK, Sharma AK (2022) Tissue culture mediated biotechnological interventions in medicinal trees: recent progress. Plant Cell Tissue Organ Cult 150:267–287

Barra-Jiménez A, Aronen TS, Alegre J, Toribio M (2015) Cryopreservation of embryogenic tissues from mature holm oak trees. Cryobiology 70:217–225

Benelli C (2021) Plant cryopreservation: a look at the present and the future. Plants 10(12):2744

Benelli C, De Carlo A, Engelmann F (2013) Recent advances in the cryopreservation of shoot-derived germplasm of economically important fruit trees of *Actinidia*, *Diospyros*, *Malus*, *Olea*, *Prunus*, *Pyrus* and *Vitis*. Biotechnol Adv 31:175–185

Bhojwani SS, Dantu PK (2013) Conservation of phytodiversity. In: Bhojwani SS, Dantu PK (eds) Plant tissue culture: an introductory text. Springer India, pp 287–298

Bonnart R, Waddell J, Haiby K, Widrlechner MP, Volk GM (2014) Cryopreservation of *Populus trichocarpa* and *Salix* dormant buds with recovery by grafting or direct rooting. CryoLetters 35:507–515

Capuana M, Di Lonardo S (2013) *In-vitro* conservation of chestnut (*Castanea sativa*) by slow growth. In Vitro Cell Dev Biol-Plant 49:605–610

Carvalho MD, Souza AD, Santos EB, Soares Filho WD, Ledo CA, Costa EM, Souza FV (2022) *In-vitro* conservation of 'Florida Rough' lemon plants. Ciência Rural 52(12):e20210276

Chauhan R, Singh V, Quraishi A (2019) *In-vitro* conservation through slow-growth storage. In: Faisal M, Alatar AA (eds) Synthetic seeds. Springer, pp 397–416

Condello E, Palombi MA, Tonelli MG (2009) Genetic stability of wild pear (*Pyrus pyraster*, Burgsd) after cryopreservation by encapsulation dehydration. Agricult Food Sci 18:136–143

Condello E, Caboni E, Andrè E, Piette B, Druart P, Swennen R, Panis B (2011) Cryopreservation of apple *in-vitro* axillary buds using droplet-vitrification. CryoLetters 32:175–185

El-Bahr MK, Abd El-Hamid A, Matter MA, Shaltout A, Bekheet SA, El-Ashry AA (2016) In-vitro conservation of embryogenic cultures of date palm using osmotic mediated growth agents. J Gen Eng Biotechnol 14:363–370

Endoh K, Hanaoka S, Matsushita M, Ubukata M, Yamada H (2023) Ex situ conservation of birch trees by cryopreservation of dormant buds adapted to subzero temperatures by extracellular freezing. New For 54:515–523

Engelmann F (1997) *In-vitro* conservation methods. In: Callow JA, Ford-Lloyd BV, Newbury HJ (eds) Biotechnology and plant genetic resources. CAB International, Oxford, pp 119–161

Engelmann F (2000) Importance of cryopreservation for the conservation of plant genetic resources. In: Engelmann F, Takagi H (eds) Cryopreservation of tropical plant germplasm: current research progress and application. IPGRI, Tsukuba, pp 8–20

Engelmann F (2004) Plant cryopreservation: progress and prospects. In Vitro Cell Dev Biol-Plant 40:427–433

Engelmann F (2011) Use of biotechnologies for the conservation of plant biodiversity. In Vitro Cell Dev Biol-Plant 47:5–16

Faisal M, Alatar A (2019) Synthetic seeds: germplasm regeneration, preservation and prospects. Springer, New York

Fernandes P, Rodriguez E, Pinto G, Roldan-Ruiz I, De Loose M, Santos C (2008) Cryopreservation of *Quercus suber* somatic embryos by encapsulation-dehydration and evaluation of genetic stability. Tree Physiol 28:1841–1850

Fki L, Bouaziz N, Chkir O, Benjemaa-Masmoudi R, Rival A, Swennen R, Drira N, Panis B (2013) Cold hardening and sucrose treatment improve cryopreservation of date palm meristems. Biol Plant 57:375–379

Gale S, Benson EE, Harding K (2013) A life cycle model to enable research of cryostorage recalcitrance in temperate woody species: the case of Sitka spruce (*Picea sitchensis*). CryoLetters 34:30–39

Gantait S, Sinniah UR, Suranthran P, Palanyandy SR, Subramaniam S (2015) Improved cryopreservation of oil palm (*Elaeis guineensis* Jacq.) polyembryoids using droplet vitrification approach and assessment of genetic fidelity. Protoplasma 252:89–101

Gomes F, Clemente M, Figueiredo P, Costa RL (2021) Medium-term *in-vitro* conservation of *Castanea* spp. hybrid clones. Vegetos 34:127–137

Guzmán-García E, Bradaï F, Sánchez-Romero C (2013) Cryopreservation of avocado embryogenic cultures using the droplet-vitrification method. Acta Physiol Plant 35:183–193

Hazubska-Przybył T, Chmielarz P, Michalak M, Bojarczuk K (2010) Cryopreservation of embryogenic tissues of *Picea omorika* (Serbian spruce). Plant Cell Tissue Organ Cult 102:35–44

Höfer M, Flachowsky H (2023) Cryopreservation of *Malus* and *Pyrus* wild species in the 'fruit Genebank'in Dresden-Pillnitz, Germany. Biology 12:200

Hung CD, Trueman SJ (2011) Encapsulation technology for short-term preservation and germplasm distribution of the African mahogany *Khaya senegalensis*. Plant Cell Tissue Organ Cult 107:397–405

Hwida MF, Abd El-Kader EM (2012) Slow growth conservation and molecular characterization of *Balanites aegyptiaca* L. Res J Agricult Biol Sci 8:179–190

Jenderek MM, Forsline P, Postman J, Stover E, Ellis D (2011) Effect of geographical location, year, and cultivar on survival of *Malus* sp. dormant buds stored in vapors of liquid nitrogen. Hort Sci 46:1230–1234

Jenderek MM, Ambruzs B, Tanner J, Holman G, Ledbetter C, Postman J, Ellis D, Leslie C (2013) Extending the dormant bud cryopreservation method to new tree species. Acta Hortic 1039:133–136

Juan-Vicedo J, Serrano-Martínez F, Cano-Castillo M, Casas JL (2022) *In-vitro* propagation, genetic assessment, and medium-term conservation of the coastal endangered species *Tetraclinis articulata* (Vahl) Masters (Cupressaceae) from adult trees. Plants 11:187

Kaya E, Alves A, Rodrigues L, Jenderek M, Hernandez-Ellis M, Ozudogru A, Ellis D (2013) Cryopreservation of *Eucalyptus* genetic resources. CryoLetters 34:608–618

Koc I, Akdemir H, Onay A, Ozden Ciftci Y (2014) Cold-induced genetic instability in micropropagated *Pistacia lentiscus* L. plantlets. Acta Physiol Plant 36:2373–2384

Kovalchuk I, Zhumagulova Z, Turdiev T, Reed BM (2014) Growth medium alterations improve *in-vitro* cold storage of pear germplasm. CryoLetters 35:197–203

Lambardi M, De Carlo A (2003) Application of tissue culture to the germplasm conservation of temperate broad-leaf trees. In: Jain SM, Ishii K (eds) Micropropagation of woody trees and fruits, Forestry sciences, vol 75. Springer, Dordrecht, pp 815–840

Latutrie M, Aronen T (2013) Long-term cryopreservation of embryogenic *Pinus sylvestris* cultures. Scand J For Res 28:103–109

Li JW, Ozudogru EA, Li J, Wang MR, Bi WL, Lambardi M, Wang QC (2018) Cryobiotechnology of forest trees: recent advances and future prospects. Biodivers Conserv 27:795–814

Marriott P, Sarasan V (2010) Novel micropropagation and weaning methods for the integrated conservation of a critically endangered tree species, *Medusagyne oppositifolia*. In Vitro Cell Dev Biol-Plant 46:516–523

Martínez MT, Ballester A, Vieitez AM (2003) Cryopreservation of embryogenic cultures of *Quercus robur* using desiccation and vitrification procedures. Cryobiology 46:182–189

Mascarello C, Sacco E, Di Silvestro D, Pamato M, Ruffoni B (2015) Reduction of the subculture frequency in three Mediterranean species during micropropagation. Acta Hortic 1155:461–466

Matsumoto T (2017) Cryopreservation of plant genetic resources: conventional and new methods. Rev Agricult Sci 5:13–20

Matsumoto T, Mochida K, Itamura H, Sakai A (2001) Cryopreservation of persimmon (*Diospyros kaki* Thunb.) by vitrification of dormant shoot tips. Plant Cell Rep 20:398–402

Matsumoto T, Yamamoto SI, Fukui K, Rafique T, Engelmann F, Niino T (2015) Cryopreservation of persimmon shoot tips from dormant buds using the D cryo-plate technique. Horticult J 84:106–110

Navya BL, Rajasekharan PE, Gutam S (2023) Pollen cryopreservation in jackfruit (*Artocarpus heterophyllus*) for crop improvement. In: Rajasekharan P, Rohini M (eds) Pollen cryopreservation protocols. Springer Protocols Handbooks. Humana, New York, pp 135–145

Niu YL, Zhang YF, Zhang QL, Luo ZR (2010) A preliminary study on cryopreservation protocol applicable to all types of *Diospyros kaki* Thunb. Biotechnol Equip 24:1960–1964

O'Brien C, Hiti-Bandaralage J, Folgado R, Hayward A, Lahmeyer S, Folsom J, Mitter N (2021) Cryopreservation of woody crops: the avocado case. Plants 10:934

Ozudogru EA, Capuana M, Kaya E, Panis B, Lambardi M (2010) Cryopreservation of *Fraxinus excelsior* L. embryogenic callus by one-step freezing and slow cooling techniques. CryoLetters 31:63–75

Ozudogru EA, Kirdok E, Kaya E, Capuana M, De Carlo A, Engelmann F (2011a) Medium-term conservation of redwood (*Sequoia sempervirens* (D. Don.) Endl.) *in-vitro* shoot cultures and encapsulated buds. Sci Hortic 127(3):431–435

Ozudogru EA, Kirdok E, Kaya E, Capuana M, Benelli C, Engelmann F (2011b) Cryopreservation of redwood (*Sequoia sempervirens* (D. Don.) Endl.) *in-vitro* buds using vitrification-based techniques. CryoLetters 32:99–110

Padilla IM, Barba-Espín G, Hernández JA, Piqueras A, Burgos L (2021) Acetylsalicylic acid improved antioxidative status and cold storage of encapsulated nodal segments of neem (*Azadirachta indica* A. Juss.). Plant Cell Tissue Organ Cult 144:261–270

Patel AK, Lodha D, Shekhawat NS (2020) An improved micropropagation protocol for the ex situ conservation of *Mitragyna parvifolia* (Roxb.) Korth. (Rubiaceae): an endangered tree of pharmaceutical importance. In Vitro Cell Dev Biol Plant 56(6):817–826

Phulwaria M, Patel AK, Rathore JS, Ram K, Shekhawat NS (2014) An improved micropropagation and assessment of genetic stability of micropropagated *Salvadora oleoides* using RAPD and ISSR markers. Acta Physiol Plant 36:1115–1122

Rai MK (2022) Plant tissue culture targeting germplasm conservation. In: Rai AC, Kumar A, Modi A, Singh M (eds) Advances in plant tissue culture: current developments and future trends. Academic, Elsevier, London, pp 205–221

Rai MK (2023) Start codon targeted (SCoT) polymorphism marker in plant genome analysis: current status and prospects. Planta 257:34

Rai MK, Jaiswal VS, Jaiswal U (2008) Encapsulation of shoot tips of guava (*Psidium guajava* L.) for short-term storage and germplasm exchange. Sci Hortic 118:33–38

Rai MK, Asthana P, Singh SK, Jaiswal VS, Jaiswal U (2009) The encapsulation technology in fruit plants—a review. Biotechnol Adv 27:671–679

Rai MK, Shekhawat NS, Harish, Gupta AK, Phulwaria M, Ram K, Jaiswal U (2011) The role of abscisic acid in plant tissue culture—a review of recent progress. Plant Cell Tissue Organ Cult 106:179–190

Rai MK, Rathour R, Behera S, Kaushik S, Naik SK (2021) Encapsulation technology: an assessment of its role in *in-vitro* conservation of medicinal and threatened plant species. In: Srivastava DK, Thakur AK, Kumar P (eds) Agricultural biotechnology: latest research and trends. Springer, Singapore, pp 103–128

Rajasekharan PE, Sahijram L (2015) *In-vitro* conservation of plant germplasm. In: Bahadur B, Rajam MV, Sahijram L, Krishnamurthy K (eds) Plant biology and biotechnology. Springer, New Delhi, pp 417–443

Rajasekharan PE, Wani SH (2020) Conservation and utilization of threatened medicinal plants. Springer, New York

Rathwell R, Popova E, Shukla MR, Saxena PK (2016) Development of cryopreservation methods for cherry birch (*Betula lenta* L.), an endangered tree species in Canada. Can J For Res 46:1284–1292

Reed BM, Sarasan V, Kane M, Bunn E, Pence VC (2011) Biodiversity conservation and conservation biotechnology tools. In Vitro Cell Dev Biol-Plant 47:1–4

Sakai A, Engelmann F (2007) Vitrification, encapsulation-vitrification and droplet-vitrification: a review. CryoLetters 28(3):151–172

Salaj T, Blehová A, Salaj J (2007) Embryogenic suspension cultures of *Pinus nigra* Arn.: growth parameters and maturation ability. Acta Physiol Plant 29:225–231

Salaj T, Matusikova I, Panis B, Swennen R, Salaj J (2010) Recovery and characterisation of hybrid firs (*Abies alba* x *A. cephalonica*, *Abies alba* x *A. numidica*) embryogenic tissues after cryopreservation. CryoLetters 31:206–217

Salaj T, Matušíková I, Fráterová L, Piršelová B, Salaj J (2011) Regrowth of embryogenic tissues of *Pinus nigra* following cryopreservation. Plant Cell Tissue Organ Cult 106:55–61

Salaj T, Panis B, Klubicova K, Salaj J (2022) Cryopreservation of *Abies alba* embryogenic tissues by slow-freezing method. Notulae Botanicae Horti Agrobotanici Cluj-Napoca 50:1–15

Sharma N, Gowthami R, Pandey R (2019) Synthetic seeds: a valuable adjunct for conservation of medicinal plants. In: Faisal M, Alatar A (eds) Synthetic seeds. Springer, Cham, pp 181–216

Shekhawat NS, Rai MK, Phulwaria M, Rathore JS, Gupta AK, Purohit M, Patel AK, Kataria V, Shekhawat S (2014) Tree biotechnology with special reference to species of fragile ecosystems and arid environments. In: Ramawat KG, Merillon JM, Ahuja MR (eds) Tree biotechnology. CRC Press, Boca Raton, pp 187–222

Sota V, Kongjika E (2014) Slow growth *in-vitro* conservation of *Zizyphus jujuba* mill. Agric For 60:27–37

Tanner JD, Chen KY, Bonnart RM, Minas IS, Volk GM (2021) Considerations for large-scale implementation of dormant budwood cryopreservation. Plant Cell Tissue Organ Cult 144:35–48

Uchendu EE, Shukla MR, Reed BM, Saxena PK (2013) Melatonin enhances the recovery of cryopreserved shoot tips of American elm (*Ulmus americana* L.). J Pineal Res 55:435–442

Välimäki S, Rusanen M, Pečínková D, Tikkinen M, Aronen T (2021) Cryopreservation and micropropagation methods for conservation of genetic resources of *Ulmus laevis* and *Ulmus glabra*. Forests 12:1121

Varshney A, Anis M (2014) Synseed conception for short-term storage, germplasm exchange and potentialities of regeneration genetically stable plantlets of desert date tree (*Balanites aegyptiaca* Del.). Agrofor Syst 88:321–329

Verzhuk V, Murashev S, Novikova L, Kiru S, Orlova S (2023) Conservation of the bird cherry (*Padus* Mill.) germplasm by cold storage and cryopreservation of winter cuttings. Biol 12:1071

Vidal N, Vieitez AM, Fernández MR, Cuenca B, Ballester A (2010) Establishment of cryopreserved gene banks of European chestnut and cork oak. Eur J For Res 129:635–643

Vujović T, Sylvestre I, Ružić D, Engelmann F (2011) Droplet-vitrification of apical shoot tips of *Rubus fruticosus* L. and *Prunus cerasifera* Ehrh. Sci Hortic 130:222–228

Vujović T, Chatelet P, Ružić Đ, Engelmann F (2015) Cryopreservation of *Prunus* spp. using aluminum cryo-plates. Sci Hortic 195:173–182

Wang MR, Lambardi M, Engelmann F, Pathirana R, Panis B, Volk GM, Wang QC (2021) Advances in cryopreservation of in vitro-derived propagules: technologies and explant sources. Plant Cell Tissue Organ Cult 144:7–20

Yamamoto SI, Rafique T, Priyantha WS, Fukui K, Matsumoto T, Niino T (2011) Development of a cryopreservation procedure using aluminum cryo-plates. CryoLetters 32:256–265

Yanli N, Zhengrong L, Yanfang Z, Qinglin Z (2012) Cryopreservation of *in-vitro* grown shoot tips of *Diospyros kaki* Thunb. using different methods. CryoLetters 33:69–74

Zhao Y, Wu Y, Chang Y, Reed BM (2008) Cryopreservation of fruit and ornamental trees. In: Reed BM (ed) Plant cryopreservation: a practical guide. Springer, New York, pp 387–420

Application of Biotechnological Tools in the *Abies* Genus: An Overview about the Application of Somatic Embryogenesis and Other Cloning Techniques

2

Nourhene Jouini, Mariana Correia, Maria Antonietta Germanà, Tiziano Caruso, and Jorge Canhoto

Abstract

In the last decades, plant cloning took a prominent place in the propagation of many tree species and proved its importance to achieve large-scale purposes. Somatic embryogenesis (SE) is a biotechnological tool, which has been applied to propagate many conifer species with many advantages. This technique is used in the intensive propagation of many conifer species and, especially, of Christmas trees because of its high demand in the market. For the genus *Abies*, this technique was initiated decades ago and still it has many challenges to improve its application to attain the commercial level. As many other conifers, diverse *Abies* species are threatened by climate change, deforestation, wildfires, livestock grazing, and genetic drift, making the improvement of its preservation and multiplication of utmost relevance. Therefore, the optimization of SE protocols could be a crucial tool to its propagation. In this review, the achievements of SE in the *Abies* genus to date are reported, with a focus on the use of the embryogenic tissue (ET) for cryopreservation, genetic advances, and other research studies.

N. Jouini (✉) · M. A. Germanà · T. Caruso
Department of Agricultural, Food and Forestry Sciences, University of Palermo, Palermo, Italy
e-mail: nourhene.jouini@unipa.it

M. Correia · J. Canhoto
Centre for Functional Ecology, Associate Laboratory Terra, Department of Life Sciences, University of Coimbra, Coimbra, Portugal

R. K. Kalia, R. Pathak (eds.), *Tree Biology and Biotechnology*,
https://doi.org/10.1007/978-981-96-0002-1_2

Keywords

Biotechnology · Christmas trees · Conifers · Cryo-conservation · Fir · Micropropagation

2.1 Introduction

The *Abies* genus commonly known as "Fir" is the second largest genus in the *Pinaceae* family. It includes 52 species, one subspecies, and 12 varieties (Xiang et al. 2007). The development of the trees from these species is very slow in the first years, then accelerates as the tree matures. Firs are most closely related to the genus *Cedrus* (cedar), typically monoecious evergreen trees with a spire-like or conic crown that often becomes flattened or rounded in old trees. Those trees are distributed all over the world, mostly located in mountains, and specifically in North and Central America, Europe, Asia, and North Africa. Many *Abies* species are endemic to specific regions, making their conservation essential for preserving biodiversity. As stated by IUCN, many fir species, viz., *Abies nebrodensis* in Italy, *Abies cephalonica* in Greece, Fraser fir in the USA, and *A. koreana* in South Korea, are subjected to extermination due to climate change (Koo et al. 2016) and thus need to be conserved urgently. The species *Abies* holds several significant roles and importance in different fields. Ecologically, it is considered as an important and functional balancer of the global forests and can serve as a crucial species for sustaining and increasing high biodiversity in different ecosystems (Caudullo and Tinner 2016). Fir trees provide habitat, food, and shelter for numerous animal species, including birds, mammals, insects, as well as to microbes. *Abies* trees are highly valued for their timber and wood products, being the wood often used as a good material for carpentry and furniture (Farjon 2010). Fir forests contribute significantly to the economy in many regions, as some species are used to produce the Christmas trees and decorative greenery, also contributing to rural economies (Caudullo and Tinner 2016).

In recent years, the propagation of many conifers species was advanced, and the use of biotechnological tools was largely applied. Somatic embryogenesis, an asexual method to propagate plants, depends on a process in which the plant somatic cells undergo a series of developmental changes that lead to the formation of embryogenic callus, which further develop into structures resembling zygotic embryos called somatic embryos and, later on, genetically uniform plantlets. This process can be induced in vitro, with the use of PGRs to induce callus formation. SE was first described in conifer species for *Picea abies* (Norway spruce) by Chalupa (1985) and Hakman and Arnold (1985). Since then, protocols for somatic embryogenesis induction have been applied to propagate a large numbers of conifer species (Salaj et al. 2019; Tretyakova and Mineev 2021). The first protocol for SE in the *Abies* was reported for silver fir [*Abies alba* (Mill.)] (Erdelský and Barancok 1986), followed by more *Abies* species and hybrids (Nørgaard and Krogstrup 1995). The technology of SE has the possibility to accomplish the aims of clonal propagation and breeding in different ways such as reducing the time required to produce plants, enabling control over genetic variation, the production of commercial hybrids for large scale cloning, and allowing the regeneration of gene edited genotypes combining desirable traits and also through the possibility of automation of some of these processes (Hazubska-Przybył et al. 2022). Moreover, SE technology offers advantages such as facilitating long-term preservation through cryopreservation and enabling species replanting if necessary (Tarraf et al. 2022). The present chapter discusses the recent insight on the utilization of somatic embryogenesis, genetic advances, organogenesis, and cryopreservation in *Abies* genus.

2.2 Somatic Embryogenesis in *Abies* Species

In the last decades, SE took a noticeable place in the propagation of many conifer species and showed the importance of applying plant biotechnological tools to attend commercial objectives and scale up the propagation of those species. SE in the *Abies* genus was first reported in *A. alba* (Erdelský and Barancok 1986); then this technique was applied in many species such as *A. alba* (Krajňáková et al. 2013), *A. balsamea* (Guevin et al. 1994), *A. cephalonica* (Krajňáková and Häggman 2018), *Abies concolor*—white fir (Kormutak and Vookova 2001), *A. fraseri* (Pullman et al. 2016), *Abies nebrodensis* (Jouini et al. 2023), *Abies koreana* (Lee et al. 2014), and other *Abies* species have been initiated before the last decade but haven't shown improvement on the protocols such as *A. lasiocarpa* (Kvaalen et al. 2005), *A. numidica* (Vookova and Kormutak 2001, Vookova et al. 2003), and hybrid firs (Korecký and Vitámvás 2011). Conventionally, SE in conifers proceeds through four major steps: initiation, proliferation, maturation, and conversion. Moreover, a supplementary step of cryopreservation could be complemented when storage of ET is desired.

2.2.1 Initiation of Embryogenic Callus

The success of the SE depends essentially on the accomplishment of the initiation phase. This step depends principally on the type of explants, basal salt medium, vitamins, and the PGRs used. The basal salts of the culture medium, the vitamins, other factors, and their interactions are vital components to induce the embryogenic callus. Choosing the right basal salt medium and vitamin complex has a significant effect in the success of the first step of SE, different culture media such as DCR, ½ Murashige and Skoog medium, Schenk and Hildebrandt medium, ½ Litvay medium, and ½ BLG medium having diverse PGRs were investigated (Table 2.1). For *A. lasiocarpa*, a most potent medium was developed for embryogenic tissue initiation from zygotic embryos isolated from the megagametophytes, named AL medium (Kvaalen et al. 2005).

For conifers, both auxin and cytokinin are majorly responsible PGRs to stimulate the callus induction, whereas in the *Abies*, cytokinin is considered as the dependable

Table 2.1 Protocols applied to induce SE from different *Abies* species

Species	Explant	Initiation medium and PGRs	Maturation medium and PGRs	References
Abies alba	Immature embryos	½ MS medium + .44 µM BA.	½ MS macronutrients maturation medium + 0.05% (w/v) casein hydrolysate and 1.7 mM L-glutamine	Krajňáková et al. (2013)
		DCR medium + one of cytokinins (BA, ZEA, TDZ, KIN, 2iP) at a concentration of 1 mg L^{-1}.	DCR basal salt + 10 mg L^{-1} ABA and supplemented with 7.5% PEG-4000	Salaj et al. (2020)
	Mature zygotic embryos	Modified MCM medium with 2.2 µM BA and 2.3 µM KIN	MCM basal medium + 20 µM ABA + 11 mM maltose	Hristoforoglu et al. (1995)
Abies alba × *Abies cephalonica*	Zygotic mature embryo	SH medium + 1 mg L^{-1} BA	DCR basal salt + 10 mg L^{-1} ABA and supplemented with 7.5% PEG-4000	Salaj and Salaj (2003)
Abies alba × *Abies numidica*	Zygotic mature embryo	SH medium supplemented with 1 mg L^{-1} BAP	DCR medium containing 1 or 10 mg L^{-1} ABA	Salajova et al. (1996)
Abies cephalonica	Immature zygotic embryos	1/2 MS + 1 mg L^{-1} (4.44 lmol/L) BA	MS + 17 mg L^{-1} (64 lM) ABA + 10% (w/v) PEG-4000 in liquid/semi-solid medium	Krajňáková and Häggman (2018)
		Solid MS, SH, and DCR culture media	DCR+ 8.5 mg L^{-1} (32 µM) ABA+10% (w/v) PEG4000, or ½ MS + 17 mg L^{-1} (64 µM) or 8.5 mg/L (32 µM) ABA, with or without 10% (w/v) PEG 4000,	Krajňáková and Häggman (2016)
Abies nebrodensis	Mature zygotic embryos	SH media +4,43 µM of BAP	SH basal salt medium+ 4,27 µM ABA + 8% PEG-4000	Jouini et al. (2023)

(continued)

Table 2.1 (continued)

Species	Explant	Initiation medium and PGRs	Maturation medium and PGRs	References
Abies cilicica and *A. cilicica* × *A. nordmanniana*	Immature zygotic embryos	SH + 5 µM BAP	MS and SH + 40 µM ABA + 4% maltose + 10% PEG 4000	Vookova and Kormutak (2003)
Abies nordmanniana	Immature embryos	½ MS + 2 or 10 µM BAP or 0.1 µM TDZ	½ MS medium containing 10 µM ABA	Nörgaard and Krogstrup (1991)
	Mature zygotic embryo	SH medium + 2.5 mg L^{-1} BA + 1 mg L^{-1} KIN + 0.04 mg L^{-1} 2,4-D	The SH + 2640 mg L^{-1} ABA + 36 mg L^{-1} picloram prepared in two different combinations: With 50 g L^{-1} PEG-4000 and without	Nawrot-Chorabik (2016)
A. Numidica	Immature embryos	SH + 1 mg L^{-1} BAP + 20 g L^{-1} sucrose	MS + 40 g L^{-1} maltose + 100 g L^{-1} PEG-4000 + 10 mg L^{-1} ABA and 500 mg L^{-1} L-glutamine and casein hydrolysate	Vookova and Kormutak (2004)
Abies cilicica	Immature embryos	SH + 1 mg L^{-1} BAP + 20 g L^{-1} sucrose	MS + 40 g L^{-1} maltose + 100 g L^{-1} polyethylene glycol-4000 + 10 mg l^{-1} ABA + 500 mg l^{-1} L-glutamine and casein hydrolysate	Vookova and Kormutak (2004)
Abies concolor	Immature embryos	SH + 1 mg L^{-1} BAP + 20 g L^{-1} sucrose	MS + 40 g L^{-1} maltose + 100 g L^{-1} polyethylene glycol-4000 + 10 mg L^{-1} ABA + 500 mg L^{-1} L-glutamine and casein hydrolysate	Vookova and Kormutak (2004)
Abies fraseri	Mature or immature seeds	AL salts, brassinolide, paclobutrazol, and abscisic acid Maintenance medium AL salts and 1 mg L^{-1} BAP	NS and AL salt medium supplemented with 5 mg L^{-1} ABA, 4% maltose, 10% PEG and 3 g L^{-1} Gelrite	Pullman et al. (2016)
Abies koreana	Mature zygotic embryos	SH +5.0 mg L^{-1} TDZ and 0.3% (w/v) gelrite	mLV medium + 60 µM ABA	Lee et al. (2014)
Abies balsamea	Mature zygotic embryos	½ BLG medium (Verhagen and Wann 1989), containing 10 µM cytokinin (TDZ, BA, or 2iP)	½ BLG basal medium + between 0 and 80 µM ABA	Guevin et al. (1994)

(continued)

Table 2.1 (continued)

Species	Explant	Initiation medium and PGRs	Maturation medium and PGRs	References
Abies lasiocarpa	Immature zygotic embryos	*Abies lasiocarpa* medium (AL) or SH medium supplemented with BA (5 μM)	Pretreatment for 8 days: AL hormone free + 10 g·L^{-1} of activated charcoal (Merck1.02186) and 58 mmol·L^{-1} sucrose. AL + ABA (40 μM)	Kvaalen et al. (2005)

PGRs to initiate SE (Krajňáková et al. 2008). For *A. alba*, *A. balsamea*, *A. cephalonica*, *A. concolor*, *A. fraseri*, *A. koreana*, *A. nebrodensis*, and *A. nordmanniana*, it was indicated that cytokinin alone is enough for induction and proliferation of the ET (Table 2.1). For the Greek fir, *A. cephalonica*, ET was induced either on solid MS, SH, or DCR culture media supplemented with 1 mg L^{-1} (4.44 μM) Benzyladenine (BA) (Krajňáková and Häggman 2016). In some cases, 0.1% (w/v) casein hydrolysate promotes cell proliferation as described by Krajňáková and Häggman (2018). Different types of cytokinins have been used to initiate SE. Salaj et al. (2020) tested diverse types of cytokinins, namely, benzyladenine, zeatin, thidiazuron, Kinetin, and/or 6-γ,γ- y-Dimethylallylamino purine, to initiate SE in *A. alba* from immature zygotic embryos, and their findings showed no significant difference on callus induction due to altered cytokinins. Moreover, in other conifer species, abscisic acid (ABA) showed the ability to promote SE in the Douglas fir in contrast to the use of auxins and cytokinin (Walther et al. 2022). Likewise, Pullman et al. (2016) have developed a favorable initiation protocol for Fraser fir comprised of a medium having abscisic acid, brassinolide, paclobutrazol, sodium thiosulfate, and aluminum salts. The embryo initiation rates recorded were between 6% and 62% depending on the composition of the medium and embryo age.

Additionally, the growth stage of the zygotic embryo during the culture is a major factor controlling the success of the SE. For *Abies*, both mature and immature zygotic embryos were utilized to induce embryogenic callus. Furthermore, secondary or repetitive SE has been tested from embryogenic cell lines of *A. alba* × *A. cephalonica* derived from immature and mature zygotic embryos and one cell line of the hybrid *A. alba* × *A. numidica* and showed that the initiation frequencies differed among genotypes and reached up to 24.3% of initiation (Salajová and Salaj 2001). Other factors such as the mother plants age, the storage period of cones, the luminosity, and the temperature could influence the initiation rate. Krajňáková et al. (2013) revealed that the storage period of cones considerably affects the induction rates in *A. alba,* where cones stored for 1 week showed the highest initiation rate than the percentages of immature embryos excised from fresh cones or from cones stored for 1 month.

2.2.2 Proliferation and Maintenance

The embryogenic callus generally appeared after some days to weeks from the establishment of the culture. Embryogenic tissues are taken from each explant independently after initiation that represents individual cell line and are shifted to a Petri dish having proliferation medium to develop fresh cell line. Then, regular sub-culturing must be performed, usually every 2 or 3 weeks on a fresh medium, to ensure a good proliferation and maintenance. ECMs are sub-cultured for months together, in the dark at 22 ± 2 °C; nevertheless, through continued proliferation, callus can lose their embryogenic capacity and turn non-embryogenic. The mechanisms associated with embryogenic capability attainment, manifestation, and maintenance in *Abies* due to somatic cell reprogramming are incomprehensive. However, suspension liquid culture was used for the ET proliferation. For *A. cephalonica,* a protocol for SE was described by Krajňáková and Häggman (2018), in which suspension culture was assessed in half-strength macro-elements MS without PGRs. Similarly, proliferation was achieved in *A. alba* using a suspension culture with half-strength macro- and microelements DCR (Salaj et al. 2020). Moreover, a trial for cultivation of the Greek fir ET with the ectomycorrhiza fungi during the proliferation phase showed a reduction in the proliferation rate; however, it improved the succeeding embryo formation (Krajňáková et al. 2012).

2.2.3 Maturation Phase

The maturation phase generally in *Abies* is promoted by ABA, polyethylene glycol (PEG), and/or the presence of different carbohydrates in the maturation medium. Maturation medium with 10 mg L^{-1} ABA is the most frequently used to achieve the cotyledonary stage of embryo development (Vookova and Kormutak 2014). Different works on many *Abies* species, viz., *A. nebrodensis, A. alba, A. balsamea, A. cephalonica, A. concolor, A. koreana, A. nordmanniana*, and *A. fraseri*, showed that the presence of ABA in the maturation medium, in combination with PEGs or without PEGs, is indispensable for the maturation of somatic embryos (Table 2.1).

The importance of carbohydrate source and concentration was earlier highlighted by Nörgaard (1997), in *A. nordmanniana*, showing that maltose was significantly better than sucrose. Studies were conducted on *A. cephalonica* to test the maturation media effect on the number of somatic embryos produced where the effect of carbohydrates source was tested. Studies suggested that the media containing maltose had more ability to form pre-cotyledonary embryos and somatic embryos than the media containing sucrose (Krajňáková et al. 2009).

Recently, studies were carried out to investigate the factors interfering on the growth and development of somatic embryos from *A. nordmanniana* and *A. bornmuelleriana* during the maturation phase (Nielsen et al. 2022). Various factors such as temperature and light conditions during germination, its time, ex vitro survival, and plant growth traits and vitality were studied. The findings indicated that almost experimental variables considerably affected plant growth in the first 3 years, nevertheless it had varied significance. The plant survival rate after nursery period was significantly affected by temperature, embryo quality, and the specific clone (Nielsen et al. 2022). The study showed and quantified intricate association within several advance phases of SE, suggesting the significance of all the aspects of cultural steps in developing protocols for SE.

2.2.4 Germination and Conversion

The phases of germination and regeneration of somatic embryos to emblings are still a bottleneck in several SE protocols for *Abies* when comparing with other conifers such as *Pinus* and *Picea* species. The conversion from SE to plantlets usually requires diverse steps. Generally, a treatment in hormone-free medium after maturation is needed. Nörgaard (1997) showed that the media composition has no significant effects on germination of SE in *A. nordmanniana*; however, about 60% germination could be attained on a half MS medium. Nevertheless, the inclusion of activated charcoal showed positive effect on the germination. Studies also reported 80–90% conversion of somatic embryos into emblings in *A. nordmanniana* (Zoglauer et al. 2012). For *A. lasiocarpa*, a desiccation treatment for 3 weeks was proceeded after maturation; then SE were transferred for germination to hormone-free AL medium supplemented with sucrose (10 g L^{-1}) and incubated under red light for 3 to 4 weeks, then transferred to a fresh medium under Fluora light with the same intensity. The conversion rate to plant was recorded to 8.4%, and only 35 plants out of 212 emblings could be grown after acclimatization (Kvaalen et al. 2005).

For *A. alba*, a partial desiccation has been done, then somatic embryos were shifted to DCR basal salt augmented with 2% maltose and 1% activated charcoal and were inoculated for 7–10 days under dark followed by culturing under dimmed light. Upon elongation of hypocotyl and cotyledons, the plantlets were transferred to light. Simultaneously, plantlets were also inoculated in Magenta jars kept in light for further development. Germination frequencies recorded were between 43 and 53% (Salaj et al. 2020). The plantlets from *A. cilicica* and its hybrid *A. cilicica* × *A. nordmanniana* having cotyledons, hypocotyls, and radicles were procured from mature somatic embryos partially desiccated and cultured on SH medium enriched by 1% sucrose and 1% activated charcoal. For *A. cilicica*, germination rate recorded was 75–94%, while for *A. cilicica* × *A. nordmanniana*, it was higher and attained 99.6% germination (Vookova and Kormutak 2003). It was observed that the higher temperatures during somatic embryogenesis may influence the early budburst in *A. nordmanniana* emblings (Lobo et al. 2022).

2.2.5 Cryopreservation

Since global forest is diminishing and climate changes are leading to more extreme environmental variations, germplasm conservation has become increasingly more significant to protect biodiversity. Cryo-conservation is a safer and economic approach for preservation of the genetic stock of non-orthodox seeds species (Engelmann 2004), wherein plant cell/tissues or other plant parts like embryos, etc., are stored under extremely low temperatures (−196 °C). The ultralow temperature stops the biological activities of the genetic material and protects it from any type of destruction (Salaj et al. 2022).

As SE is the most practiced methodology for conifers in vitro propagation, the established cryo-conservation protocols generally involve active proliferating embryogenic masses. Within the *Abies* genus, there are established protocols for *A. nordmanniana* (Nörgaard et al. 1993), *A. alba* (Salaj et al. 2022), *A. fraseri* (Pullman et al. 2016), hybrids *A. alba* × *A. cephalonica*, *A. alba* × *A. numidica* (Salaj et al. 2016), and *A. nebrodensis* (Benelli et al. 2024). Most of the protocols for embryogenic cultures consist of the conventional slow cooling and fast thawing wherein samples are given a prior treatment of sucrose or sorbitol followed by application of a cryoprotectant. Generally 5–15% DMSO are used as cryoprotectant. After that, the samples should be pre-frozen and lastly immersed into liquid nitrogen. For regeneration, samples are swiftly thawed at a higher temperature (30–40 °C) to evade the recrystallization and cell recovery. In addition to the advantages mentioned prior, cryo-conservation has greater importance in the extensive multiplication of conifers using SE (Cyr 2000). The challenges that this technique presents are often related to crystallization that causes cell damage and to genetic fidelity of the samples through long-term cryopreservation (Krajňáková et al. 2011).

2.3 Organogenesis

Organogenesis refers to the procedure in which meristems, roots, and shoots are developed through an explant tissue under controlled conditions (Chawla 2011). De novo meristems are inducted in this method under the presence of auxins and cytokinins in a balanced medium culture (Bhat et al. 2014). There are two distinct processes in organogenesis. The commonly practiced methodology is the indirect organogenesis wherein callus induction is achieved from de novo meristems as in *A. pindrow* (Bhat et al. 2014). Usually, the best pathway to produce callus was the supplementation of MS media with NAA (1-naphthaleneacetic acid), and maximum shoot regeneration was obtained by supplementing the media with a combination of BAP (6-benzylaminopurine) and IBA (indole-3-butyric acid) (Bhat et al. 2014). For *A. balsamea*, organogenesis was directly achieved by obtaining first an embryogenic shoot from a terminal whorl of buds on twigs (Bonga 1977). Then, the embryogenic shoots were placed in a medium supplemented with IBA that led to the regeneration of shoots, without the middle step of callus formation (Bonga 1977). Direct organogenesis was also achieved for *A. concolor* × *A. grandis* hybrid, with cotyledons as initial explants (Vookova and Gajdošová 1992). Organogenesis can be used for conservation and plant breeding purposes, enabling rapid multiplication of elite *Abies* genotypes with desirable traits where traditional methods might not be applicable (Krajňáková et al. 2011). It has great cloning prospective and a single explant may develop several plantlets. Some of its disadvantages are related to genetic instability that may occur in the cultured cells, as well as the fact that long-term storage is not an option, as it is for somatic embryogenesis. Nonetheless, different *Abies* species might respond differently to these techniques, so the consideration of the biology and growth patterns of the species is important.

2.4 Applications of In Vitro Culture and Other Biotechnological Tools for Conservation and Breeding

Advances in molecular and genetic analyses on the *Abies* genus have been carried out for different purposes. One of them is conservation and population genetics with the aim of understanding and protecting endangered and rare *Abies* species. For *A. cilicica* and *A. koreana*, genetic assessments were performed to prepare a management plan (Awad et al. 2014; Kwak et al. 2017). Molecular markers, such as microsatellites, were used to describe the demographic history of populations and subsequently addressed the issues regarding evolutionary ecology of the species. For *A. koreana*, no significant reduction in genetic diversity, despite population fragmentation and small population size, was observed (Kwak et al. 2017). The study revealed identification of two genetic clusters and recommended the conservation strategy to maintain these two unique clusters to avoid outbreeding depression (Kwak et al. 2017). For *A. cilicica*, there was also no evidence of bottleneck, but the populations showed low genetic variation (Awad et al. 2014). The results showed a better understanding of the evolutionary history of the species in Lebanon and recommended that regeneration should be prioritized to conserve the genetic diversity. Introduction of seeds or seedlings from exotic locations should be avoided (Awad et al. 2014).

Conservation genetics was also used to link isolation inheritances and existing habitat disintegration in *A. pinsapo* (Cobo-Simón et al. 2020), as well as to assess the conservation value of *A. alba* in the Italian northern Apennines (Piovani et al. 2010). Molecular research was also applied to establish a phylogenetic tree of the *Abies* genus with a strong statistical basis (Semerikova et al. 2018). Additionally, it aids in estimating the time of divergence for its primary clades, offering insights into the geographical distribution of fir trees and the historical aspects of their intercontinental migration(s) (Semerikova et al. 2018). Further research is required to comprehend the evolutionary mechanisms operating within populations, especially to develop and execute forest management practices that align with genetic conservation efforts (Sagnard et al. 2002).

Genomic studies such as genomic sequencing are also significant to understand the genetic make-up of different *Abies* species. The genome of *A. alba* was drafted, assembled, and implemented through a community-based interventions (Mosca et al. 2019). The aim of this type of work is the prediction of species' future distribution and survival. The study also gathered convened chloroplast genome on a solitary platform that exhibited higher linearity with the chloroplast genomes of *A. sibirica* and *A. koreana* (Mosca et al. 2019). The mitochondrial genome of the *A. alba* was also sequenced, opening new possibilities for comparative studies and allowing the answer to questions related to structural, phylogenetic, and evolutionary fields (Kersten et al. 2022).

Investigations into the molecular mechanisms underlying responses to environmental stressors such as climate change, pathogens, and pollutants are also vital to understand the gene-environment interactions. The complementary transcriptomic model for *A. pinsapo* revealed a genomic basis for drought resilience (Cobo-Simón et al. 2023). Gas-exchange parameters, transcriptomic analysis, and single nucleotide polymorphisms were studied in after drought period (24 h) and extended drought period (22 days). Throughout the extended drought genes related to cellular injury and homeostasis triumphed, resilient and sensitive phenotypes were defined (Cobo-Simón et al. 2023).

Breeding programs also benefit from genetic advances using SNP genotyping via high-throughput targeted sequencing to breed *A. procera* considered a genetically depauperate landrace (Hansen et al. 2020) or a combination of DNA markers and quantitative genetics for production of Christmas tree *A. bornmülleriana* (Xu et al. 2018).

Moreover, embryogenic callus from *Abies* species were used to apply genetic transformation. In Korean fir, an efficient *Agrobacterium*-mediated transformation system was evaluated through the somatic embryogenesis technique wherein embryogenic masses were obtained using *Agrobacterium tumefaciens* (Lee et al. 2014). The study observed the appearance of the GUS reporter gene in the somatic embryo and suggested that the technique may have potential to develop breeding and conservation planning in Korean fir (Lee et al. 2014).

The environmental pollution with heavy metals has increased due to industrial development, and researchers are using the embryogenic tissue to study the stress caused with heavy metals. Nawrot-Chorabik (2017) examined the consequences of lead, cadmium, and copper using on the callus cells of *A. nordmanniana* and reported that lead had the most toxic effects on the tissue while accumulation of copper was higher in the cells. Moreover, results showed that the tolerance for metal toxicity was different between genotypes. Likewise, the response of embryogenic tissue from in vitro hybrid firs was tested to assess the effect of cadmium ions, and it was observed that cadmium had adverse effect on the development with difference between cell lines (Matusova et al. 2020). Thus, the research highlighted that the embryogenic tissue is an appropriate plant material to assess the heavy metals stress on the cultured cells and can be applied to select the most tolerant cell lines for a mass production.

2.5 Conclusion

Somatic embryogenesis technique is being widely used for propagation of diverse species of conifers. The technique has many advantages over other propagation methods for many conifer trees including the *Abies* species for different purposes including genetic transformation, molecular analysis, cryo-conservation, and even to study the plant cells response to stress. Despite all the advantages of the SE in the

Abies genus, the technique is still less exploited in several species and the information on the procedures to convert somatic embryos into plantlets is meagre.

References

Awad L, Fady B, Khater C, Roig A, Cheddadi R (2014) Genetic structure and diversity of the endangered fir tree of Lebanon (*Abies cilicica* Carr.): implications for conservation. PLoS One 9(2):e90086

Benelli C, Tarraf W, İzgü T, Anichini M, Faraloni C, Salvatici MC, Jouini N, Germanà MA, Lambardi M (2024) Long-term conservation for the safeguard of *Abies nebrodensis*: an endemic and endangered species of Sicily. Plan Theory 13(12):1682

Bhat SJA, Gangoo SA, Geelani SM, Qasba SS, Parray AA (2014) Callus culture and organogenesis in fir (*Abies pindrow* Royle). J Cell Tissue Res 14(3):4653

Bonga JM (1977) Organogenesis *in vitro* cultures of embryonic shoots of *Abies balsamea* (balsam fir). In Vitro 13:41–48

Caudullo G, Tinner W (2016) *Abies*–Circum-Mediterranean firs in Europe: distribution, habitat, usage, and threats. In: San-Miguel-Ayanz J, de Rigo D, Caudullo G, Houston Durrant T, Mauri A (eds) European atlas of forest tree species. Publication Office of the European Union, Luxembourg, pp 50–51

Chalupa V (1985) Somatic embryogenesis and plant regeneration from cultured immature and mature embryos of *Picea abies* (L.) Karst. Commun Inst For Cech 14:57–63

Chawla HS (2011) Introduction to plant biotechnology. CRC Press, Pantnagar

Cobo-Simón I, Méndez-Cea B, Jump AS, Seco J, Gallego FJ, Linares JC (2020) Understanding genetic diversity of relict forests. Linking long-term isolation legacies and current habitat fragmentation in *Abies pinsapo* Boiss. For Ecol Manag 461:117947

Cobo-Simón I, Maloof JN, Li R, Amini H, Méndez-Cea B, García-García I, Gómez-Garrido J, Esteve-Codina A, Dabad M, Alioto T, Wegrzyn JL, Gallego FJ (2023) Contrasting transcriptomic patterns reveal a genomic basis for drought resilience in the relict fir *Abies pinsapo* Boiss. Tree Physiol 43(2):315–334

Cyr DR (2000) Cryopreservation: roles in clonal propagation and germplasm conservation of conifers. In: Cryopreservation of tropical plant germplasm: current research progress and application. Proceedings of an international workshop, Tsukuba, Japan, October, 1998. International Plant Genetic Resources Institute (IPGRI), pp 261–268

Engelmann F (2004) Plant cryopreservation: progress and prospects. In Vitro Cell Dev Biol-Plant 40:427–433

Erdelský K, Barancok P (1986) Cultivating embryos of silver fir (*Abies alba* Mill.) *in vitro*. Acta Fac Rerum Nat Univ Comen Physiol Plant 23:25–29

Farjon A (2010) A handbook of the World's conifers, vol I, II. Brill, Leiden/Boston

Guevin TG, Micah V, Kirby EG (1994) Somatic embryogenesis in cultured mature zygotic embryos of *Abies balsamea*. Plant Cell Tissue Org Cult 37:205–208

Hakman I, Arnold SV (1985) Plantlet regeneration through somatic embryogenesis in *Picea abies* (Norway spruce). J Plant Physiol 121(2):149–158

Hansen OK, Lamour K, Whetten R, Xu J, Cuenca A, Islam MS, Nielsen UB (2020) Ad hoc breeding of a genetically depauperate landrace of noble fir (*Abies procera* Rehder) using SNP genotyping via high-throughput targeted sequencing. Tree Genet Genomes 16(5):63

Hazubska-Przybył T, Wawrzyniak MK, Kijowska-Oberc J, Staszak AM, Ratajczak E (2022) Somatic embryogenesis of Norway spruce and scots pine: possibility of application in modern forestry. Forests 13(2):155

Hristoforoglu K, Schmidt J, Bolhar-Nordenkampf H (1995) Development and germination of *Abies alba* somatic embryos. Plant Cell Tissue Org Cult 40:277–284

Jouini N, Yahyaoui E, Tarraf W, İzgü T, Benelli C, Lambardi M, Germanà MA (2023) Somatic embryogenesis in *Abies nebrodensis*, an endangered Sicilian fir. Plant Cell Tissue Org Cult 152(2):393–404

Kersten B, Rellstab C, Schroeder H, Brodbeck S, Fladung M, Krutovsky KV, Gugerli F (2022) The mitochondrial genome sequence of *Abies alba* Mill. reveals a high structural and combinatorial variation. BMC Genomics 23(1):1–15

Koo KA, Kim J, Kong WS, Jung H, Kim G (2016) Projecting the potential distribution of *Abies koreana* in Korea under the climate change based on RCP scenarios. J Korean Soc Environ Restor Technol 19(6):19–30

Korecký J, Vitámvás J (2011) Somatic embryogenesis in the hybrid *Abies cilicica* x *Abies cephalonica*. J For Sci 57:401–408

Kormutak A, Vookova B (2001) Peroxidase activity in non-embryogenic and embryognie calli and in developing somatic embryos of white fir (*Abies concolor* gord, et glend). Plant Biosyst 135(1):101–105

Krajňáková J, Häggman H (2016) Somatic embryogenesis of *Abies cephalonica* Loud. In: Germana M, Lambardi M (eds) In vitro embryogenesis in higher plants. Methods in molecular biology, vol 1359. Humana Press, New York, pp 417–430

Krajňáková J, Häggman H (2018) Somatic embryogenesis of Greek Fir (*Abies cephalonica* Loud.). In: Jain S, Gupta P (eds) Step wise protocols for somatic embryogenesis of important woody plants. Forestry sciences, vol 84. Springer, Cham

Krajňáková J, Gömöry D, Häggman H (2008) Somatic embryogenesis in Greek fir. Can J For Res 38:760–769

Krajňáková J, Häggman H, Gömöry D (2009) Effect of sucrose concentration, polyethylene glycol and activated charcoal on maturation and regeneration of *Abies cephalonica* somatic embryos. Plant Cell Tissue Org Cult 96:251–262

Krajňáková J, Sutela S, Aronen T, Gömöry D, Vianello A, Häggman H (2011) Long-term cryopreservation of Greek fir embryogenic cell lines: recovery, maturation and genetic fidelity. Cryobiology 63(1):17–25

Krajňáková J, Niemi K, Gomory D, Haggman H (2012) Effects of different ectomycorrhizal fungi on somatic embryogenesis of *Abies cephalonica* Loud. Plant Cell Tissue Org Cult 109:353–361

Krajňáková J, Bertolini A, Gömöry D, Vianello A, Häggman H (2013) Initiation, long-term cryopreservation, and recovery of *Abies alba* mill. Embryogenic cell line. In Vitro Cell Dev Biol-Plant 49:560–571

Kvaalen H, Daehlen OG, Rognstad AT, Grønstad B, Egertsdotter U (2005) Somatic embryogenesis for plant production of *Abies lasiocarpa*. Canadian J For Res 35(5):1053–1060

Kwak M, Hong JK, Park JH, Lee BY, Suh MH, Kim CS (2017) Genetic assessment of *Abies koreana* (Pinaceae), the endangered Korean fir, and conservation implications. Conserv Genet 18:1165–1176

Lee H, Moo HK, Park SY (2014) Agrobacterium-mediated transformation via somatic embryogenesis system in Korean fir (*Abies koreana* Wil.), a Korean native conifer. Korean J Plant Resour 27(3):242–248

Lobo A, Find JI, Hansen JK, Ræbild A, Kjaer ED (2022) Effect of temperature and osmotic stress during somatic embryogenesis on phenology and physiology of *Abies nordmanniana* emblings. For Ecol Manag 514:120212

Matusova R, Cárach M, Labun P, Salaj T (2020) Physiological and structural responses of hybrid firs embryogenic tissue under cadmium stress. S Afr J Bot 131:240–249

Mosca E, Cruz F, Gómez-Garrido J, Bianco L, Rellstab C, Brodbeck S, Csilléry K, Fady B, Fladung M, Fussi B, Gömöry D, Neale DB (2019) A reference genome sequence for the European silver fir (*Abies alba* Mill.): a community-generated genomic resource. G3 Genes Genomes Genetics 9(7):2039–2049

Nawrot-Chorabik K (2016) Plantlet regeneration through somatic embryogenesis in Nordmann's fir (*Abies nordmanniana*). J For Res 27(6):1219–1228

Nawrot-Chorabik K (2017) Response of the callus cells of fir (*Abies nordmanniana*) to *in vitro* heavy metal stress. Folia Forestalia Polonica Series A Forestry 59(1):25–33

Nielsen UB, Hansen CB, Hansen U, Johansen VK, Egertsdotter U (2022) Accumulated effects of factors determining plant development from somatic embryos of *Abies nordmanniana* and *Abies bornmuelleriana*. Front Plant Sci 13:989484

Nörgaard JV (1997) Somatic embryo maturation and plant regeneration in *Abies nordmanniana* Lk. Plant Sci 124(2):211–221

Nörgaard J, Krogstrup P (1991) Cytokinin induced somatic embryogenesis from immature embryos of *Abies nordmanniana* Lk. Plant Cell Rep 9:509–513

Nørgaard JV, Krogstrup P (1995) Somatic embryogenesis in *Abies* spp. In: Somatic embryogenesis in woody plants, vol 3-Gymnosperms. Springer, Dordrecht, pp 341–355

Nörgaard JV, Baldursson S, Krogstrup P (1993) Genotypic differences in ability of embryogenic *Abies nordmanniana* cultures to survive cryopreservation. Silvae Genetica 42:93–97

Piovani P, Leonardi S, Piotti A, Menozzi P (2010) Conservation genetics of small relic populations of silver fir (*Abies alba* Mill.) in the northern Apennines. Plant Biosyst 144(3):683–691

Pullman GS, Olson K, Fischer T, Egertsdotter U, Frampton J, Bucalo K (2016) Fraser fir somatic embryogenesis: high frequency initiation, maintenance, embryo development, germination and cryopreservation. New For 47:453–480

Sagnard F, Barberot C, Fady B (2002) Structure of genetic diversity in *Abies alba* Mill. from southwestern Alps: multivariate analysis of adaptive and non-adaptive traits for conservation in France. For Ecol Manag 157(1–3):175–189

Salaj T, Salaj J (2003) Somatic embryo formation on mature *Abies alba* × *Abies cephalonica* zygotic embryo explants. Biol Plant 47:7–11

Salaj T, Matusova R, Panis B, Swennen R, Salaj J (2016) Tissue regeneration of *Abies* embryogenic cell lines after 1 year storage in liquid nitrogen. Biologia 71:93–99

Salaj T, Klubicová K, Matusova R, Salaj J (2019) Somatic embryogenesis in selected conifer trees *Pinus* nigra Arn. and *Abies* hybrids. Frontiers. Plant Sci 10:408324

Salaj T, Klubicová K, Panis B, Swennen R, Salaj J (2020) Physiological and structural aspects of in vitro somatic embryogenesis in *Abies alba* mill. Forests 11(11):1–15

Salaj T, Panis B, Klubicova K, Salaj J (2022) Cryopreservation of *Abies alba* embryogenic tissues by slow-freezing method. Notulae Botanicae Horti Agrobotanici Cluj-Napoca 50(4):12770

Salajová T, Salaj J (2001) Somatic embryogenesis and plantlet regeneration from cotyledon explants isolated from emblings and seedlings of hybrid fir. J Plant Physiol 158:747–755

Salajova T, Jasik J, Kormutak A, Salaj J, Hakman I (1996) Embryogenic culture initiation and somatic embryo development in hybrid firs (*Abies alba* x *Abies cephalonica*, and *Abies alba* x *Abies numidica*). Plant Cell Rep 15:527–530

Semerikova SA, Khrunyk YY, Lascoux M, Semerikov VL (2018) From America to Eurasia: a multigenomes history of the genus *Abies*. Mol Phylogenet Evol 125:14–28

Tarraf W, Izgu T, Jouini N (2022) Strategies for the conservation by biotechnological approaches of *Abies nebrodensis*, a relict conifer of Sicily. In: XXXI International Horticultural Congress (IHC2022): international symposium on in vitro technology and micropropagated plants 1359, pp 215–222

Tretyakova IN, Mineev VV (2021) Reproductive potential of conifers, somatic embryogenesis and apomixis. Russ J Dev Biol 52(2):75–86

Verhagen SA, Wann SR (1989) Norway spruce somatic embryogenesis: high-frequency initiation from light cultured mature embryos. Plant Cell Tissue Org Cult 16:103–111

Vookova B, Gajdošová A (1992) Induction of adventitious buds on the cotyledons of *Abies concolor* x *Abies grandis* hybrid seedlings. Biol Plant 34:23–29

Vookova B, Kormutak A (2001) Effect of sucrose concentration, charcoal, and indole-3-butyric acid on germination of *Abies numidica* somatic embryos. Biol Plant 44:181–184

Vookova B, Kormutak A (2003) Plantlet regeneration in *Abies cilicica* Carr. and *Abies cilicica* x *Abies nordmanniana* hybrid via somatic embryogenesis. Turk J Bot 27(2):71–76

Vookova B, Kormutak A (2004) Propagation of some *Abies* species by somatic embryogenesis. Acta Univ Latv 676:257–676

Vookova B, Kormutak A (2014) Study of *Abies* somatic embryogenesis and its application. Dendrobiology 71:149–157

Vookova B, Matúšová R, Kormutak A (2003) Secondary somatic embryogenesis in *Abies numidica*. Biol Plant 46:513–517

Walther M, Wagner I, Raschke J, Zoglauer K, Rupps A (2022) Abscisic acid induces somatic embryogenesis and enables the capture of high-value genotypes in Douglas fir (*Pseudotsuga menziesii* [MIRB.] Franco). Plant Cell Tiss Organ Cult 148(1):45–59

Xiang X, Cao M, Zhou Z (2007) Fossil history and modern distribution of the genus *Abies* (*Pinaceae*). Front For China 2(4):355–365

Xu J, Nielsen UB, Hansen OK (2018) Ad hoc breeding of *Abies bornmülleriana* for Christmas tree production using a combination of DNA markers and quantitative genetics – a case study. Tree Genet Genomes 14:1–15

Zoglauer K, Aurich C, Uehre P, Herrmann S (2012) Somatic embryogenesis in *Abies nordmanniana*: present status and future application. In: Proceedings of the S1-2 (Oral presentation), 2nd international conference IUFRO working party 2.09.02, Brno, Czech Republic, 25–28 June 2012, pp S1–S2

Application of Tissue Culture in Major Dye Yielding Agroforestry Trees in India

3

S. Kala, Suresh Ramanan S., Ambika Rajendran, S. Vennila, K. Kumaran, V. Subbulakshmi, and K. Aswini

Abstract

Natural dyes utilize 75% of the country's total dye production of 48,000 tons. The market for natural dyestuffs is limited, but rising. Natural dyes frequently lack consistency and reproducibility of hue. Given the World Health Organization's prohibitions on the use of synthetic dyes, there is a growing interest in natural dye-based goods, and demand for these products is also expanding. The main issue is the availability of natural dyes in huge quantities. There is a scarcity of scientific information on enhanced and high-yielding selection, planting stock quality, colour extraction uniformity, raw material availability, standardized dye extraction procedures, cost-effective methods of dyeing with natural dyes, and raising awareness among locals about the cultivation of natural dye yielding species. In this review, the status of natural dyes utilization, production and prospective of major dye yielding agroforestry tree species in India,

S. Kala (✉)
ICAR-IISWC-Research Centre, Kota, Rajasthan, India

S. Ramanan S.
ICAR-Central Agroforestry Research Institute (CAFRI), Jhansi, Uttar Pradesh, India

A. Rajendran
Division of Genetics, ICAR-IARI-Pusa Campus, New Delhi, India

S. Vennila
ACRI-TNAU, Trivannnamalai, Tamil Nadu, India

K. Kumaran
Department of Agronomy, ADAC&RI, TNAU, Trichy, Tamil Nadu, India

V. Subbulakshmi
ICAR-CAZRI-Regional Research Station, Bikaner, Rajasthan, India

K. Aswini
Division of Microbiology, ICAR-IARI-Pusa Campus, New Delhi, India

R. K. Kalia, R. Pathak (eds.), *Tree Biology and Biotechnology*, https://doi.org/10.1007/978-981-96-0002-1_3

techniques for dye extraction, importance of tissue culture, and standardization of tissue culture approaches in dye yielding species have been discussed thoroughly.

Keywords

Colours · Dye extraction · Eco-friendly · Edible dye · Genetic improvement · Micropropagation · Nodal explants · Species conservation

3.1 Introduction

The practice of natural dyeing has long been credited to Indians. Textiles, medications, cosmetics, and other items can be coloured with natural dyes. They are also used to colour a variety of food goods because of their harmless properties. Over 450 different plants can be used to make colours in India. Some of these plants have medical benefit in addition to their ability to yield dye. Despite the vastness of the plant resource base, not much has been used up to this point. It has not achieved the same level of commercial success as synthetic dyes due to a lack of exact scientific expertise on the extraction and colouring process. India possesses around 39.24 million hectares of wastelands (Sheikh and Abhilash 2016); the sustainable use of this land will yield multiple benefits, for example, biomass production, the generation of bio-dye material, the sequestration of carbon in the soil, and the rehabilitation of ecosystem services. Proper cultivation of potential dye-producing plants on these lands could lead to the revival and use of vegetable dyes. It might be beneficial to do agronomic research on different plants and shrubs and to evaluate and purify dyes chemically. Using natural colours is the safest option because they do not harm the environment and are also environmentally beneficial. Because natural dyes are non-toxic, biodegradable, and visually appealing, they might be a superior option for creating jobs and making use of wastelands.

Though dyes also made from insects, including cochineal, kermes, and lac, are also mentioned, plant sources are the major sources of the natural dyes. Several plants, shrubs, trees, insects, animals, microorganisms, and minerals have been recognized for extraction of colour compounds. Naturally collected dyes are utilized in textile industry for a variety of purposes, including dying yarns that are subsequently woven into cloth or any other useable form. Even while it is doubtful that all dyestuffs will be made entirely of plants, it is an intriguing and thrilling possibility that someday a portion of common colours may come from natural sources.

Natural dye yielding plant-based agroforestry systems may be useful in dryland and rainfed agriculture, and they are a potential source of improving land productivity and addressing several associated issues including biodiversity loss and land degradation, etc. These problems coupled with open grazing and overuse under changing climatic conditions make it more difficult for dryland crop production to be sustainable. Thus, a multi-tier, biodiverse planting strategy for woody perennials may be the solution. This approach envisions combining of pulses, oilseeds crop with multipurpose tree species along with livestock husbandry. The primary goal of this system's survival is the effective conservation and maximum utilization of the resources. The majority of farming systems include tree crops, which give the system some stability in regions with extremely fluctuating rainfall.

The foundation for the study of tissue culture was established over 230 years ago when research on plant wound healing revealed that elm plants' decorticated areas spontaneously formed calluses. The idea of in vitro cell cultivation was invented by German botanist Haberlandt in 1902 wherein fully differentiated cells could be cultivated and separated on a nutritional medium that included peptone, glucose, and salt. Over the past 20 years, tissue culture techniques witnessed significant advancements and their potential applications in forestry, despite certain inherent constraints could be realized (Bonga and Durzan 1987). Even though tissue culture didn't start in India till the 1960s, but there has been a tremendous rise through in vitro commercialization in the country during recent years. In vitro culture techniques are becoming more and more useful in the biotechnology era for plant propagation, breeding, unfavourable elimination, and germplasm conservation. Cutting, budding, grafting, air layering, and other traditional methods of multiplication are either very sluggish or altogether unsuccessful. Under this circumstance, the ability to cultivate superior clones and identify valuable variants through tissue culture technologies has shown to be a flexible instrument. The present chapter discusses the recent insight related to status of natural dyes utilization, production, and prospective of major dye yielding agroforestry tree species in India, techniques for dye extraction, importance of tissue culture, and standardization of tissue culture approaches in dye yielding species.

3.2 Status of Natural Dyes

Utilization of natural dyes in India is a very old tradition. Traditional skilled artisans in the nation have been creating naturally dyed textiles for centuries in numerous areas. When used alone, natural dyes have a number of drawbacks, such as (i) it is challenging to replicate manners when employing natural dyes or colorants because these agro products differ across the species, between crop seasons, between locations, and between harvesting periods, among other factors, (ii) standardizing the production of natural dyes is challenging since the development of colour in natural dyes depends on colour constituent and on the materials used, (iii) research is still needed to fully understand the scientific basis for most of the science underlying with natural dyeing, (iv) the absence of accurate technical understanding on dying and extraction methods, (v) when the coloured material is exposed to air, sweat, and sunlight, it may change colour, and (vi) compared to synthetic dye, natural dye and dye products are more expensive.

A natural dye is any pigment, dye, or other material that comes from plants, animals, or minerals; these are sustainable, renewable bio-resource goods that have little to no negative effect on the environment. These are made of organic materials that are harvested or obtained from natural resources. In comparison to synthetic dyes, natural dyes are typically less poisonous, non-allergic, biodegradable, and

Table 3.1 Commonly used natural dyes

Species name	Colour of dye
Eugenia jambalana, Tamarindus indica, Hibiscus sinensis	Red dye
Bixa orellana, Lawsonia inermis, Nerium indicum	Orange dye
Annona squamosa, Butea monsperma, Terminalia chebula, Cassia auriculata	Yellow dye
Ocimum sanctum, Eupotorium spp., *Bougainvillea* spp.	Green dye
Indigofera tinctoria, Wrightia tinctoria, Anogeioiss latifolia	Indigo dye
Calotropis procera, Clitoria ternatea, Acacia nilotica	Blue dye
Ficus religiosa, Rubia cordifolia, Rubia tinctoria	Pink dye
Eucalyptus Spp., *Emblica officinalis, Azadirachta indica*	Brown dye

Source: Vennila et al. (2023)

environmental friendly. The majority of plant dyes are safe to wear, and since they degrade naturally, spilling natural dyes into water bodies doesn't affect the ecosystem. A wide range of plant, shrub, and tree species, insects, animals, microorganisms, and minerals have been recognized to extract dyes. These dyes can produce different colours such as red, yellow, brown, blue, black, green, and orange (Table 3.1). The natural dyes tend to fade quickly, and dye fixatives, such as starch, kelp, alum, table salt, and vinegar, are employed to make the dyes stay to the fabric; however, the natural dyed fabrics need to be handled carefully and should not be dried in direct sunlight because they are more sensitive. People have been aware of utilization of natural dyes in colouring food ingredients, clothing, silk, cotton, wool, leather, and cosmetics since ancient times.

3.3 Types of Dye Extraction

Natural dyes from renewable sources are not suitable for direct usage. There are numerous restrictions when using raw ingredients to dye products. The most crucial thing is to safely and affordably extract the primary colouring component without compromising the extraction conditions or introducing any contaminants into different extraction methods. A number of economical extraction methodologies for natural dye that abide by consumer preference and regulatory regulation are gaining favour. These are a few techniques such as ultrasonic microwave extraction, fermentation, alkaline, acidic, enzymatic, solvent, supercritical fluid, and/or aqueous extraction used for the production of natural colorants.

3.4 Production and Prospective of Major Dye Yielding Agroforestry Tree Species

People working in the agricultural farming sector should be well-versed in land, soil types, irrigation facilities, water availability, seeds, rainfall amount and distribution, human resources, livestock, marketing, credit institutions, input supply systems, and processing facilities in the area, all of which can lead to better results for individuals, states, and the country as a whole. Plants are the most prevalent source of dyes; however, these may also be obtained from animal, mineral, and microbes. On the basis of this, natural dyes can also be generally categorized in to plant, animal, mineral, or microbiological colours. Nonetheless, the majority of natural colours are derived from plants. Many plant parts, for example, twigs, bark, branches, leaves, roots, heartwood, flowers, fruits, rinds, hulls, husks, can be used to make natural colours. Some of the most important dyes are listed below (Table 3.2 and Fig. 3.1).

A brief detail on some of the most important dyes is summarized below.

3.4.1 Annatto Orange Dye

Annatto dye is essentially a red orange pigment called Bixin (natural carotenoid) derived from the seed coat of *Bixa orellana*. It is an edible dye yielding tree, native to the Neotropics, and belongs to Bixaceae family. Since it is an edible dye, it is frequently utilized to colour several edibles including butter, cheese. However, cotton, wool, and silk are also subjected to colour using this dye to obtain reddish-orange hues. Boiling is employed to extract dye using the aqueous technique or alkali extraction. The carotenoid, norbixin and bixin, is the main colouring agent. Seed-specific pigment known as bixin has been utilized in cosmetics and food since pre-Columbian times. After caramel, Anatto seeds are the second most important natural colorant in the world, generating colours ranging from yellow to red

Table 3.2 Source of natural dye from plants

Species name	Plant part used
Wrightia tinctoria, Ailanthus excelsa, Tectona grandis, Mangifera indica, Indigofera tinctoria	Leaves
Hibiscus sabdariffa, Clitoria ternatea, Carthamus tinctorius, Kigelia pinnata, Spathodea companulata,	Flowers
Terminalia chebula, Syzygium cuminii	Fruits
Pterocarpus santalinus, Biancea sappan, Pterocarpus dalbergioides	Wood
Eucalyptus tereticornis, Casuarina equsetifolia	Bark
Bixa orellana	Seed
Sterculia foetida	Pod

Source: Vennila et al. (2023)

Fig. 3.1 Potential eco-dye yielding multipurpose tree species

(Mercadante and Pfander 1998). Bixin is a highly unsaturated chemical and its amount in the seed ranges from 0.73% to 1.30%.

3.4.2 Eucalyptus Red Amber Dye

Eucalyptus is found all over India and grows on a wide range of soil types. Due to the vast range of climatic zones, this foreign species has primarily overrun in the country. It is an important source of natural colorants, producing numerous orange-yellowish dyes. It grows quickly and can be collected in 3–4 years after planting. The dye is retrieved from its bark after boiling the bark in water that provides the fabric material including cotton, silk, and wool a vivid orange hue.

3.4.3 Tropical Chestnut Dye

The tropical Skunk tree, known by its Latin name Sterculia and a member of the Sterculiaceae family, has a straight trunk and can reach to the heights of 50–70 feet. The fruit shell was utilized as a raw material for the dye extraction. The fruits shells are oven dried at 80 °C and pulverized into a fine powder in a mixer. The coloured solution used for dyeing is obtained by the refluxing procedure. After that, the extracted dye is centrifuged for 20 min at 4000 rpm to settle any remaining particles. Textile dyeing was done using the supernatant solution.

3.4.4 Mango Carmine Dye

Mango leaves were thoroughly cleaned with water and dried and grounded into extremely small pieces through grinding machine. The fine powder is passed through a sieve to remove the coarse particles. The colorant is isolated through an aqueous extraction method in which grounded mango leaves are boiled for 60 min in a water bath in a material to liquor ratio of 1:10, at pH 10, and temperature of 100 °C in order to extract the colour. After that heated dye solution is allowed to cool down and dye extract is carefully filtered. When it is used for dyeing in textile industries, it gives different brown colour shade.

3.4.5 Teak Leaf Brown Dye

Although the teak tree is mainly used as timber, it is also recognized for its orange-red dye. The young leaves and the bark of the roots yield a yellowish brown-reddish dye that is used to make colourful matting, paper, food, and clothing. Anthocyanin, the pigment responsible for the brown hue in teak leaves, can be extracted through macerations, soxhlet and reflux extraction. Solvent extraction technique is more convenient technique utilized in dye extraction.

3.4.6 Terminalia Black Dye

The fruit of *Terminalia chebula*, generally well-known as black- or chebulic myrobalan, is a species of genus Terminalia found in Pakistan, India, Nepal, China, Sri Lanka, Malaysia, and Vietnam. The fruits of Myrobalan are source of tannins and customarily utilized as natural dye to colour cotton before its dyeing with another colorant in India. It provides yellow or bronze colour that can also be solely used as dye.

3.4.7 Dalbergia Dark Blue Dye

Dalbergia is a genus of trees and shrubs in the legume family, known for many species that are valued for their high-quality timber. Some species of Dalbergia are indeed known for their dye-yielding properties. For instance, besides being a valued timber tree, *Dalbergia sissoo* (Shisham tree) can also be used to produce dyes, although this is less common compared to its use in woodworking. The heartwood of this species can yield a deep, rich dye. These trees are significant in various cultures for their economic and practical uses, including their role in creating beautiful, durable dyes. Dalbergia wood is collected and chopped into small pieces or shavings and soaked in water to soften it. Then, the wood is simmered in water for several hours to release the dye. Longer simmering may give more intense colour. Its dye is used to colour fabrics or other materials.

3.4.8 Pterocarpus Santolin Dye

Pterocarpus, sometimes known as East Indian mahogany, Andaman padauk, or Andaman redwood, is a member of the Fabaceae family. This tree grows best on well-drained banks in undulating terrain, on or near riverbanks, and occasionally forms pure stands. It is primarily found in deciduous and semi-moist deciduous woods. While the bark of the Andaman redwood tree is used medicinally and as a source of tannins, other portions of the plant, like blossoms and young leaves, are delicious. The wood is strong and impervious to termite damage. It is utilized in building, furniture, panelling, cabinetry, farming tools, bridges, piles, etc. After soaking in alkaline or aqueous solution, the wood dust or powders are cooked in water for 60 min at 600 °C to obtain red to yellowish colour of dye.

3.5 Importance of Tissue Culture in Dye Yielding Species

Tree species has a large geographic range, which allows for significant genetic variety. Seeds, the primary mechanism of propagation, vary greatly from one origin to the next in terms of morphological and physiological characteristics (Mathur et al. 1984). Understanding variation among species is essential for designing efficient tree improvement and/or breeding strategies (Vakshasya et al. 1992). Although plants are traditionally grown from open pollinated seeds, they have a high level of genetic variety. Genetic variety is a critical component of adaptability and, by extension, forest ecosystem stability. Variation is the primary resource to be investigated for genetic improvement in any species and hence plays an important part in plant improvement programs (Zobel and Talbert 1984; Tiwari 1992). A genetic characterization of natural resources is a necessary step toward a better understanding of genetic resources and the execution of in situ and ex situ conservation efforts. As a result, enough plant knowledge is required for planning sustainable development in any region, such as India, where the flora is diverse and endemism rich (Siva 2003).

The degree of variety provides breeders with more options for selection. Once discovered, the promising genotypes can be used to create varieties with a combination of more desired traits. Choosing trees with desired phenotypic traits aids in the generation of progeny for breeding and reforestation plans; this in turn makes tree improvement programs easier (Kedharnath 1982; Zobel and Talbert 1984). A large-scale plantation requires a lot of planting material, and high-quality planting material is essential to achieving higher yields. Therefore, it is imperative to build plantations using high-yielding material that, once discovered, can be massively duplicated via clonal propagation. In light of the aforementioned factors, progeny evaluations of superior phenotypes of this species were conducted in nurseries in order to gather data on genetic variation and correlations between different traits. If a crop is chosen for its yield and dye content, micropropagation should be simple and enable quick advancements. Despite possessing a broad range of adaptability and potential uses, this Annatto species has not garnered much attention in terms of clonal improvement research. As a result, super individual clonal propagation is sometimes recommended as a substitute for conventional breeding techniques.

Over the past 20 years, significant advancements have been made in the development of tree tissue culture techniques and their potential applications in forestry, despite certain inherent constraints (Bonga and Durzan 1987). Even though tissue culture was introduced to India in the 1960s, the pulp and paper industries, in particular, have shown a growing interest in commercializing the in vitro propagation of forest trees in recent years (Mascarenhas and Muralidharan 1989). Compared to traditional approaches, tissue culture micropropagation offers far more control and manipulation over the development of tissues inside the culture tube (Parthiban et al. 2000). Plantlets are directly regenerated from explants such as buds and meristems, ensuring genetic stock cloning. Macropropagation is typically regarded as a significant aspect of tree improvement in terms of selected trees, enabling rapid genetic gain through the multiplication of selected genotypes using low-cost, simple processes and assisting in the establishment of clonal seed orchards. Micropropagation would be advantageous since it allows for virus-free stocks (Quak 1977), micro-grafting of desired rootstocks (Jonard et al. 1983), and the creation of true-to-type progenies. Explants should be collected from phenotypically superior trees in order to limit genetic variation with cautious care.

Tissue culture would be advantageous since it allows for the development of true-to-type progenies, micro-grafting of desired rootstocks, and the acquisition of virus-free stocks (Quak 1977; Jonard et al. 1983). For this reason, careful consideration should be taken while collecting transplants from phenotypically superior trees in order to reduce genetic variance. In addition, grafts rather than seedlings can be planted to shorten the gestation period to 3–4 years (Rao et al. 2000). These plants, which were grown from a single clone in the field, have the potential to replace the present diverse plantations since they may produce bixin steadily (Parimalan et al. 2008). The method aids in the development of clonal seed orchards, which may yield a sufficient supply of high-quality planting stock with significant genetic values for commercial plantations and running afforestation initiatives. Early flowering and fruiting on clonal propagules may indicate a reduction in the tree's juvenile period and an acceleration of the breeding program. To multiply the desired genotypes/chosen trees producing high quantity and quality products, clonal propagation is the urgent need of the hour.

3.6 Standardization of Tissue Culture Approaches in Dye Yielding Trees

The in vitro culture techniques find tremendous improvement and usage in plant propagation of high valued and difficult trees, breeding, adverse elimination, and germplasm conservation. The ability to cultivate superior clones and identify valuable variants through tissue culture technologies has shown to be a flexible instrument (Parthiban et al. 2000). Forest tree tissue culture technology is not as advanced as it is for many agricultural crops. The presence of endogenous microbial pollutants, particularly in the tissues of material produced in fields, is the primary cause of this delay (Bonga and Durzan 1987). Numerous plant organs have been found to be contaminated by bacteria (Bastiaens 1983), with woody plants being more frequently affected (Young et al. 1984). Over the past 20 years, tree tissue culture techniques witnessed significant advancements and their potential applications in forestry, despite certain inherent constraints (Bonga and Durzan 1987). The tissue culture technique has got tremendous interest in India for in vitro micropropagation of forest trees due to increasing demand by several industries including pulp and paper industry. Gupta et al. (1980, 1983) observed rapid propagation of teak and Eucalyptus trees through tissue culture technique accomplished through axillary buds of 100 and 20 years old trees, respectively.

3.6.1 Standardize the Concentration of Suitable Sterilants

The tissue culture technique in forest trees is lagging behind that of many agricultural crops. The main reason for this lagging is the occurrence of endogenous microbial contaminants, especially, in tissues of field grown material (Bonga and Durzan 1987). Bacterial contamination is very common in several plant organs (Bastiaens 1983) and was more common in woody plants (Young et al. 1984). Sterilization is carried out through 3 steps including 'pre-treatment', 'treatment', and the 'post-treatment' (Sita et al. 1986). 'Pre-treatment' involves washing with soap solution and rinsing in running water for few hours. 'Treatment' involves the rinsing of explants in suitable surface sterilants. 'Post-treatment' involves 3–5 rinses in sterile distilled water. Ethyl alcohol, calcium hypochlorite, mercuric chloride, chlorine/bromine water, and silver nitrate are commonly used sterilizing agents in tissue culture (Hendre 1991). Successive surface sterilization of mature seeds of *D. sissoo* in 70% ethanol for 10 s and 0.1% $HgCl_2$ for 8 min followed by four times washing in distilled water gave good results (Tewari et al. 1993).

The explants from young shoots of 65 years old sandal plus tree, after preliminary washings in running water for 30 min, were subjected to the following sterilization treatments, that is, soaking in the fungicide, Bavistin (0.1–0.2%), for 15 min and two washing with distilled water and 0.1% $HgCl_2$ (5 min), twice washing with sterile distilled water and 50% v/v $NAOCl_2$, and finally thorough washing with sterile distilled water. After the surface sterilization, the explants were cultured aseptically in a culture tubes having MS medium (Reddy and Subramanian 1998). Shoot tips and newly formed young leaves or phyllodes obtained from five-month-old seedlings were washed under tap water for 20 min followed by Tween-20 solution for 10 min. The explants were later surface-sterilized using 0.1% $HgCl_2$ solution for 5 min and washed under distilled water (Sumana et al. 1998).

3.6.2 Optimization of Growth Regulators for Shoot and Root Proliferation

Cell division is induced by auxins [indole-3-acetic acid (IAA), indole-3-butryic acid (IBA), 2,4-D (dichlorophenoxyacetic acid), naphthalene acetic acid (NAA)] that promotes the root initiation as well. Adenine derivatives called cytokinins play a crucial part in the induction of shoots. The four most often used cytokinins are iso-pentenyl adenine, 6-(4-Hydroxy-3-methyl-Butyl-2 enylamino) purine (Zeatin), N-6 benzyl adenine (BA), and 6-furfurylaminopurine (Kinetin). Benzyl adenine is one of the efficient growth regulators used for the culture of meristem, shoot tip, and buds (Kitto and Young 1981). Hasegawa (1980) reported high callus induction when auxin, especially 2,4-D, was used at high concentration. Explants must be necessarily sterilized before culture inoculation (George and Sherrington 1984). NAA has been reported to be the most successful root promoter (Sood et al. 1994); however, in order to give roots with enough stimulation without causing an excessive amount of callus to grow, the concentration of auxin needed is crucial. Auxin combinations can occasionally result in a better response (Gupta et al. 1980). Additionally, IBA functions well (Sood et al. 1994).

3.6.3 Screening of Basal Medium for Tissue Culture

Success of plant tissue culture mainly depends on the choice of suitable culture media used. Semi-solid media are typically used for plant tissue culture, to which gelling agents like agar or gelrite are added. A suitable number of major and minor nutrients, a source of carbohydrates (sucrose) and trace levels of organic components (vitamins, amino acids, water, fruit juices, yeast extract casein hydrolysate also works well), should be present in a basal medium. In 1962, Murashighe and Skoog (MS) created the most widely utilized medium for all tissue culture applications. The medium described by Nitsch and Nitsch (1969) is widely utilized in anther culture. Several workers reported the accessibility for the cultivation of cells and protoplasts, the B5 medium, or its several derivatives (Gamborg et al. 1968; Kao 1977). For the purpose of anther and cereal culture, the N6 medium was created (Chu 1978). Woody Plant Medium (WPM) was especially developed for in vitro culturing of tree species (Lloyd and McCown 1980). It has lower amount of ammonium nitrate and potassium chloride while it has higher amount of sulphate in comparison to the MS medium. Gamborg (1984) found E1 medium as most supportive for the rapid growth of cell for embryogenesis and for the culture of protoplasts, while MS medium was found to be optimal for the nodal segment explant in teak compared to other media (Parthiban 2001).

3.7 Exemplary Research Findings of Tissue Culture Development in Major Dye Yielding Species

Comparatively to traditional approaches, tissue culture micropropagation offers far more control and manipulation over the development of tissues inside a culture tube (Parthiban et al. 2000) because it can lead to a significant level of plant variability and sexual reproduction that is unreliable through seeds. Tissue culture is crucial for finding beneficial variations and for quickly multiplying genotypes in vitro. The MS + 1 mg/l IBA treatment gave better results with respect to root and shoot development across a number of different tree species as compared to the other treatments.

3.7.1 Annatto (*Bixa orellana*)

The WHO's prohibitions on the use of synthetic dyes have drawn attention to plant-based edible dye and dye-based products. Genetic diversity, propagation techniques, and other biochemical characteristics, as well as their extraction in species that create dyes, particularly annatto plants, have been the focus of increased research in recent times. Research on in vitro propagation revealed that nodal segments from mother trees made the finest explants and that the best medium for growing shoot tips was the MS medium. The in vitro regeneration of *B. orellana* has the potential to speed up true-breeding initiatives and provide genotypes with higher yields. Studies on plant regeneration through seedling axillary buds (D'Souza and Sharon 2001) and by intervening seed callus (Sha-Valli-Khan et al. 2002) are available in the literature. Kala et al. (2022) reported that the most effective treatment for sterilizing *B. orellana* explants was 0.1% $HgCl_2$ for 5 min and exhibited maximum survival percentage (65.67%). The study reports better shoot elongation in the presence of BA (1.5 mg/l) and IBA (1 mg/l) while higher morphogenic responses were observed by nodal segments (25.88%) than shoot tip (13.0%). The combination of MS + BAP 2.5 mg/l + Kinetin 3 mg/l was better for greater shoot induction and number of shoots per explants (Kala et al. 2022). In order to meet the demand for edible colour, it would be beneficial to popularize superior Candidate Plus Trees (CPTs) with higher bixin concentration among commercial producers and encourage further improvement programs. Many farmers and nursery owners are in urgent need of better Annatto planting stock.

3.7.2 Neem (*Azadirachta indica*)

The endophytic contamination and systemic infection can be relatively easy to eradicate, and surface sterilization of explants using hydrogen peroxide, mercuric chloride, and nascent chlorine may be more challenging (Thorpe et al. 1991). According to Reddy and Subramanian (1998), soaking in Bavistin (0.1–0.2%) and then washing with 0.1% $HgCl_2$ was an effective therapy for removing fungal contamination from neem explants, but it was not effective for bacterial contamination. *Eucalyptus tereticornis* cultures treated with 2% sodium hypochloride produced contamination-free cultures (Parthiban et al. 1999). The induced shoots gave rooting upon transfer to MS medium enriched by 2% sucrose supplemented with 4 μM IAA (George and Kulkarni 1997). Arya and Arya (1998) reported that in *A. indica* multiple shoots differentiated from hypocotyls when cultured on MS medium supplemented with 2 mg/l BA + 50 mg/l adenine sulphate. Axillary buds inoculated on MS medium having 2–10 mg/l BA resulted in multiple shoot proliferation. In *A. indica,* best results (60%) were obtained from rooting of excised shoots with 1–2 mg/l IBA supplemented with MS medium.

3.7.3 Eucalyptus (*Eucalyptus tereticornis*) and Dalbergia (*Dalbergia lanceolata*)

The superiority of MS medium gives better results through nodal segments in *Eucalyptus*, for example, F1 hybrid FRI-5 (Bisht et al. 2000); cotyledons in *D. sissoo* (Tewari et al. 1993) and *D. lanceolata* (Anand and Bir 1984). Among the four distinct basal media, that is, MS, B5, modified MS, and WPM-MS medium, the WPM-MS medium outperformed (25.69%) the others in terms of the morphogenic response in eucalyptus.

It has also been previously well proven that MS medium is superior to other media for successful propagation in several tree species, including Eucalyptus (Chauhan et al. 1996; Parthiban et al. 1999), Eucalyptus hybrids (Bisht et al. 2000), *D. sissoo* (Tewari et al. 1993). Based on research conducted by Joshi et al. (2003), it was determined that MS medium produced a higher rate of shoot proliferation in *D. sissoo* than B5 medium, out of the two nutritional media that were tried. Auxin and cytokinin have been shown to work synergistically in *D. sissoo* (Rao 1986). Explants from mature trees had larger concentrations of auxins for roots and cytokinin for shoot initiation than did explants from seedlings (Gupta et al. 1983).

Shoot buds formed in MS having BAP (2 mg/l) at cotyledon tip of *D. lanceolata*, and rooting took place in MS having IBA (Anand and Bir 1984). There are comparable reports for *D. latifolia* (Sita et al. 1986). The MS and B5 nutrient media were found to be most appropriate in the culturing of nodal explants taken from ramets of a 60-year-old *D. sissoo* trees and media combinations supplemented with BAP (0.10–1 mg/l) alone as well as in combinations with IAA or NAA (0.10–0.5 mg/l) showed 100% bud breaks (Joshi et al. 2003).

3.7.4 Mahua (*Madhuca longifolia*) and Sterculia

Propagation of *Madhuca longifolia* carried out on MS medium enriched with IBA or NAA through apical and auxiliary meristems taken from 10 days old seedlings induced shoot multiplication and root induction (Rout and Das, 1993). Large scale production of *Sterculia* was realized through cotyledonary nodal segments inoculated on MS medium coupled with BAP and IBA (Purohit and Dave 1996). The study also reported improved rooting response and minimized callusing with this treatment. Jagatram et al. (2003) reported that in *M. longifolia*, nodal segments exhibited superiority over shoot tips in terms of per cent culture response. In the case of nodal segments, the MS medium having BAP showed maximum (79.25%) culture response while MS medium enriched with activated charcoal proved to be the best for shoot bud proliferation due to its activity on absorption of phenolic substances (Jagatram et al. 2003). Cytokinins were known to promote cell division and exerted other growth regulating function (Mothes 1967).

3.7.5 Teak (*Tectona grandis*) and Terminalia

Axillary meristem explants of teak were subjected to sterilization by dipping in 0.1% $HgCl_2$ for 10 min (Khatri et al. 2001). Studies on micropropagation in teak revealed that nodal explants of old trees grown on MS media enriched with BAP and NAA were capable of in vitro bud break and shoot multiplication (Gupta et al. 1980). The combination of 6-Benzylaminopurine and kinetin was superior in terms of percent establishment (62.50%) in teak (Khatri et al. (2001). It was discovered that teak shoots grown in vitro could successfully root when treated with IAA and IBA on ½ strength MS media. Parthiban (2001) observed maximum shoot induction and shoot numbers in the MS medium encompassing BAP and Kinetin in *Tectona grandis*. For the production of numerous shoots, *Terminalia arjuna* nodal explants (80% culture response) outperformed terminal buds. BA concentration of 0.5 mg/l was found to be ideal for establishing a maximal culture. On WPM containing BA, nodal explants developed more multiple shoots than terminal buds. Concentration of IAA at the tune of 1 mg/l was shown to be superior to IBA and NAA for rooting of numerous shoots in *Terminalia arjuna* (Thomas et al. 2003).

3.7.6 Pterocarpus

Increasing the concentration of the sterilants, viz., $HgCl_2$ and NaOCl, there was an increase in the drying percentage and it was 100% with 0.4% $HgCl_2$ because of the scorching effect while the contamination and survival percentage was zero at 0.4%. This may be the reason for poor survivability at higher concentrations. Higher concentration of BAP and kinetin reduced the multiplication rate of Pterocarpus marsupium, whereas amalgamation of lower amount of IAA or NAA along with BAP initiated shoot bud initiation (Anis et al. 2005). According to the literatures, the level of heterogeneity in discovered CPTs provided greater possibility for subsequent genetic enhancement of dye yielding species including Bixa, Pterocarpus, Teak, Eucalyptus, Dalbergia, Terminalia, Sterculia, Neem, etc. The identified superior CPTs with higher dye content might be supported for further improvement, and their widespread adoption among commercial growers would be extremely beneficial in meeting the demand for edible and non-edible colouring.

3.8 Conclusion

Natural dyes are non-poisonous, non-carcinogenic, less harmful to health, and less polluting. Furthermore, they have soothing colours that are soft, delicate, and harmonious, which has a calming impact. Most importantly, they may be recycled after use and are safe for the environment. Micropropagation technique of biotechnology plays important roles in the cultivation and improvement of dye-yielding tree species, such as those used for natural dyes like indigo, henna, and other plant-based colorants. Tissue culture enables rapid plant propagation using a little amount of tissue, such as shoot tips or meristems. This is critical for ensuring genetic purity and growing huge quantities of homogenous plants. In the case of dye-yielding trees, tissue culture can assure that favourable dye production features are regularly reproduced. Biotechnological techniques such as genetic engineering can be used to improve the dye-yielding capacity of tree species or alter the metabolic pathways to improve dye yield and quality. It may help in preservation of genetic variety and minimization of the loss of key genetic resources. Besides these, identification of novel dye compounds having commercial significance may be carried out through metabolomics and genetic profiling. Hence, tissue culture and biotechnology are invaluable instruments for the long-term cultivation, genetic enhancement, conservation, and commercialization of dye-producing tree species. They provide a solution to address rising demand for natural dyes while maintaining environmental sustainability and economic viability.

References

Anand M, Bir SS (1984) Organogenesis differentiation in tissue cultures of *Dalbergia lanceolata* L. Curr Sci 53:1305–1307

Anis M, Husain MK, Shahzad A (2005) *In vitro* plantlet regeneration of *Pterocarpus marsupium* Roxb., an endangered leguminous tree. Curr Sci 88(6):861–863

Arya S, Arya ID (1998) Micropropagation of neem (*Azadirachta indica*) by adventitious organogenesis. Ann For 6(2):123–129

Bastiaens L (1983) Endogenous bacteria in plants and their implications in tissue culture, a review. Med Fac Landbouw Rijkuniversiteit Genet 48:1–11

Bisht P, Joshi I, Chaunan JMS, Sharma VK (2000) *In vitro* clonal propagation of mature *Eucalyptus* F1 hybrid FRI-5 (*E. camaldulensis* Dehn. x *E. tereticornis* Sm.). Indian J For 23(1):28–32

Bonga JM, Durzan DJ (1987) Cell and tissue culture in forestry, vol 1, 2 & 3. Martinas Nijhoff, The Hague

Chauhan JMS, Bisht P, Kapoor ML, Rawat MS (1996) *In vitro* clonal propagation of *Eucalyptus* hybrid FRI-4 (*Etereticornis* Sm. x *E. camaldulensis* Dehn.). Ann For 4(2):186–191

Chu CC (1978) The MS medium and its application to anther culture of cereal crops. In: Proceedings of symposium on plant tissue culture. Science Press, Perkins, pp 43–50

D'Souza MC, Sharon M (2001) *In vitro* clonal propagation of annatto (*Bixa orellana*). In vitro Cell Dev Plant 37:168–172

Gamborg OL (1984) Cell cultures: nutrition and media. In: Cell culture and somatic cell genetics of plants, vol II. Academic, New York, pp 18–26

Gamborg OLC, Miller RA, Ojima K (1968) Nutrient requirements of suspension cultures of soybean root cells. Exp Cell Res 50:151–158

George I, Kulkarni AR (1997) Micropropagation of neem – *Azadirachta indica* A. Juss. Ann For 5(1):50–54

George EF, Sherrington PD (1984) Plant propagation by tissue culture. Handbook and Directory of Commercial Laboratories, Besignstoke

Gupta PK, Nadgir AL, Mascarenhas AF, Jagannathan V (1980) Tissue culture of forest trees -clonal multiplication of *Tectona grandis* (teak) by tissue culture. Plant Sci Lett 17:259–268

Gupta PK, Mehta UJ, Mascarenhas UF (1983) A tissue culture method of rapid propagation of mature trees of *Eucalyptus torelliana* and *Eucalyptus camaldulensis*. Plant Cell Rep 2:296–299

Hasegawa PM (1980) Factors affecting shoot and root initiation from cultured rose shoot tips. J Am Soc Hortic Sci 105:216–220

Hendre RR (1991) Nutritional requirements of plants tissue cultures. In: Mascacentron AF (ed) Handbook of plant tissue culture. ICAR, New Delhi, pp 26–32

Jagatram M, Surendran C, Paramathma M, Parthiban KT, Sasikumar K (2003) Micropropagation of *Madhuca latifolia*. Indian J For 26(4):445–448

Jonard R, Hugard JJ, Macheix J, Martinez L, Mosella-Chancel JL, Villemur P (1983) *In vitro* micrografting and its application to fruit science. Sci Hortic 20:147–159

Joshi IB, Sharma P, Uniyal VK (2003) Studies on effect of nutrient media for clonal propagation of superior phenotypes of *Dalbergia sissoo* Roxb. through tissue culture. Silvae Genetica 52(3/4):143–147

Kala S, Kumaran K, Reeja S, Vennila S, Subbulakshmi V, Smitha GN (2022) Micropropagation studies in annatto (*Bixa orellana* L). In: Choudhary M et al (eds) Innovative biotechnological approaches in plant sciences. Vital Biotech Publications-An International Publishers. ISBN:978-933-92953-29-3

Kao KN (1977) Chromosomal behaviour in somatic hybrids of soybean—*Nicotiana glauca*. Mol Gen Genet 100:225–230

Kedharnath S (1982) Genetic variation and heritability of juvenile height growth of *Eucalyptus grandis*. J Tree Sci 1(1–2):46–49

Khatri JH, Kukadia MV, Singh RR (2001) Micropropagation of teak (*Tectona grandis* Linn.). Indian J For 24(3):368–371

Kitto SL, Young MJ (1981) *In vitro* propagation of *Carrizo citrange*. Hort Sci 16:305–306

Lloyd G, McCown B (1980) Commercial feasible, micropropagation of mountain laurel (*Kamila latifolia*) by use of shoot tip culture. Proc Int Pt Prog Soc 30:421–427

Mascarenhas AF, Muralidharan EM (1989) Tissue culture of forest trees in India. Curr Sci 58:606–613

Mathur RS, Sharma KK, Rawat MMS (1984) Germination behaviour of various provenances of *Acacia nilotica* spp *indica*. Indian For 111:435–449

Mercadante AZ, Pfander H (1998) Carotenoids from annatto: a review. Recent research developments in agriculture and food chemistry 2(1):79–91

Mothes H (1967) Hormonal control of root growth. Plant Cell Physiol 8:132–137

Murashige T, Skoog F (1962) A revised medium for rapid growth and bioassays with tobacco tissue culture. Physiol Plant 15:473–497

Nitsch JP, Nitsch G (1969) Haploid plants from pollen grains. Science 163:85–87

Parimalan R, Giridhar P, Gururaj HB, Ravishankar GA (2008) Mass multiplication of *Bixa orellana* L, through tissue culture for commercial propagation. Ind Crop Prod 28:122–127

Parthiban KT (2001) Seed source variations, molecular characterization and clonal propagation in teak (*Tectona grandis* Linn f.). Ph.D. Thesis, Tamil Nadu Agricultural University, Coimbatore

Parthiban KT, Vanangamudi M, Surendran C, Balaji S (1999) Callogenesis and organogenesis in *Eucalyptus tereticornis*. Indian J For 22(2):101–105

Parthiban KT, Surendran C, Balaji S (2000) Tissue culture techniques for tree species. In: Surendran C, Parthiban KT, Vanangamudi K, Balaji S (eds) Vegetative propagation of trees—principles and practices. Kaisar Graphics Ltd, Coimbatore, pp 110–124

Purohit SD, Dave A (1996) Micropropagation of *Sterculia urens* Roxb. – an endangered tree species. Plant Cell Rep 15:704–706

Quak F (1977) Meristem culture and the virus free plants. In: Reinert J, YPS B (eds) Applied and fundamental aspects of plant cell, tissue and organ culture, pp 598–615

Rao KS (1986) Plantlets from somatic callus tissues of the Indian rosewood (*Dalbergia latifolia* Roxb.). Plant Cell Rep 5:199–202

Rao GMR, Basavaraja TB, Joshi S (2000) Sel-1: an early flowering variant in seedling progeny of *Simarouba glauca*. Indian J For 23(4):444–445

Reddy MM, Subramanian S (1998) Bacterial contamination in micropropagation of matured sandal (*Santalum album* L.) plus tree. Indian J For 21(2):108–110

Rout GR, Das P (1993) Micropropagation of *Madhuca longifolia* (Koenig) MacBride var. *latifolia* Roxb. Plant Cell Rep 13(9):513–516

Sha-Valli-Khan PS, Prakash E, Rao KR (2002) Callus induction and plantlet regeneration in Bixa orellana L., an Annatto-yielding tree. In Vitro Cell Dev Biol Plant 38:186–190

Sheikh AE, Abhilash PC (2016) Exploring marginal and degraded lands for biomass and bioenergy production: an Indian scenario. Renew Sust Energ Rev 54:1537–1551

Sita GL, Chattopadhyay S, Tejavathi DB (1986) Plant regeneration from shoot callus of rosewood. Plant Cell Rep 5:266–268

Siva R (2003) Assessment of genetic variations in some dye yielding plants using isozyme data. Ph.D. Thesis, Bharathidasan University, Thiruchirappalli, India

Sood A, Palni LMS, Sharma M, Sharma OP (1994) Micropropagation of *Dendrocalamus hamiltonii*. Munro using single node cuttings taken from elite seedling plants. In: Yeoman MM (ed) Proc. 4th international bamboo workshop in Asia and the Pacific Chiangima, Thailand. Plant cell culture technology. Blackwell Scientific Publication, Melbourne, p 33

Sumana KR, Kaveriappa KM, Krishna Bhat CH (1998) Micropropagation of *Acacia mangium* Willd., a commercially important tree species. Indian J For 21(3):181–184

Tewari SK, Tewari L, Lavania S, Garg AK (1993) Callus induction from different explants of *Dalbergia sissoo*. J Tree Sci 12(1):55–56

Thomas TV, Rema Shree AB, Nabeesa E, Neelakandan N, Nandakumar S (2003) *In vitro* propagation of *Terminalia arjuna* Roxb.: a multipurpose tree. Plant cell biotechnology and molecular biology 4(1&2):95–98

Thorpe TA, Harry IS, Kumar PP (1991) Application of micropropagation to forestry. In: Debergh PC, Zimmerman RH (eds) Micropropagation. Kulwer Academic Publishers, Dordrecht, pp 311–336

Tiwari DN (1992) A monograph on teak. International Distributors, Dehra Dun

Vakshasya RK, Rajora OP, Rawat MMS (1992) Seed and seedling traits of *Dalbergia sissoo* Roxb. Seed source variation studies among ten sources in India. For Ecol Manag 48:265–275

Vennila S, Kala S, Fernandaz C (2023) Scope for adaptation of potential natural dye yielding woody perennials on farm bunds and marginal lands. Kisan World 50(1):36–38

Young PM, Hutchins AS, Canfield ML (1984) Use of antibiotics to control bacteria in shoot cultures of woody plants. Plant Sci Lett 34:203–209

Zobel BJ, Talbert J (1984) Applied tree improvement. Wiley, New York, p 503

Tissue Culture of Threatened African Hardwood Tree Species

4

Priscilla Kimani, Constantin Dushimimana, and Joseph Machua

Abstract

Many African hardwood tree species provide valuable products and services. However, compared to introduced conifers, they are considered slow-growing. Hence, natural regeneration alone cannot sustain their production, necessitating the development of managed tree stands that can continually feed the ever-increasing demand. To generate adequate planting stocks, propagation difficulties including low fecundity, slow fruit and seed development, low seed viability, seed dormancy, and recalcitrance to vegetative propagation such as rooting of cuttings and grafting need to be overcome. Tissue culture offers a solution to propagation difficulties through rapid production of quality germplasm and a high multiplication rate of selected clones. During tissue culture, in-vitro rooting and subsequent acclimatization are the most challenging stages, resulting in poor survival rates at the protocol development stage in some instances. Results of some numerous successful cases show that the massive demand for commercially crucial African tree species could potentially be met through tissue culture. The present chapter discusses the utilization of tissue culture techniques in the multiplication of African tropical hardwood tree species including methods that enhance success, challenges encountered, and potential for augmentation of these applications for improved productivity at a time of increasing demand for tree products and services.

P. Kimani (✉) · J. Machua
Department of Biosciences, Kenya Forestry Research Institute, Nairobi, Kenya
e-mail: pkimani@kefri.org

C. Dushimimana
Independent Consultant, Kigali, Rwanda

R. K. Kalia, R. Pathak (eds.), *Tree Biology and Biotechnology*,
https://doi.org/10.1007/978-981-96-0002-1_4

Keywords

African tree species · Threatened hardwoods · Conservation · Tissue culture

4.1 Introduction

Africa is home to iconic tropical timber species that have long been traded in domestic and international markets for their superior wood qualities (Jenkins et al. 2002; Uzu et al. 2022). Apart from the wood, these tree species provide non-wood products, the most important of which probably remain the bioactive phytochemical extracts used in industrial, pharmaceutical, agricultural, and cosmetics industries. Natural gums and resins, essential oils, natural dyes, rubber, and a myriad of complex aromatic compounds extracted from these trees have found diverse applications in the production and preservation of human food, manufacture of human and veterinary medicines, insecticides, pesticides, and herbicides, all due to their antimicrobial, antiparasitic, antibiotic, antioxidant, and anti-inflammatory properties among others (Bolouri et al. 2022). These species also provide aesthetic, ecological, and environmental services ranging from shade, wildlife habitats, climate regulation, soil nutrition and stabilization, and socio-economic activities to local communities (Belsky et al. 1989).

Demand for the products of these species has led to excessive, often illegal, extraction from their natural habitats, leaving them threatened and endangered (World Conservation Monitoring Centre 1998; Barstow 2020). Many tropical hardwood trees have very long generational cycles, taking over 50 years to reach harvestable maturity. Reproductive maturity equally takes time, and so do the seed maturation periods. The seeds frequently experience dormancy, low viability, and germination challenges, requiring elaborate pre-treatment techniques to enhance seedling recovery. Vegetative multiplication approaches, for example, rooting and grafting of cuttings often result in low success rates (Amri et al. 2009; Ky-Dembele et al. 2011). Natural regeneration of these precious hardwoods cannot keep up with the high deforestation rates, which are aggravated by the conversion of natural landscapes to agriculture and settlements.

Tissue culture is an essential alternative propagation technique that can provide large quantities of regenerative material, especially for species with difficulties in seed availability and propagation. Some applications of tissue culture include rapid multiplication of selected germplasm from a limited stock source through adventitious shoot initiation and multiplication, shoot tip and nodal culture for clonal propagation, callus culture, in-vitro synthesis of bioactive compounds from secondary metabolites, genetic transformation, micrografting, protoplast culture, somatic embryogenesis, and cryopreservation of in-vitro germplasm (FAO 2011; Pijut et al. 2012). Although tissue culture technology is still largely underdeveloped in Africa for large-scale forest regeneration (FAO 2011), efforts have been made to determine the amenability of some economically important tree species to the technique. This chapter reviews the utilization of tissue culture techniques in multiplication of

African tropical hardwood tree species including methods that enhance success, challenges encountered, and potential for augmentation of these applications for improved productivity at the time of increasing demand for tree products and services. Case studies are given on *Allanblackia stuhlmannii, Melia volkensii, Khaya senegalensis, Dalbergia melanoxylon* and *Balanites aegyptiaca,* being renowned African tree species of high economic potential.

4.1.1 *Allanblackia stuhlmannii*

Allanblackia stuhlmannii, of the Clusiaceae family, is an endangered edible-oil-producing tree native only to Tanzania. It grows well at altitudes of 500–1600 meters above sea level, maximum annual temperatures of 25–33 °C and average annual rainfall of 1200–2400 mm. The tree can grow up to a height of 45–50 m at maturity and possesses large, brown fruits which contain the oil-producing seeds (Alvum-Toll 2013; Crockett 2015). *A. stuhlmanni* seed oil is edible. It remains solid at room temperature owing to a high melting point of ~34 °C and contains high contents of stearic and oleic acids, which makes it a more nutritiously and functionally preferred oil for human consumption compared to oils with high contents of palmitic, myristic and/or lauric acids which have been shown to affect blood cholesterol levels (Crockett 2015). The oil is traditionally used as cooking oil and substitute for butter in spreads, as traditional medicine for coughs, chest pain, rheumatism, sore joints, wounds, hypertension, and impotence. More recently, its oil has been utilized for soap manufacturing and also in the food and cosmetics industries. The protein-rich seed cake that remains after extraction of the seed oil is used as animal feed (Crockett 2015). The increasing demand for seed oil and other *A. stuhlmannii* tree products such as fuelwood, construction timber, medicinal extracts from the roots, bark and leaves, and animal fodder from the leaves led to overexploitation of the wild populations of this species, leaving it vulnerable to genetic erosion, which may consequently diminish the quality of harvested products. However, there have recently been successful efforts to domesticate the species for increased production and commercialization and to protect the wild populations, with manufacturing companies and non-governmental stakeholders supporting farmers and farmer groups to grow and market the trees (Alvum-Toll 2013).

Naturally, *A. stuhlmannii* trees start flowering at the age of 15–20 years, and fruiting commences one year thereafter (Mathew et al. 2009; Munjuga et al. 2013; Alvum-Toll 2013). However, full fruit production occurs after 40–50 years (Munjuga et al. 2013). Poor seed germination, a long juvenile phase, slow natural regeneration, and a general recalcitrance of the species to vegetative propagation methods reduce the available planting stock and impede commercial utilization (Neondo et al. 2011; Munjuga et al. 2013). To enhance and accelerate germplasm production, Neondo et al. (2011) evaluated the amenability of *A. stuhlmannii* to micropropagation and reported promising results for explant sterilization, direct shoot induction, and callus induction. Further investigations on explant sterilization and callus induction were reported by Kahia et al. (2015). Both studies could not record the recovery of rooted plantlets and ended at the preliminary stages. Besides these two reports, we could not get any other publication on micropropagation of *A. stuhlmannii.*

4.1.2 *Melia volkensii*

Melia volkensii, of the Meliaceae family, is an endemic tree species to East Africa (Kenya, Somalia, Tanzania and Ethiopia) having drought tolerant, rapid growth potential (Stewart and Blomley 1994; Orwa et al. 2009). The tree can be harvested for its hardwood, termite-resistant timber in a short period of 10–15 years in some environments (Kariuki et al. 2012). *M. volkensii* is popular for its attractive, easily-worked and durable wood used in construction and in tool and furniture making. It is a multipurpose tree that provides fuelwood, leaf fodder for livestock, shade in the drylands, and bee forage for honey production (Ndathi et al. 2012). Its leaf extracts, fruit pulp, seeds, roots, and bark have medicinal properties and are used traditionally to cure pain, diarrhea, skin rashes, and eczema owing to their antifungal, antibacterial, and antiplasmodial properties (Mwangi and Mukiama 1988; Mwangi and Rembold 1988; Kamau et al. 2016). The extracts also contain active biological compounds that exhibit pesticidal and insecticidal activity, making them potentially useful for pest control in agricultural systems (Jaoko et al. 2020, 2021). *M. volkensii* enhances the soils through mulching, intercropping, soil stability, and erosion control (Githae and Mutiga 2021).

Despite its desirable qualities and potential for commercialization in the timber industry, the propagation of *M. volkensii* faces significant challenges. The tree reaches reproductive maturity in 2–3 years (Orwa et al. 2009). The seeds thereafter take 11–13 months to mature (Milimo 1986). At maturity, the seeds are encased in a woody nut that requires manual cracking and opening to extract the seeds. Cracking nuts is arduous, requiring a significant amount of time and effort and often resulting in the loss of many seeds. As a result, this decreases the amount of seeds that can be collected and the availability of germplasm for planting initiatives. Seed viability is low and extracted seeds are highly susceptible to fungal infection (Maundu and Tengnas 2005). The seeds must undergo elaborate pre-germination treatments to enhance germination percentages (Muok et al. 2010). Vegetative propagation techniques for *M. volkensii* have not been very successful either. Rooting of cuttings was reported to be hindered by heavy fungal contaminations, rotting of the cuttings, and callusing instead of rooting when planted in media treated with auxins (Indieka 2007).

Several studies have attempted tissue culture of *M. volkensii* using different explant sources. Indieka (2007) directly multiplied and rooted young shoot explants and initiated direct somatic embryos from cotyledonary explants. Mulanda et al. (2012) regenerated indirect somatic embryos through mature zygotic embryos and cotyledonary explants; Werbrouck et al. (2017) developed shoots through leaf explants while Dushimimana et al. (2022a, b) regenerated Melia shoots into whole plantlets. Kimani et al. (2024) regenerated immature zygotic embryos from 60-day-old fruits, effectively accelerating germplasm production.

4.1.3 *Khaya senegalensis*

Commonly known as the African mahogany, *K. senegalensis* is a valuable West African timber tree in the family Meliaceae. Its attractive and durable wood is among the hardest African mahoganies and is the hardest among the Khaya species. The species is also used to produce wood fuel (Olagbende et al. 2021), gums, and resins useful in industrial applications as binders, film coatings and drug carriers (Ozoude et al. 2020), leaf fodder for domestic animals, edible seed oil used for cooking in West Africa, and medicinal extracts for treatment of ailments such as malaria, fever, headache, diarrhea, skin conditions, jaundice, allergies, and wounds. Decoctions are used as anthelmintic and laxative (Orwa et al. 2009). Various plant extracts of *K. senegalensis* have also been reported to have insecticidal, pesticidal, and nematicidal activity against crop pests (Bamaiyi et al. 2007; Liman et al. 2010; Yakubu and Nda 2012; Loko et al. 2017; Fabiyi 2022;), with potential for application as biopesticides in agriculture. Because of its many uses and timber trade, the species has largely been overexploited in its natural habitats and has been enlisted as a susceptible species (World Conservation Monitoring Centre 1998).

Natural seed regeneration of *K. senegalensis* is poor owing to the seed's short viability of only 2–3 months under natural storage conditions (Danthu et al. 1999; Ky-Dembele et al. 2011). However, the trees readily produce suckers, making them suitable for reforestation purposes (Orwa et al. 2009). Rooting of cuttings, especially from young juvenile plants has generally been successful (Nikles et al. 2008; Ky-Dembele et al. 2011; Hung and Trueman 2011a; de Vasconcelos et al. 2016). Promising in-vitro propagation has been attempted in several studies (Danthu et al. 2003; Hung and Trueman 2011a, b; Darwesh et al. 2017) with the potential for enhanced germplasm production under improvement programs of the species.

4.1.4 *Dalbergia melanoxylon*

Dalbergia melanoxylon is a tree species in the Fabaceae family. The species occurs in Senegal and Nigeria in West Africa and is widespread in most tropical South-East Africa, including Kenya, Tanzania, Mozambique, Angola, and South Africa (Orwa et al. 2009). It is commonly known as the African blackwood due to the deeply dark color of its heartwood (Alkadri et al. 2020). Its wood is strong and durable and has among the highest densities of all the commercially traded timbers in the world (Alkadri et al. 2020). The wood is fine-textured and attractive, making it famous for high-end furniture, artwork, and musical instruments such as clarinets and guitars (Zadro 1975; Alkadri et al. 2020; Liu et al. 2020). Many active chemical compounds such as neoflavones, flavanones, benzofurans, N-cinnamoyl, quinone pigments, sesquiterpene, and phenanthrenedione have been extracted from *D. melanoxylon* and are possibly responsible for its effective use to cure human ailments, for example, pain, headache, skin conditions, venereal diseases, respiratory diseases, and heart conditions (Yin et al. 2018; Orwa et al. 2009; Nakai and Yoshimura 2020; Xu et al. 2021). The potential cytotoxic activity of the active constituents against colorectal cancer, breast cancer, and melanoma cell lines has also been investigated (Chung 2022). The active extractives have also been associated with the durability and pest and decay resistance of *D. melanoxylon* heartwood (Alkadri et al. 2020; Nakai and Yoshimura 2020).

Its numerous benefits notwithstanding, *D. melanoxylon* is a slow-growing tree, taking between 70 and 100 years to reach maturity (Malimbwi and Luoga 2000). Fruit development after pollination takes 6–8 months, and seed viability is low under natural conditions. Regeneration is usually achieved through seedlings, root suckers, and cuttings (Malimbwi and Luoga 2000; Washa 2021). Over time, the species has been overexploited for its quality timber and the overexploitation has led it to be enlisted as near threatened species (Barstow 2020). The species conservation programs have been initiated, requiring rapid multiplication of germplasm. Successful attempts at micropropagation of *D. melanoxylon* have been reported by Kiondo et al. (2014a, b) and Washa and Nyomora (2014). Successful rooting of cuttings was also reported (Amri et al. 2009).

4.1.5 *Balanites aegyptiaca*

The species also named as the desert date is a small, multi-purpose, and drought-tolerant tree in the family Balanitaceae, or Zygophyllaceae (Abdelaziz et al. 2020). It is widespread in most of the African continent across various ecological conditions and soil types. The species does well in the arid and semi-arid regions. Its wood produces good firewood and charcoal that burn well with little smoke, but the tree is rarely harvested for timber (Orwa et al. 2009). The seeds produce edible cooking oil and biofuel, and the seed cake is fed to livestock as a protein source (Aldeen et al. 2020; Abou-Elfadl et al. 2023; Nahed et al. 2023). The leaves, shoots, and sprouts are used as vegetables and fodder for livestock. The fruits are edible, raw, or ripe in many recipes across their native range (Orwa et al. 2009). The fruits are nutritious and contain health-promoting phytochemicals used to treat diabetes and hypertension (Abdelaziz et al. 2020; Zaky et al. 2022). Indeed, all parts of the tree are reported to contain bioactive medicinal compounds used in treating abdominal pain, skin diseases, asthma, malaria, jaundice, syphilis, bilharzia, epilepsy, and gastric ulcers. Promising anticancer properties have also been reported (Ibrahim et al. 2022; Murthy et al. 2021). The phytochemicals exhibited insecticidal and pesticide activity, particularly against mosquitoes and freshwater snails (Abdelaziz et al. 2020; Murthy et al. 2021).

B. aegyptiaca is a slow-growing tree, reaching reproductive maturity at 5–8 years. The fruits thereafter take about one year to mature (Orwa et al. 2009). The species regenerates naturally through seed and coppices and pollards well (Ndoye et al. 2003). However, the trees are being overharvested due to product demand, and domestication and reforestation are recommended to maintain adequate stocks. Seeds are encased in a nut that requires manual extraction and present physical dormancy requiring pre-treatment for successful germination (Nahed et al. 2023). Alternative vegetative propagation techniques such as rooting of stem cuttings, suckering and air-layering have been trialed, with considerable success rates (Massaoudou et al. 2022). Tissue and callus culture studies and their associated applications for *B. aegyptiaca* have also been reported by several studies (Ndoye et al. 2003; Aldeen et al. 2020; Ishaku et al. 2020; Abdel-Kareem et al. 2021; Abou-Elfadl et al. 2023), particularly in the synthesis of secondary metabolites.

4.2 Tissue Culture of African Hardwood Tree Species

4.2.1 Direct Shoot Initiation and Proliferation

In-vitro shoot multiplication was not established for *A. stuhlmannii.* However, best shoot elongation (1.44 ± 0.066 cm) from shoot explants was recorded on Modified Lloyd and McCown's Woody plant medium enriched by 1.2 mg/l kinetin in a single PGR treatment, while amalgamated PGR treatment of 6 mg/l BAP and 2 mg/l of 2, 4–D was highly successful for shoot elongation at 1.35 ± 0.041 cm after 12 weeks (Neondo et al. 2011).

For *M. volkensii*, Indieka (2007) reported the highest average of only 1.8 shoots per nodal explant inoculated on MS media accompanied with 2 mg/l BAP alone after 30 days of culture. However, the amalgamation of 0.5 mg/l BAP and 0.2 mg/l IBA enhanced the number of shoots up to 5.4 per explants after 30 days. Direct shoot regeneration was documented from mature cotyledons (Mulanda 2016; Dushimimana et al. 2023). Oryzalin application at 10 mg/l to cotyledons promoted the development of polyploidy in *M. volkensii* plants (Dushimimana et al. 2023). The transfer of callus derived from immature fruits to a growth medium without PGRs contributed in the formation of microshoots (Kimani et al. 2024).

In *K. senegalensis*, epicotyls of 5-week-old seedlings were utilized as explants by Darwesh et al. (2017). Explants inoculated on modified MS media containing 0.8 mg/l BAP produced maximum number (3.3) of shootlets per explant after three subcultures, subsequently by 0.6 mg/l BAP, which had also produced higher (12.67) nodes per explant. The length of shootlets decreased as the BAP concentration increased from 0.4 to 1.0 mg/l. Moreover, explants cultured on ½ MS media yielded the maximum number of shootlets as well as number of nodes per explant (12.11 and 38.11, respectively). In contrast, full-strength WPM and B5 media yielded a lesser number of shoots (5.56 and 4.22, respectively) and number of nodes (12.67 and 9.44 respectively) per explant (Darwesh et al. 2017). The maximum number of shoots (2.6 ± 0.2) was obtained on MS media enriched with 8.9 μM BAP + 0.3 μM NAA after four-weeks of culture (Hung and Trueman 2011a). Cumulatively, media with 4.4 μM BAP alone had the highest shoot proliferation (52.9 ± 4.9 shoots per clone) over a 15-weeks culture period (Darwesh et al., 2017). Danthu et al. (2003) reported best shoot proliferation for *K. senegalensis* on to MS media complemented with 2.2 μM BAP + 0.26 μM IBA (2.3 ± 0.4 shoots), or with 8.8 μM BAP alone (2.4 ± 0.5 shoots) after 1 month in culture (Danthu et al. 2003).

In *D. melanoxylon*, MS media supplemented with 1.5 mg/l BAP alone gave maximum shoots per explant (3.56 ± 0.19) of all single BAP treatments, while the mixture of 2 mg/l BAP + 1 mg/l NAA gave the highest number (6.12 ± 0.13) of shoots per explant among all PGR treatments after 4 weeks in culture (Kiondo et al. 2014a). More shoot length (5.84 ± 0.08 cm) was realized on media containing 2 mg/l BAP + 0.25 mg/l NAA. In B. *aegyptiaca*, MS media supplemented with 5 mg/l BAP + 0.1 mg/l NAA induced more number of shoots (4.30) per nodal explant after 4 weeks in culture, whereas, 2.5 mg/l BAP + 0.1 mg/l NAA gave a combination of high shoot multiplication with long shoots (Ndoye et al. 2003).

4.2.2 Callus Induction

Callus induction is the most important method practiced in the field of plant tissue culture for several objectives, including large-scale proliferation, genetic improvements, and preservation of edible crops and hardwood species. The process of callus induction in African hardwoods is affected by various factors, including explant type, species, composition of the growth media, and experimental surroundings (Long et al. 2022).

Limited studies have been conducted to induce callus in African hardwood species. Attempt has been reported for callus formation using leaf explants of *Allanblackia stuhlmannii* cultured into MS medium having TDZ (Munjuga et al. 2015). Further studies on callus induction are needed to enhance micropropagation protocol of *Allanblackia.* Callus induction in *Melia volkensii* was effectively achieved using various explants, for example, leaves, roots, cotyledons, rachis, and petiole. The leaves were inoculated in B5 medium coupled either with 0.5 mg/l of BAP or a combination of 2 mg/l of kinetin and 0.05 mg/l of 1-Naphthaleneacetic acid and callusing was reported after 10 days in the culture medium (Mulanda 2016). The cotyledon culture in MS medium enriched with TDZ or 2,4-D formed either soft white yellow or compact green callus (Dushimimana 2018). Thidiazuron was the most successful hormone in induction of *M. volkensii* callus from cotyledon explants (Mulanda et al. 2012; Mulanda 2016). *Melia volkensii* explants induce callus quickly and easily, even during multiplication and rooting. Callus induction from young fruits has been observed in media containing BAP, TDZ, IAA, or their mixtures (Kimani et al. 2024). Callus induction of *Balanites aegyptiaca* was achieved by culturing root and fruit explants in MS medium having 2.5 mg/l NAA and 2 mg/l 2,4-D over 3 weeks (Gour and Kant 2006). Callus has been successfully generated from apical buds, cotyledon portions, and root segments and juvenile thorn explants grown on MS media enriched with 2.23 μM 2,4-D (Gour et al. 2007). Higher amount of TDZ was reported to induce callus at the base of the shoot (Khamis et al. 2016). Callus formation from mature cotyledons of *B. aegyptiaca*, buds and roots section was achieved after 8 weeks in MS medium having 2 μM 2,4-D (Gour et al. 2007). While, Ishaku et al. (2020) observed that the *B. aegyptiaca* callus induction from cotyledon is successful using MS plus 0.5 mg/l NAA, 1 mg/l 2,4-D and 0.5 mg/l of BAP.

4.2.3 Rooting of African Hardwood Species

Successful rooting of *Allanblackia* single-node foliage and stem cuttings was achieved after 38 days through treatment with 50 µg of auxins (IBA, IAA and NAA) dissolved in 10 µl of alcohol (Atangana et al. 2006). Generally, information regarding the process of in-vitro rooting of *Allanblackia* is scanty. *K. senegalensis* microshoots were reported to induce roots after 4 weeks when inoculated on half-strength MS supplemented with 19.6 µM IBA (Hung and Trueman 2011a, b). Pre-conditioning of *K. senegalensis* plantlets with 4-mg/l IBA in darkness for a duration of 1 week and subsequently transferring them onto a culture medium of half-strength WPM or MS was reported to be an efficient method for encouraging rooting (Darwesh et al. 2017). Danthu et al. (2003) achieved rooting of *K. senegalensis* shoots on a half-strength medium with a low concentration of IBA.

Rooting in *M. volkensii* plantlets were successful using modified WPM with half-strength salt enriched with 2 µM IBA and 0.02 M silver thiosulfate (Dushimimana et al. 2022a, b). In contrast, during rooting of *M. volkensii* micro-shoots cultured full- or half-strength MS medium supplemented with 0.1–2 mg/l IBA reported to produce abundant callus at the base of the shoot but in some cases root callused (Indieka et al. 2007; Mulanda 2016; Werbrouck et al. 2017). Rooting *D. melanoxylon* shoots was reported to be efficient when MS medium was added with 0.1–1 mg/l NAA compared to the media with IBA (Kiondo et al. 2014b). Amri et al. (2009) found that *D. melanoxylon* cuttings treated with IBA solution showed excellent results, with a rooting percentage of 71% and longer roots were observed as compared to the untreated cuttings. The better rooting of *B. aegyptiaca* shoots was attained on half-strength MS medium added with 1.0 µM IBA (Anis et al. 2010). Shoots of *B. aegyptiaca* inoculated on MS medium containing a combination of IBA and NAA were reported to have comparable rooting results to the shoots inoculated on MS media having 2 mg/l IBA (Mohamed and Bekhit 2007). Shoots of *B. aegyptiaca* successfully rooted well after 4 weeks, when cultured onto Gamborg (B5) medium supplemented with 9.84 µM IBA (Gour et al. 2007). Rooting of two *B. aegyptiaca* shoots from two provenances in Egypt was achieved with a maximum rate of 60% on MS media having 1.2–4.8 µM IBA (Khamis et al. 2016).

4.2.4 Somatic Embryogenesis

Induction of somatic embryogenesis in Allanblackia stuhlmannii was reported using leaf explants inoculated into MS media enhanced with TDZ (Munjuga et al. 2015). While Neondo et al. (2011) could not obtain somatic embryos from callus cultured into WPM containing Gamborg's vitamins. *Melia volkensii* SE induction was realized from cotyledons and mature zygotic embryos cultured into MS medium having 0.5 mg/l BAP and 0.2 mg/l 2,4-D (Indieka et al. 2007). The germination of somatic embryos of *M. volkensii* was also reported on MS-free medium (Werbrouck et al. 2017). Sub-culturing callus regenerated through zygotic embryos inoculated on Gamborg's B5 media enriched with 4 mg/l TDZ into hormone-free B5 medium produced multiple somatic embryos (Mulanda et al. 2012). Somatic embryo regeneration from callus of immature zygotic *M. volkensii* embryos was reported as an alternative to accelerated breeding program by avoiding protracted juvenile period and slow seed maturity (Kimani et al. 2024). The somatic embryo regeneration from immature embryos cultured into media with 0.25 mg/l TDZ could be realized in the study. *B. aegyptiaca* root explants cultured on semisolid MS media supplemented with 2.5 mg/l 2,4-D and 0.5 mg/l meta-topolin induced embryogenic callus (Saharan et al. 2011).

4.2.5 Inhibitory Compounds

During the tissue culture process, production and oxidation of phenolic compounds occur, hindering the action of PGRs and subsequent browning of explants. Exogenous substances are added to the growth media to hinder excess production and oxidation of the phenolic compounds. During the process, the oxidation of phenolic chemicals is reduced, which affects the shoot regeneration. The commonly used inhibitory compounds are 2-aminoindane2-phosphonic acid (Alagarsamy et al. 2018), activated charcoal (Malik et al. 2005; Thomas 2008; Mona 2012), ascorbic acid (Elmore et al. 1990; Chikezie 2012), silver nitrate and its nanoparticles (Mahendran et al. 2019). Based on inhibitory processes Permadi et al. (2024) grouped inhibitory substances into six categories viz., antioxidants, acidulants, adsorbents, chelating agents, nanoparticles, and mixed-type inhibitors.

Activated charcoal added into *M. volkensii* rooting medium was reported to be effective in reducing shoots callusing and defoliation (Mulanda 2016). The addition of 10% coconut water into medium having 0.1 mg/l BAP enhanced shoot regeneration from somatic embryos (Mulanda et al. 2012). Silver thiosulfate containing rooting media increased *M. volkensii* shoot rooting up to 100% and lowered callusing (Dushimimana et al. 2022a). Inclusion of activated charcoal (0.5%) to MS medium supplemented with 1 μM IBA influenced the rooting of *B. aegyptiaca* microshoots positively (Siddique and Anis 2009).

4.3 Conclusions

This review evaluated the use of tissue culture methods in propagating five threatened and hard-to-propagate African hardwood tree species and revealed various levels of success. The review did not find any uniformity in the methods or protocols used to regenerate plantlets either through induction of direct organogenesis, callus induction or somatic embryogenesis among the species under review. In all the five threatened hardwood species, the review did not find any successful programs of cultivation of the said species with the materials generated from tissue culture. This indicates that tissue culture programs for these species are still at the developmental stages. Nevertheless, the use of biotechnological approaches remains an important alternative for the mass production and conservation of threatened African hardwood tree species. The results from the studies reviewed contain very useful information that can be utilized by breeding programs and conservation strategies for the threatened African hardwood tree species.

References

Abdelaziz SM, Lemine FMM, Tfeil HO, Filali-Maltouf A, Boukhary AOMS (2020) Phytochemicals, antioxidant activity and ethnobotanical uses of *Balanites aegyptiaca* (L.) Del fruits from the arid zone of Mauritania, Northwest Africa. Plants 9:401

Abdel-Kareem AAA, El-Shamy HA, Dawh AK, Gwiefel SG (2021) Influence of auxin on *in vitro* secondary metabolites production from *Balanites aegyptiaca* callus cultures. Plant Arch 21(2):89–93

Abou-Elfadl RE, Ahmad ME, Abdel-Magiud MR, Mahdi AA (2023) Comparative analysis of bioactive compounds of Balanites aegyptiaca L Callus. Egypt J Chem 66(3):339–349

Alagarsamy K, Shamala LF, Wei S (2018) Influence of media supplements on inhibition of oxidative browning and bacterial endophytes of *Camellia sinensis* var. Dermatol Sin 3 Biotech 8:356

Aldeen AMT, El-Sadek ME, Abdul-hakim AS, Hussein EA, Al-kaf AG (2020) Saponins, glycosides and flavonoids in cells and tissues of *Balanites aegyptiaca* cultured on solid and liquid culture media. Univers J Pharm Res 5(4):16–20

Alkadri A, Jullien D, Arnould O, Rosenkrantz E, Langbour P, Hovasse L, Gril J (2020) Hygromechanical properties of grenadilla wood (*Dalbergia melanoxylon*). Wood Sci Technol, Springer 54:1269–1297

Alvum-Toll K (2013) *Allanblackia stuhlmannii*—a tree under current domestication: what are the soil requirements? Master's Thesis, Swedish University of Agricultural Sciences

Amri E, Lyaruu HVM, Nyomora AS, Kanyeka ZL (2009) Vegetative propagation of African Blackwood (*Dalbergia melanoxylon* Guill. & Perr.): effects of age of donor plant, IBA treatment and cutting position on rooting ability of stem cuttings. New For 39:183–194

Anis M, Varshney A, Siddique I (2010) *In vitro* clonal propagation of *Balanites aegyptiaca* (L.) Del. Agroforest Syst 78(2):151–158

Atangana AR, Tchoundjeu Z, Asaah EK, Simons AJ, Khasa DP (2006) Domestication of *Allanblackia floribunda*: amenability to vegetative propagation. For Ecol Manag 237(1–3):246–251

Bamaiyi LJ, Adams LS, Toro WA, Odekina S (2007) Laboratory evaluation of mahogany (*Khaya senegalensis* (Desv.) seed oil and powder for the control of *Callosobruchus maculatus* (Fab.) (Coleoptera: Bruchidae) on stored cowpea. J Entomol 4(3):237–242

Barstow M (2020) *Dalbergia melanoxylon*. The IUCN red list of threatened species 2020: e.T32504A67798379

Belsky AJ, Amundson RG, Duxbury JM, Riha SJ, Ali AR, Mwonga SM (1989) The effects of trees on their physical, chemical, and biological environments in a semi-arid Savanna in Kenya. J Appl Ecol 26:1005–1024

Bolouri P, Salami R, Kouhi S, Kordi M, Lajayer BA, Hadian J, Astatkie T (2022) Applications of essential oils and plant extracts in different industries. Molecules 27:8999

Chikezie NYU (2012) Effect of ascorbic acid on blackening and sprouting of Musa spp shoot tips. ISABB J Biotechnol Bioinform 2(2):11–17

Chung HS (2022) Cytotoxic neoflavonoids and chalcones from the heartwood of *Dalbergia melanoxylon*. Nat Prod Sci 28(3):115–120

Crockett SL (2015) Allanblackia oil: Phytochemistry and use as a functional food. Int J Mol Sci 16:22333–22349

Danthu P, Gaye A, Sarr A (1999) Long term storage of *Khaya senegalensis* seeds. Int Tree Crops J 10:93–100

Danthu P, Diaite-Sanogo D, Sagna M, Sagna P, Dia-Gassama YK (2003) Micropropagation of *Khaya senegalensis*, an African mahogany from dry tropical zones. J Trop For Sci 15(1):164–175

Darwesh MA, Naser AA, Habba EEA, Taha LS, Gabr AMM, El-Assaly RMB (2017) *In vitro* propagation protocol for improving African mahogany (*Khaya Senegalensis*) endangered tree. J Biol Sci 17(5):235–246

de Vasconcelos RT, Valeri SV, Martins ABG, Biagiotti G, Perez BAP (2016) Rooting of African mahogany (*Khaya senegalensis* A. Juss.) leafy stem cuttings under different concentrations of indole-3-butyric acid. Afr J Agric Res 11(23):2050–2057

Dushimimana C (2018) Development of *in vitro* multiplication and acclimatization protocols for *Melia volkensii*. MSc Thesis, University of Nairobi

Dushimimana C, Magomere T, Mulatya J, Vandenabeele J, Olubayo F, Smagghe G, Werbrouck SPO (2022a) Variation of morphological traits and quality indices of micropropagated *Melia volkensii* Gürke clones before field planting. Forests 13:337

Dushimimana C, Sakha MA, Korir MJ, Jefwa JM, Vandenabeele J, Magomere T, Mutitu EW, Mulatya J, Olubayo F, Smagghe G, Werbrouck SPO (2022b) Early growth performance of in vitro raised *Melia volkensii* Gürke plantlets in response to beneficial microorganisms under semi-arid conditions. Plan Theory 11:1300

Dushimimana C, Van Laere K, Magomere T, Smagghe G, Werbrouck SPO (2023) Production of tetraploid plants from cotyledons of diploid *Melia volkensii* Gürke. Horticulturae 9:791

Elmore HW, Samples B, Sharma S, Harrison M (1990) Influence of cultural and physiochemical factors on ascorbate stability in plant tissue culture media. Plant Cell Tissue Org Cult 20:131–135

Fabiyi OA (2022) Cytotoxicity and nematicidal potential of leaf extracts of *Adansonia digitata* and *Khaya senegalensis* on root knot nematode (*Meloidogyne incognita*) associated with cabbage. J Agric Scie Sri Lanka 17:3

FAO (2011) Agricultural biotechnologies in developing countries: options and opportunities in crops, forestry, livestock, fisheries and agro-industry to face the challenges of food insecurity and climate change (ABDC-10). Guadalajara, Mexico

Githae EW, Mutiga IM (2021) Ecological restoration of pastoral landscapes in the drylands of East Africa. J Dryland Agric 7:34–41

Gour VS, Kant T (2006) Optimized callus production in *Balanites Aegyptiaca* (L.) Del.: a potential source of diosgenin. J Phytol Res 19(1):53–57

Gour VS, Sharma SK, Emmanuel CJSK, Kant T (2007) A rapid in vitro morphogenesis and acclimatization protocol for *Balanites aegyptiaca* (L) Del – a medicinally important xerophytic tree. J Plant Biochem Biotechnol 16(2):151–153

Hung CD, Trueman SJ (2011a) In vitro propagation of the African mahogany *Khaya senegalensis*. New For 42(1):117–130

Hung CD, Trueman SJ (2011b) Encapsulation technology for short-term preservation and germplasm distribution of the African mahogany *Khaya senegalensis*. Plant Cell Tissue Org Cult 107:397–405

Ibrahim OHM, Al-Qurashi AD, Asiry KA, Mousa MAA, Alhakamy NA, Abo-Elyousr KAM (2022) Investigation of potential *in vitro* anticancer and antimicrobial activities of *Balanites aegyptiaca* (L.) Del fruit extract and its phytochemical components. Plan Theory 11:2621

Indieka AS (2007) Macro- and micro-propagation of *Melia volkensii* Gurke (Meliaceae): an indigenous multipurpose drylands tree species. MSc Thesis, Kenyatta University, pp 87

Indieka SA, Odee DW, Muluvi GM, Rao KN, Machuka J (2007) Regeneration of *Melia volkensii* Gürke (Meliaceae) through direct somatic embryogenesis. New For 34:73–81

Ishaku GA, Haruna A, Kalum AA, Vargas-De-La-Cruz C, Solórzano-Acosta R (2020) Callogenesis and antibacterial activity of *Balanites aegyptiaca*. J Biosci Med 08(10):157–168

Jaoko V, Taning CNT, Backx S, Mulatya J, Van den Abeele J, Magomere T, Olubayo F, Mangelinckx S, Werbrouck SPO, Smagghe G (2020) The phytochemical composition of *Melia volkensii* and its potential for insect pest management. Plants (Basel) 9(2):143

Jaoko V, Taning CNT, Backx S, Motti P, Mulatya J, Vandenabeele J, Magomere T, Olubayo F, Mangelinckx S, Werbrouck SPO, Smagghe G (2021, 1994) Laboratory and greenhouse evaluation of *Melia volkensii* extracts for potency against African sweet potato weevil, *Cylas puncticollis*, and fall armyworm, *Spodoptera frugiperda*. Agronomy 11(10)

Jenkins M, Oldfield S, Aylett T (2002) International trade in African blackwood. Fauna & Flora International, Cambridge

Kahia J, Kouame C, Ofori DA, Jamnadass R (2015) In vitro propagation of *Allanblackia*. In: Anjarwalla P, Musili PM, Gachuiri A, Ofori DA (eds) Allanblackia meeting. Workshop report. World Agroforestry Centre (ICRAF), Nairobi

Kamau RW, Juma BF, Baraza LD (2016) Antimicrobial compounds from root, stem bark and seeds of *Melia volkensii*. Nat Prod Res 30:1984–1987

Kariuki JG, Mulatya J, Kimondo JM, Ogawa S, Muchiri D (2012) Genetic improvement of *Melia volkensii*: evaluation of candidate plus trees. KEFRI

Khamis G, Winkelmann T, Schaarschmidt F, Papenbrock J (2016) Establishment of an in vitro propagation and transformation system of *Balanites aegyptiaca*. Plant Cell Tissue Org Cult 125(3):457–470

Kimani PN, Wamalwa LN, Ngugi K, Omondi SF, Machua JM, Magomere T, Mulatya J, Werbrouck SPO (2024) In vitro regeneration of immature zygotic embryos of *Melia Volkensii* Gürke for accelerated breeding. Plant Cell Tissue Org Cult 156:27

Kiondo F, Feyissa T, Ndakidemi PA, Seth M, Lema K (2014a) Micropropagation of *Dalbergia melanoxylon* Guill. & Perr.: a threatened tree species. Am J Res Commun 2(12):43–52

Kiondo F, Feyissa T, Ndakidemi PA, Seth M (2014b) In vitro propagation of *Dalbergia melanoxylon* Guill. & Perr.: a multipurpose tree. Am J Res Commun 2(11):181–194

Ky-Dembele C, Tigabu M, Bayala J, Savadogo P, Boussim IJ, Oden PC (2011) Clonal propagation of *Khaya senegalensis*: the effects of stem length, leaf area, auxins, smoke solution, and stockplant age. Int J For Res 2011:281269

Liman B, Ibrahim M, Ibrahim NT, Rabah AB (2010) Effect of Mahogany (*Khaya senegalensis* L) leaf extract on root-knot nematode of tomatoes (*Lycopersicum esculentum* L.). Niger J Basic Appl Sci 18(2):272–276

Liu M, Peng L, Lyu S, Lyu J (2020) Properties of common tropical hardwoods for fretboard of string instruments. J Wood Sci 66:14

Loko LY, Alagbe O, Dannon EA, Datinon B, Orobiyi A, Thomas-Odjo A, Dansi A, Tamò M (2017) Repellent effect and insecticidal activities of *Bridelia ferruginea*, *Blighia sapida*, and *Khaya senegalensis* leaves powders and extracts against *Dinoderus porcellus* in infested dried yam chips. Psyche J Entomol 2017:1–18

Long Y, Yang Y, Pan G, Shen Y (2022) New insights into tissue culture plant-regeneration mechanisms. Front Plant Sci 13:926752

Mahendran D, Geetha N, Venkatachalam P (2019) Role of silver nitrate and silver nanoparticles on tissue culture medium and enhanced the plant growth and development. In: Kumar M, Muthusamy A, Kumar V, Bhalla-Sarin N (eds) In vitro plant breeding towards novel agronomic traits: biotic and abiotic stress tolerance. Springer, Singapore, pp 1–273

Malik SK, Chaudhury R, Kalia RK (2005) Rapid in vitro multiplication and conservation of *Garcinia indica*: a tropical medicinal tree species. Sci Hortic 106:539–553

Malimbwi RE, Luoga EJ (2000) Prevalence and standing volume of *Dalbergia melanoxylon* in coastal and inland sites of southern Tanzania. J Trop For Sci 12(2):336–347

Massaoudou M, Habou R, Felix ZCS, Tougiani A, Damme PV (2022) Development of vegetative propagation strategies for *Balanites aegyptiaca* in the Sahel. Niger Int J For Res 2022:5110018

Mathew MM, Munjuga MR, Ndangalasi HJ, Cordeiro NJ (2009) Aspects of the floral and fruit biology of *Allanblackia stuhlmannii* (Clusiaceae), an endemic Tanzanian tree. J East Afr Nat Hist 98(1):79–93

Maundu P, Tengnas B (2005) Useful trees and shrubs for Kenya, Technical handbook No.35. Nairobi, World Agroforestry Centre–Eastern and Central Africa Regional Programme (ICRAF-ECA)

Milimo PBW (1986) The control of germination in *Melia volkensii* Gurke seeds. MSc Thesis University of Alberta, Canada

Mohamed A, Bekhit M (2007) In vitro propagation of *Balanites aegyptiaca* (L) Del., an endangered medicinal plant. J Plant Prod 32(4):2907–2915

Mona AA (2012) In vitro propagation of *Swietenia macrophylla* King. Res J Agric Biol Sci 8(2):282–287

Mulanda ES (2016) In vitro regeneration, morphological and molecular characterization of *Melia volkensii* (Gurke). PhD Thesis, University of Nairobi

Mulanda ES, Adero MO, Amugune NO, Akunda E, Kinyamario JI (2012) High-frequency regeneration of the drought-tolerant tree *Melia volkensii* Gurke using low-cost agrochemical Thidiazuron. Biotechnol Res Int 2012:818472

Munjuga M, Kariuki W, Njoroge JBM, Ofori D, Jamnadass R (2013) Effect of rootstock type, scion source and grafting methods on the healing of *Allanblackia stuhlmannii* grafts under two nursery conditions. Afr J Hortic Sci 7:1–10

Munjuga M, Ofori D, Asaah E, Peprah T, Tsobeng A, Schmidt L, Henneh S, Ofori E, Mwanaidi K, Kahia J, Sesiwa H, Mpanda M, Simon T, Tchoundjeu Z, Jamnadass R (2015) Tissue culture. In: Allanblackia propagation protocol. World Agroforestry Centre (ICRAF), pp 81–85

Muok B, Mwamburi A, Kyalo E, Auka S (2010) Growing *Melia volkensii*: a guide for farmers and tree growers in the drylands. KEFRI Information Bulletin No 3

Murthy HN, Yadav GG, Dewir YH, Ibrahim A (2021) Phytochemicals and biological activity of desert date (*Balanites aegyptiaca* (L.) Delile). Plan Theory 10:32

Mwangi RW, Mukiama TK (1988) Evaluation of *Melia volkensii* extract fractions as mosquito larvicides. J Am Mosq Control Assoc 4:442–447

Mwangi RW, Rembold H (1988) Growth-inhibiting and larvicidal effects of *Melia volkensii* extracts on *Aedes aegypti* larvae. Entomol Exp Appl 46:103–108

Nahed ND, El-Azazi EMA, Ali MAM, El-Mekawy MA (2023) Enhancing *Balanites aegyptiaca* seed germination in Egyptian deserts gene bank by breaking dormancy treatments. Int J Dev 12(1):7–15

Nakai K, Yoshimura T (2020) African blackwood (*Dalbergia melanoxylon*) and other local Tanzanian tree species' biological performance against subterranean termites and wood decay fungi. Bioresources 15(2):2994–3005

Ndathi, AJ, Nyangito MM, Musimba NK, Mitaru BN (2012) Farmers' preference and nutritive value of selected indigenous plant feed materials for cattle in drylands of South-Eastern Kenya. Livest Res Rural Dev 24, Article #28. Retrieved September 4, 2024, from http://www.lrrd.org/lrrd24/2/ndat24028.htm

Ndoye M, Diallo I, Gassama Dia YK (2003) In vitro multiplication of the semi-arid forest tree, *Balanites aegyptiaca* (L.) Del. Afr J Biotechnol 2(11):421–424

Neondo J, Machua J, Muigai A, Nyende AB, Munjuga M, Jamnadass R, Muchugi A (2011) Micropropagation of *Allanblackia stuhlmannii*: amenability to tissue culture technique. Int J Biotechnol Mol Biol Res 2(11):185–194

Nikles DG, Bevege DI, Dickinson GR, Griffiths MW, Reilly DF, Lee DJ (2008) Developing African mahogany (*Khaya senegalensis*) germplasm and its management for a sustainable forest plantation industry in northern Australia: progress and needs. Aust For 71(1):33–47

Olagbende OH, Falowo OA, Latinwo LM, Betiku E (2021) Esterification of *Khaya senegalensis* seed oil with a solid heterogeneous acid catalyst: modeling, optimization, kinetic and thermodynamic studies. Clean Eng Technol 4:100200. https://doi.org/10.1016/j.clet.2021.100200

Orwa C, Mutua A, Kindt R, Jamnadass R, Anthony S (2009) Agroforestree database: a tree reference and selection guide, version 4.0

Ozoude CH, Azubuike CP, Ologunagba MO, Tonuewa SS, Igwilo CI (2020) Formulation and development of metformin-loaded microspheres using *Khaya senegalensis* (Meliaceae) gum as co-polymer. Future J Pharm Sci 6:120

Permadi N, Akbari SI, Prismantoro D, Indriyani NN, Nurzaman M, Alhasnawi AN, Doni F, Julaeha E (2024) Traditional and next-generation methods for browning control in plant. Curr Plant Biol 38:100339

Pijut PM, Beasley RR, Lawson SS, Palla KJ, Stevens ME, Wang Y (2012) In vitro propagation of tropical hardwood tree species—a review (2001–2011). Propag Ornam Plants 12(1):25–51

Saharan V, Yadav RC, Yadav NR, Wiesman Z (2011) Somatic embryogenesis and plant regeneration of *Balanites Aegyptiaca* Del (L.): an industrial important arid tree. Journal of. Cell Tissue Res 11(1):2529–2534

Siddique I, Anis M (2009) Direct plant regeneration from nodal explants of *Balanites aegyptiaca* L. (Del.): a valuable medicinal tree. New For 37:53–62

Stewart M, Blomley T (1994) Use of *Melia volkensii* in a semi-arid agroforestry system in Kenya. Commonw For Rev 73:128–131

Thomas TD (2008) The role of activated charcoal in plant tissue culture. Biotechnol Adv 26:618–631

Uzu J, Bettinger P, Siry J, Mei B (2022) Timber business in West Africa: a review and outlook. Int For Rev 24(2):240–256

Washa BW (2021) The African Blackwood (*Dalbergia melanoxylon*) seedling initiation technicality, (germinability and rootability). Recent Res Adv Biol 5:13–23

Washa WB, Nyomora AMS (2014) Advances in *Dalbergia melanoxylon* investigations towards tissue culture: problems and limitations. Am J Plant Sci 5:3501–3509

Werbrouck S, Bhogar N, Magomere T, Omondi S (2017) In vitro biotechnology of *Melia volkensii*, a high potential forestry tree from eastern Africa. Proceedings of the fourth international conference of the IUFRO Unit 2.09.02 on 'development and application of vegetative propagation technologies in plantation forestry to cope with a changing climate and environment'. La Plata, Argentina

World Conservation Monitoring Centre (1998) *Khaya senegalensis*. The IUCN Red List of Threatened Species 1998: e.T32171A9684583

Xu Z, Liu Y, Meng X, Yang L, Shao F, Liu R, Chen L (2021) Neoflavonoids from the Heartwood of *Dalbergia melanoxylon*. Rec Nat Prod, p 200

Yakubu BL, Nda AJ (2012) Evaluation of (*Khaya senegallensis*) for the control of *Tribolium confusum* (Duval) (Coleoptera: Tenebrionidae) on stored pearl millet. J Biol Agric Healthc 2:2

Yin X, Huang A, Zhang S, Liu R, Ma F (2018) Identification of three *Dalbergia* species based on differences in extractive components. Molecules 23:2163

Zadro MG (1975) Woods used for woodwind since the 16th century 2: a descriptive dictionary of the principal woods mentioned. Early Music 3(3):249–251

Zaky AS, Kandeil M, Abdel-Gabbar M, Fahmy EM, Almehmadi MM, Ali TM, Ahmed OM (2022) The antidiabetic effects and modes of action of the *Balanites aegyptiaca* fruit and seed aqueous extracts in NA/STZ-induced diabetic rats. Pharmaceutics 14:263

Forest Alchemists: Microbe-mediated Nutrient Recycling

5

V. Rathna, Siddu Lakshmi Prasanna, Jayashree Anandakumar, B. Swathi, Swathi S. Patil, and Ruby Susana Kujur

Abstract

Forests hold significant influence on global biogeochemical cycles by sequestering carbon and serving as reservoirs of biodiversity. Microbes are vital to the ecological functioning in which an intricate network of nutrient recycling occurs in the forest ecosystem. Microbial communities, encompassing bacteria, fungi, archaea, and viruses, exhibit remarkable diversity and functional versatility in forest soils. These microbes decompose the organic matter, engaging in symbiotic relationships with plants, and releases several nutrients and micronutrients including carbon, nitrogen, and phosphorus. Various enzymatic activities, including cellulases, ligninases, and nitrogenases, are carried out by microbes, contributing to the breakdown of recalcitrant organic materials. Furthermore, mycorrhizal fungi form symbiotic association with tree roots to enhance nutrient

V. Rathna
ICFRE – Institute of Forest Productivity, Ranchi, Jharkhand, India

S. L. Prasanna
Agricultural College, Bapatla Acharya N G Ranga Agricultural University, Guntur, Andhra Pradesh, India

J. Anandakumar
ICAR-National Institute of Veterinary Epidemiology and Disease Informatics, Bengaluru, Karnataka, India

B. Swathi
ICAR-Krishi Vigyan Kendra, Chikkaballapur, Karnataka, India

S. S. Patil
University of Agricultural Science, GKVK, Bengaluru, Karnataka, India

R. S. Kujur (✉)
ICFRE – Institute of Wood Science and Technology, Bengaluru, Karnataka, India
e-mail: kujurrs@icfre.org

R. K. Kalia, R. Pathak (eds.), *Tree Biology and Biotechnology*,
https://doi.org/10.1007/978-981-96-0002-1_5

absorption and transfer. This mutualistic relationship fosters the acquisition of mineral nutrients and shapes the overall health and resilience of the forests. Microbes also influence soil pH and mineral weathering processes, impacting nutrient availability to vegetation. Forests, acting as carbon sinks, rely on the carbon captured by primary producers and transformed by microorganisms, affecting geochemical cycles and climate dynamics. Human-induced disturbances, such as deforestation, land-use changes, and climate change, can disrupt microbial communities and their functions, leading to nutrient imbalances and ecosystem degradation. The chapter aims to explore the vital contributions of soil microorganisms to nutrient recycling in forest ecosystems, highlighting their role in transforming organic matter, cycling essential nutrients, and sustaining the intricate web of life within these verdant landscapes in forest management and conservation.

Keywords

Biogeochemical cycles · Nutrient recycling · Decomposition · Symbiosis · Climate change

5.1 Introduction

Forests are the most complex and bio-diverse ecosystems available on the earth, spreading over 40 million km^2 along with accounting for the 30% area of the total global land (Keenan et al. 2015). It relies on intricate processes to sustain their health and productivity. In the intricate tapestry of forest ecosystems, microorganisms act as the unsung heroes, or forest alchemists orchestrating the transformation and recycling of essential nutrients. These microscopic entities, including bacteria, fungi, and other decomposers, are main players in maintaining health and productivity of ecosystem by facilitating the flow of nutrients through various biological processes. Forests, with their rich and diverse flora, provide a dynamic environment for microorganisms. These forests are not just extensive reservoirs of biodiversity but also complex systems where every component from towering trees to the smallest soil organisms interacts in a delicate balance. The role of microbes in enhancing forest ecosystem resilience and productivity encompassing soil microbial community, effects of climate change, symbiotic relationships of mycorrhizal fungi, and flow of nutrients on the forest soil are depicted in Fig. 5.1.

Cycling of nutrients under forest soil is closely associated with the nutrients obtained through the degradation of mineral weathering processes and available organic matter. Mineral-weathering is influenced by a range of hydrological, chemical, physical, and biological factors, in addition to soil characteristics itself. Soil minerals undergo mechanical erosion and are influenced by thermal fluctuations and hydration. The speed of mineral dissolution is strongly determined by the presence of weathering agents, including organic and inorganic acids, chelating compounds (e.g., siderophores), along with pH of the soil. Biotic factors, including soil

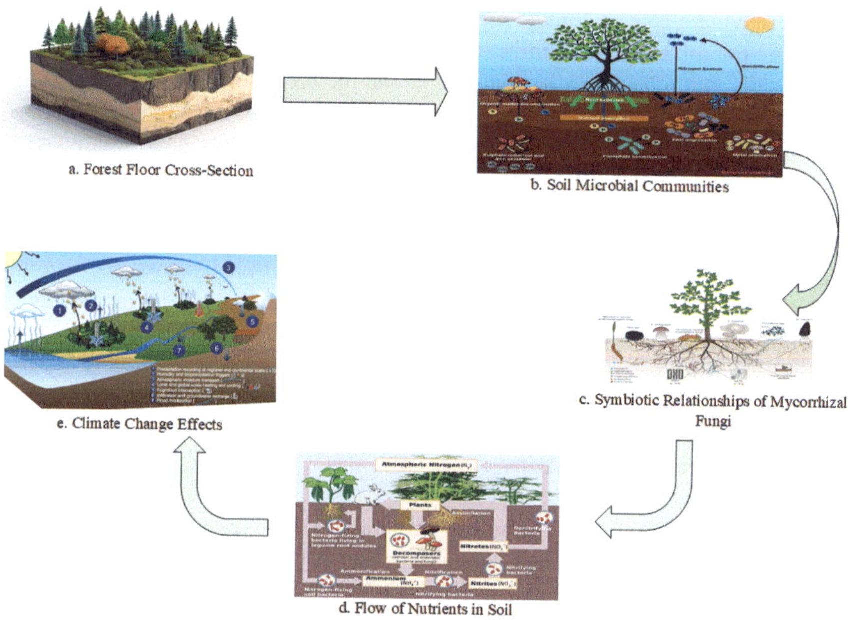

Fig. 5.1 Role of microbes in enhancing forest ecosystem resilience and productivity

microorganisms, can initiate and accelerate mineral-weathering processes (Drever 2005). These biological components act as a key factor in releasing nutrients from minerals, which influences overall nutrient cycling in forest ecosystems.

Eukaryotes and prokaryotic microbes decompose the complex material to simple forms, discharging vital nutrients, viz., carbon, nitrogen, and phosphorus, back to the soil. The decomposition is not only fundamental for nutrient recycling but also affects soil structure and fertility (Bardgett et al. 2008). Microbes also mediate key biochemical processes such as nitrogen fixation, nitrification, and denitrification, each of which acts as a major contributor in maintaining nutrient accessibility and ecosystem function. Tree roots enhance mineral weathering by producing organic acids and exerting physical forces on soil minerals. Additionally, the mycorrhizal fungi associated with the root also mobilize inorganic nutrients from minerals and transfer the essential nutrients to trees and thereby supporting nutrition (Courty et al. 2010).

For decades, mineral-weathering by soil microorganisms has primarily been credited to mycorrhizal fungi (Landeweert et al. 2001). In natural ecosystems, plants are a major reservoir for various microorganisms comprising of bacteria and fungi and also broadly known as the phytomicrobiome or plant-microbiome (Quiza et al. 2015). The temperate forests are reservoir of Ectomycorrhizal fungi wherein several soil-borne fungi colonize in the roots of various tree species such as Pine and Oak (Sebastiana et al. 2017). Under this symbiotic association, the fungi obtain photosynthetically sequestered carbon from the host plant and gives nutrients taken from the rhizosphere soil. Microbial processes influence greenhouse gas emissions, thereby playing a role in climate regulation (Davidson and Janssens 2006). Plants influence microbial communities through microhabitats (Wang et al. 2023). Additionally, microbes have a pivotal impact in the breakdown of pollutants, helping to mitigate environmental contamination (Mohn and Stewart 2000).

The aim of this book chapter is to explore the vital contributions of soil microorganisms to nutrient recycling in forest ecosystems, highlighting their role in transforming organic matter, cycling essential nutrients, and sustaining the intricate web of life within these verdant landscapes in forest management and conservation. As forests encounters rising impacts of climate change and habitat loss, the insights gained from studying microbial nutrient recycling can help in developing strategies to sustain forest ecosystems and their diverse functions.

5.2 Importance of Microbial Communities in Forest Ecosystem

Microbes are omnipresent, existing everywhere and capable of performing numerous functions. Without microbes, human life would not be possible. In forest ecosystems, microbes play varietal ecological processes that maintain health and functionality. It has a wide range of habitats including soil, plant, surfaces, streams, rocks, and others for microbes. Microbes are involved in nutrient cycling and recycling process to enhance soil fertility and improves plant growth and health, which leads to disease suppression. Forests serve as major carbon pool, where the carbon sequestered by major producers surpasses carbon destruction through inhalation by 7–25% (Malhi et al. 1999). Additionally, forests influence the geochemical rotations of various elements (Perry et al. 2008). Certain microbes remove and degrade pollutants and toxins in the soil.

Soil organic matter accumulates through various activities including litter decay, deadwood, microbial biomass, etc., and is stored within the soil (Clemmensen et al. 2013). The conversion of organic matter, coupled with various other routes, depends on the microbial activity. Diverse activities of microbes support overall ecosystem biodiversity by maintaining soil health and creating a conducive environment for various plant and animal species (Bardgett and van der Putten 2014). However, forests face numerous disturbances including insect outbursts, wind throws, fires, and human-induced factors, such as climate change, and environmental pollution, which can disrupt the balance of nutrient cycling processes and health of forest.

5.3 Nutrient Cycle

Microbes play a critical role in nutrient cycle and recycling within ecosystems, facilitating the continual availability of fundamental elements for plant nourishment and ecosystem health.

5.3.1 Organic Matter Decomposition

Microbes, including bacteria and fungi, break down dead and decayed plant and animal matter through decomposition. This process converts composite organic material to simple forms, discharging vital nutrients, viz., carbon, nitrogen, phosphorus, and sulfur back to the soil. Fungi and bacteria colonize and decompose the organic material. Fungi primarily produce enzymes that decompose complex plant materials, such as lignin and cellulose (Hattenschwiler et al. 2005), into simpler organic molecules. Subsequently, bacteria break down these simpler organic molecules into even smaller compounds, such as carbon dioxide and inorganic nitrogen, which become available for plant uptake (Bardgett and van der Putten 2014). The decomposed organic material is transformed into humus, which enhances soil structure and fertility (Swift et al. 1998). This process is essential for breaking down woody materials and forming humus.

The main fungal decomposers are saprotrophic fungi, which perform decomposition of lignin and cellulose to form humus (Bardgett et al. 2008), even though mycorrhizal fungi is having mutualistic association with roots, that mycorrhizal fungi also involve in the breakdown of organic matter in the rhizosphere (Smith and Read 2008). The cellulolytic bacteria decompose cellulose into simpler sugars, which are further decomposed by other microorganisms (Gupta et al. 2012) and proteolytic bacteria decompose proteins from dead organisms, releasing proteins and nitrogen compounds to the soil (Ojima et al. 1993). Additionally, invertebrates like insects and arthropods also accord to the physical breakdown of organic material and facilitate microbial decomposition by shredding and mixing litter (Hansen and Coleman 1998); similarly, the earthworms consume organic matter and excrete it in a more decomposed form, improving soil structure and nutrient content (Brown and Doube 2004).

5.3.2 Nitrogen Cycle

Nitrogen recycling is a fundamental process in forest ecosystems for sustaining soil fertility and supporting plant growth. Nitrogen, an essential nutrient, is an important constituents of proteins and nucleic acids. The recycling of nitrogen in forests involves a chain of complex biological and chemical processes that ensure its availability to plants and its movement through the ecosystem.

5.3.2.1 Process of Nitrogen Cycle and Recycle

Nitrogen Fixation Certain microorganisms transform the atmospheric nitrogen in ammonia or related compounds. Bacteria like Rhizobium found in the root nodules of legumes and Frankia found in non-leguminous plants fix nitrogen for plant use (Giller 2001). Abiotic nitrogen fixation happens through lightning, which transforms atmospheric nitrogen into nitrate (NO_3^-) through high-energy reactions (Schlesinger and Andrews 2000).

Nitrification Nitrifying bacteria *Nitrosomonas* (ammonia-oxidizing bacteria) convert ammonia (NH_3) or ammonium (NH_4^+) to nitrite (NO_2^-) called as ammonia oxidation (Prosser 2002). Later, nitrite (NO_2^-) is further oxidized to nitrate (NO_3^-) by nitrite-oxidizing bacteria, that is, *Nitrobacter* (Schimel and Bennett 2004); these nitrates are readily absorbed by plants for growth.

Assimilation Plants absorb NO_3^- and NH_4^+ from the soil and integrate them in organic material including nucleic acids and proteins (Fowler et al. 2013).

Ammonification Organic matter from dead plants, animals, and decomposed by microbes, releases ammonia (NH_3) or ammonium ions (NH_4^+) into the soil (Swift et al. 1998), which is known as ammonification or mineralization.

Denitrification Readily absorbed form of nitrates is reduced to nitrite (NO_2^-) and then to N_2 gas or nitrous oxide through denitrifying bacteria, that is, *Pseudomonas* and *Paracoccus*, and returns nitrogen to the atmosphere (Firestone and Davidson 1989). The leaching and runoff of nitrate (NO_3^-) can percolate into groundwater or be transported through runoff into surface water, potentially leading to nutrient pollution and eutrophication (Galloway et al. 2003).

Decomposition Nitrogen re-enters into the soil through the decomposition of plant litter, animal remains, and microbial biomass. Soil microbes break down organic matter, returning nitrogen to the soil in forms that plants can reabsorb them (Swift et al. 1998).

Soil organic nitrogen A significant portion of soil nitrogen is stored in organic forms, such as soil organic matter. Microbial activity continually mineralizes this organic nitrogen, adding to the nitrogen reserves, which is accessible by plants (Schimel and Bennett 2004).

5.3.3 Phosphorus Cycle and Recycling in Forest

Phosphorus cycle contributes in sustaining soil fertility and promoting the plant growth. Phosphorus is an essential nutrient for plants, influencing several processes such as photosynthesis, energy transfer, and nucleic acid synthesis. Phosphorus recycling in forest ecosystems is a critical process that sustains soil fertility and supports plant growth. Unlike nitrogen, phosphorus lacks a significant gaseous phase and primarily cycles through biological and geological processes.

5.3.3.1 Process of Phosphorus Cycle and Recycle

Phosphorus inputs and redistribution Phosphorus enters forest ecosystems mainly through the weathering of phosphate-rich rocks, which releases phosphate ions (PO_4^{3-}) into the soil; this process is slow but provides a long-term source of phosphorus to the forest floor. Atmospheric deposition, including dust and particulate matter, can also contribute to phosphorus inputs (Vitousek et al. 2010).

Phosphorus in the soil Phosphorus in forest soils exists both in organic and inorganic forms. Inorganic phosphorus, present in form of phosphate ions, is readily available to plants and is influenced by soil pH, moisture, and microbial activity (Frossard et al. 2000). Organic phosphorus is mineralized into inorganic forms by soil microbes which is accessible to plants (Turner et al. 2014). Additionally, microbial decomposition of organic matter, including leaf litter and dead plant material, releases phosphorus into the soil.

Phosphorus uptake by plants Many forest plants establish mutualistic relationships with mycorrhizal fungi, which extend their hyphae into the soil and enhance the surface area for phosphorus absorption. This symbiosis is crucial for accessing phosphorus from the soil, particularly in phosphorus-poor environments (Smith and Read 2008). Plants absorb phosphorus in the form of phosphate ions from the soil. This phosphorus is engaged in wide physiological processes, such as cell membrane formation energy transfer, nucleic acid synthesis, and energy transfer. Mycorrhizal fungi form 97% symbiotic relationships with terrestrial plants. Ectomycorrhiza (EcM) is formed when fungi infect the roots of about 2% of vascular plants, primarily within families such as Pinaceae, Fagaceae, Dipterocarpaceae, and Myrtaceae. This type of mycorrhiza is distinguished by the presence of a fungal mantle and a Hartig net, with hyphae that remain outside the plant cells. Globally, approximately 7750 species of fungi are known to form EcM. However, the actual species richness of ectomycorrhizal fungi (EcMF) is estimated to be between 20,000 and 25,000, considering both known and yet-to-be-discovered macromycete diversity. These EcMF encompass more than 80 independently evolved lineages and belong to over 250 genera, predominantly from the Basidiomycota and Ascomycota phyla (Liu et al. 2020).

Phosphorus solubilization Phosphorus-solubilizing bacteria transform insoluble phosphorus molecules into soluble forms, enhancing phosphorus availability to plants (Rodriguez and Fraga 1999).

5.3.3.2 Decomposition and Recycling

Dead plants, animals, and microbial biomass contribute to the phosphorus pool in forest soils through decomposition. Forest ecosystems exhibits high nutrient retention because of dense vegetation and organic matter that trap and recycle the available nutrients. This retention helps in maintaining the soil fertility to reduce phosphorus losses through leaching and runoff (Gosz et al. 1972).

5.3.4 Sulfur Cycle and Recycling in Forest

Sulfur is an essential element for all living organisms and contributes the structure and function of proteins and enzymes. In forest ecosystems, the sulfur cycle is a complex process involving the transformation of sulfur through various chemical forms, driven by both biological and abiotic factors.

5.3.4.1 Process of Sulfur Cycle and Recycle

Atmospheric Deposition Sulfur enters into forest ecosystems primarily through atmospheric deposition, like wet (rainfall) and dry deposition (gaseous and particulate sulfur compounds). Major sources of atmospheric sulfur include volcanic emissions, sea spray, and anthropogenic activities such as fossil fuel combustion (Schlesinger and Bernhardt 2013). Sulfur dioxide (SO_2) and hydrogen sulfide (H_2S) in the atmosphere can be oxidized to sulfate (SO_4^{2-}), which is deposited onto forest soils and vegetation for the plant use (Likens et al. 2002).

Soil sulfur transformations In the soil, organic compounds of sulfur from plant litter and microbial biomass are mineralized by soil microbes into inorganic sulfate. Conversely, sulfate can be immobilized back into organic forms by microbial activity (Schindler et al. 1986). These sulfate ions can be adsorbed onto soil particles, particularly clays and organic matter, or desorbed back into the soil solution (David et al. 1982).

Plant uptake and assimilation Plants absorb sulfur primarily in sulfate form from the soil. Sulfate is used by roots and transported to the shoots, where it is assimilated into organic sulfur compounds, viz., proteins (e.g., cysteine and methionine) and nucleic acids (Koprivova and Kopriva 2014). It helps synthesize chlorophyll, vitamins, and coenzymes and also contributes to plant defense mechanisms (Marschner 2012).

Decomposition and recycling The decomposition of plant litter and organic matter into organic sulfur compounds, later subsequently mineralized into sulfate, contributing to soil sulfur pool (Swift et al. 1979) by soil microbes, including bacteria and fungi by mediating processes such as sulfate reduction and sulfur oxidation (Bosatta and Berendse 1984).

Sulfur losses In soil, sulfate is relatively mobile and can be leached from the soil profile, particularly in sandy soils with low organic matter content, leading to losses of sulfur from the ecosystem (Likens et al. 2002). Some sulfur is evaporating to the atmosphere as volatile organic sulfur compounds, includes dimethyl sulfide produced by microbial activity (Schlesinger and Bernhardt 2013).

5.3.5 Carbon Cycle and Recycling in Forest

The carbon cycle in forest ecosystems is a complex and dynamic process that encompasses the movement and transformation of carbon across the varied components of the ecosystem. Microbes facilitate the breakdown of organic matter, contribute to carbon sequestration, and release carbon dioxide (CO_2) through respiration.

5.3.5.1 Process of Carbon Cycle and Recycle

Photosynthesis Plants absorb CO_2 from the atmosphere and convert it into organic compounds through photosynthesis. This process forms the basis of the forest carbon cycle, providing the base for organic carbon in the ecosystem called as carbon fixation. This fixed carbon is used to build plant biomass, including leaves, stems, and roots, which serve as carbon reservoirs (Chapin et al. 2011).

Litter fall and organic matter input Dead plant material, such as leaves, stems, and branches, falls to the forest floor, contributing to the organic matter pool (Prescott 2010) and roots release organic compounds into the soil, offering carbon pool for soil microbes (Jones et al. 2009).

Decomposition by microbes Bacteria and fungi decompose complex organic compounds in dead plant and animal material into simpler compounds, releasing nutrients and CO_2 (Swift et al. 1979). These microbes produce enzymes such as cellulases and ligninases that degrade cellulose and lignin, the primary structural components of plant biomass (Baldrian 2017).

Soil organic carbon formation Portion of decomposed organic matter converts into humus that remains in soil for longer time (Schmidt et al. 2011). Forest soils serve as significant carbon sinks, storing carbon in organic matter and mitigating atmospheric CO_2 levels (Lal 2005) called as carbon sequestration.

Respiration and carbon release Microbes metabolize organic carbon for energy, releasing CO_2 as a byproduct of respiration (Schimel 2001). The combined respiration of plant roots, soil fauna, and microbes leads to the release of CO_2 from the soil to the atmosphere (Bond-Lamberty and Thomson 2010).

5.3.5.2 Carbon Recycling

Microbes transform carbon through various metabolic pathways, influencing the stability and persistence of soil organic carbon (Fontaine et al. 2003). Microbial activity helps to form stable carbon pools in the soil, which are essential for long-term carbon storage (Schmidt et al. 2011).

5.3.5.3 Carbon-Climate Effect

Microbial activity is sensitive to temperature variations, which can affect carbon release and climate feedbacks. Warming may accelerate microbial respiration, potentially increasing CO_2 emissions from soils (Davidson and Janssens 2006). Understanding microbial processes in the carbon cycle is essential for developing strategies to mitigate climate change through carbon management in forests (Lal 2005).

5.3.6 Humification Cycling and Recycle in Forest

Humification is a critical process in forest ecosystems that involves the conversion of organic matter into humus, enhancing soil structure, maintaining the soil fertility, and storing carbon.

5.3.6.1 Processes in Humification

The humification process begins with the breakdown of plant litter (leaves, stems, and roots) and animal remains by bacteria, fungi, and actinomycetes (Swift et al. 1979). Microorganisms decompose complex organic compounds such as cellulose, hemicellulose, and lignin into simpler substances (Prescott 2010). The end products are humic substances grouped in humic acids, fulvic acids, and humin (Stevenson 1994). These substances are formed through biochemical processes, including condensation, polymerization, and oxidation-reduction reactions. Humic substances are characterized by their complex structure and high molecular weight (Kögel-Knabner 2002). These humic substances interact with soil minerals and form stable complexes, contributing to long-term carbon storage in soils (Schmidt et al. 2011).

Physical protection within soil aggregates the chemical stabilization by binding to minerals and helps protect humic substances from rapid decomposition (Six et al. 2002). Humus enhances the cation exchange capacity of the soil that increases the ability of soil to hold and exchange nutrients (Stevenson 1994) and also the development of soil aggregates. Well-aggregated soils have better aeration, water infiltration, and root penetration. The formation and decomposition of humus are integral to the carbon cycle in forest ecosystems, balancing carbon inputs and outputs (Six et al. 2002).

5.3.7 Biotransformation

Biotransformation is the process where organic and inorganic substances are chemically altered by living organisms, primarily microbes. In forest ecosystems, biotransformation is essential process of cycling and recycling of nutrients, detoxification of pollutants, and maintenance of ecosystem health. This process involves a complex interaction of biological, chemical, and physical factors that are beneficial in maintaining the dynamic equilibrium of forest soils. Processes in biotransformation involve breakdown of organic matter by microbe decomposer, followed by conversion of organic compounds into inorganic forms (e.g., ammonium, nitrate, phosphate) that can be readily taken up by plants (Paul and Clark 1996).

5.4 Role of Microbes in Maintaining the Forest Health

Microbes restore forest health by participating in numerous ecological processes that promote plant growth, nutrient cycling, and overall ecosystem resilience. Their activities include recovery of degraded forests, enhancing soil fertility, and maintaining ecosystem stability. Below are some of the important functions and roles of microbes in forest health restoration.

5.4.1 Nutrient Cycling

Microbes decompose organic matter to nitrogen, phosphorus, and sulfur back into the soil, to make them available for the plant (Swift et al. 1979). Rhizobium and Frankia convert atmospheric nitrogen that can be utilized by the plants, thereby improving soil nitrogen levels and supporting their growth (Vitousek et al. 2002). Phosphate-solubilizing bacteria and fungi transform insoluble phosphorus into soluble forms so that the plants can utilize it, thereby enhancing phosphorus availability (Rodriguez and Fraga 1999).

5.4.2 Soil Health and Structure

Microbes generate extracellular polysaccharides and other substances that help aggregate soil particles, thereby enhancing soil structure and porosity. This leads to improved water retention and root growth (Tisdall and Oades 1982). Additionally, the microbes decompose the organic matter to create humus that enhances soil structure and fertility (Stevenson 1994).

5.4.3 Stress Buffering and Sustainable Productivity

Humus helps to buffer the forest ecosystems against environmental stresses such as drought, nutrient depletion, and soil erosion (Tisdall and Oades 1982). The continuous cycling and recycling of humus ensure the sustainable productivity of forest ecosystems, maintaining their health and resilience over time (Swift et al. 1979).

5.4.4 Plant Growth Promotion

Mycorrhizal mutualistic interaction improves plant nutrient uptake and increases resistance to environmental stressors (Smith and Read 2008). Ground vegetation forms symbioses with arbuscular mycorrhizal (AM) or ericoid mycorrhizal (ERM) fungi (Clemmensen et al. 2013), which often concur in the soils with ectomycorrhizal fungi associated with trees. As compared to well-studied ECM fungi of trees, the microbiomes associated with ground-dwelling forest plants are less understood. Additionally, some soil microbes give rise to plant growth-promoting hormones (auxins, gibberellins, and cytokinins), which stimulates the plant growth and its development (Vessey 2003).

5.4.5 Disease Suppression

Some soil microbes function as biocontrol agents by outcompeting other organisms or inhibiting the progression of plant pathogens. *Trichoderma* spp. produce antifungal compounds that safeguard plants from root diseases (Harman et al. 2004). Furthermore, favorable microbes also initiate systemic resistance in plants, boosting its capacity to resist pathogen attacks (Pieterse et al. 2014).

5.4.6 Ecosystem Resilience

Enhanced nutrient cycling and pollutant detoxification through biotransformation processes increase the resilience of forest ecosystems against drought, nutrient depletion, and contamination (Swift et al. 1979). Diverse microbial communities engaged in biotransformation sustain overall biodiversity by preserving soil health and supplying essential nutrients for various plant and animal species (Brussaard 1997).

5.4.7 Detoxification of Pollutants

Microorganisms degrade toxic substances, such as polycyclic aromatic hydrocarbons (PAHs) and pesticides, into less harmful compounds through metabolic processes (Alexander 1999). Plants, in association with microbe, uptake, transform, and sequester pollutants from the soil and water (Pilon-Smits 2005).

5.4.8 Carbon Sequestration

Microbial decomposition of organic matter aids in sequestering the soil carbon by transforming it into stable forms such as humus, which helps in combating climate change (Lal 2005).

5.5 Interactions with Other Nutrient Cycles

5.5.1 Nitrogen Cycle

Nitrogen cycling is linked to the carbon cycle. Nitrogen availability influences the plant growth and carbon sequestration in forests (Holland et al. 1999). Soil pH influences microbial activity and nitrogen transformations, with acidic soils potentially enhancing nitrification but also increasing nitrate leaching (Johnson and Matson 2009).

5.5.2 Carbon Cycle

The phosphorus cycle is meticulously linked with the carbon cycle. Soil organic matter, including phosphorus, is influenced by carbon inputs from plant growth and decomposition. The cycling of these elements is interconnected, affecting overall soil fertility and ecosystem health (Vitousek et al. 2010).

5.5.3 Sulfur Cycle

The sulfur cycle is closely associated with other nutrient cycles, such as nitrogen and carbon cycles. Interactions between these cycles influence overall nutrient availability and ecosystem functioning (Schindler et al. 1986). Sulfur deficiencies can lead to decreased plant productivity and compromised ecosystem health (Marschner 2012). Conversely, excessive deposition of sulfur from human intervention can cause soil and water acidification, having a detrimental effect on forests ecosystems by causing nutrient imbalances and reducing biodiversity (Driscoll et al. 2001).

5.6 Factors Affecting the Microbial Activity

Microbial activity is essential for the health and functioning of forest ecosystems. Various factors influence microbial activity in forest soils, including environmental conditions, soil properties, and biotic interactions.

5.6.1 Soil Temperature

Temperature influences microbial metabolism and enzyme activity, affecting the rates of decomposition and nutrient cycling. Seasonal changes in temperature can lead to fluctuations in microbial activity, with higher activity generally observed during warmer periods (Davidson and Janssens 2006).

5.6.2 Soil Moisture

Soil moisture is crucial for microbial activity as water is necessary for microbial metabolism and nutrient transport (Manzoni et al. 2012). Prolonged dry conditions can reduce microbial activity and biomass, while rewetting can cause a temporary spike in activity (Birch 1958).

5.6.3 Soil pH

Most soil microbes survives better under neutral to slightly acidic pH (6–7) conditions, and deviations can inhibit microbial activity and diversity (Rousk et al. 2010). Soil pH affects the solubility of nutrients and toxins, influencing microbial growth and functions (Fierer and Jackson 2006).

5.6.4 Organic Matter Availability

The availability of organic carbon sources, such as plant litter and root exudates, are the primary driver of microbial activity (Fontaine et al. 2003). The chemical composition of organic matter (e.g., lignin content) influences its decomposition rate and involvement of the microbial communities (Berg and McClaugherty 2008).

5.6.5 Nutrient Availability

The availability of necessary nutrients like nitrogen and phosphorus can limit or enhance microbial activity and growth (Cleveland and Liptzin 2007). Imbalances or deficiencies in key nutrients can impact the composition and function of microbial communities (Treseder 2008).

5.6.6 Soil Texture and Structure

Well-aggregated soils provide a stable habitat for microbes, protecting them from environmental fluctuations and predation (Tisdall and Oades 1982). Soil texture and structure influence the aeration, water retention, and nutrient availability, affecting microbial habitat and activity (Young and Crawford 2004).

5.6.7 Vegetation Type and Root Activity

Plant-microbe interactions in rhizosphere soil shape the constituent and activity of microbial communities (Berendsen et al. 2012). Additionally, the type and quality of plant litter input can influence microbial activity and decomposition rates.

5.6.8 Environmental Disturbances

Water logging, fire, and other management practices can alter soil conditions and microbial communities (Hartmann et al. 2012). Heavy metals, pesticides, and other pollutants also have detrimental effect on microbial activity and diversity.

5.7 Impact of Climate Change on Microbial Activity

Climate change is significantly impacting forest ecosystems globally, affecting not just plant and animal species but also the microbial communities essential to ecosystem functions. Forests face numerous climate-related challenges like elevated carbon dioxide levels, rising temperatures, increased frequency of droughts and fires, pest outbursts, etc. Reaction of forest to these fluctuations is mainly arbitrated by microbes. Influence of climate fluctuations varies among boreal, temperate, and tropical forests. Microbial activities, including that of fungal symbionts, saprotrophic fungi, bacteria, and fungal plant pathogens, hold significant influence in shaping forest ecosystems.

5.7.1 Increase in Temperature

Usually higher temperature enhances the microbial metabolic rates, resulting in to faster decay and nutrient cycling (Davidson and Janssens 2006). Different microbial species have diverse temperature requirements, so warming may modify the microbial community composition, favoring thermophilic (heat-loving) species (Allison and Treseder 2008). Increased microbial respiration rates can lead to higher carbon dioxide emissions from soil, contributing to a positive feedback loop in global warming (Crowther et al. 2016).

5.7.2 Changes in Soil Moisture

Climate change can result in more frequent and severe droughts, which reduce soil moisture and hinder microbial activity and growth (Schimel et al. 2007). When dry soils are rewet, there can be a temporary spike in microbial activity and nutrient release, known as the Birch effect (Birch 1958). Both excess moisture (waterlogging) and drought can slow down decomposition rates, affecting soil organic matter dynamics (Manzoni et al. 2012).

5.7.3 Increased Frequency of Extreme Events

Increased frequency and intensity of wildfires can disrupt microbial communities by directly killing microbes and altering soil properties. Post-fire recovery involves shifts in microbial diversity structure and function (Hart et al. 2005). Extreme weather events result in soil erosion, nutrient leaching, and changes in soil structure, all of which impact microbial communities (Wang et al. 2017).

5.7.4 Elevated Atmospheric CO_2

Higher CO_2 levels can enhance plant growth and root exudation, providing more substrates for soil microbes. This can elevate microbial biomass and activity (Phillips et al. 2011). Increased plant growth can lead to greater organic matter inputs into the soil, which may enhance soil carbon sequestration, depending on the rates of microbial decomposition (Norby et al. 2010).

5.7.5 Nutrient Imbalances

Increased atmospheric nitrogen deposition, often a consequence of fossil fuel combustion and agricultural activities, can alter soil nutrient dynamics and microbial activity. Excess nitrogen can favor fast-growing, nitrogen-tolerant microbial species (Treseder 2008). Changes in nutrient cycling can lead to phosphorus limitation, affecting microbial activity and community composition (Vitousek et al. 2010).

5.8 Conclusion

In the intricate tapestry of forest ecosystems, microorganisms are the unseen alchemists, tirelessly working to transform and recycle essential nutrients. These microscopic entities, which include bacteria, fungi, and other decomposers, have indispensable role in sustaining the health and productivity of forests. Their activities ensure the continuous flow of nutrients through various biological processes, supporting the complex interactions exist between plants, animals, and other organisms within the forest. Forests, with their rich and diverse flora, provide a dynamic and conducive environment for microorganisms. The decomposition of organic matter, driven largely by microbial activity, releases essential nutrients back into the soil, which makes them accessible for plants to absorb. Additionally, microbes mediate crucial biochemical processes such as nitrogen fixation, nitrification, and denitrification, which are important for maintaining nutrient availability and ecosystem function.

As forests face rising impacts of climate change, pollution, and habitat loss, understanding and appreciating the role of these microbial alchemists becomes ever more critical. Their activities not only underpin forest productivity but also have broader environmental implications, including influencing greenhouse gas emissions and pollutant degradation. By recognizing and harnessing the power of microbial processes, we can better manage and conserve our forest ecosystems, ensuring their health, resilience, and the continued delivery of ecosystem services. This study underscores the importance of considering both above-ground and below-ground components in ecosystem management and conservation strategies. Effective soil biodiversity conservation and thoughtful landscape design in urban ecosystems should incorporate an understanding of microbial influences. By doing so, we can create sustainable and resilient environments that support both plant and microbial diversity, ultimately benefiting the broader ecosystem and human well-being.

5.9 Future Prospects

The present review may summarize the following future prospects:

- Advances in microbial ecology and genomics are deepening our understanding of how different microbes contribute to nutrient cycling, which microbes are most effective in specific nutrient transformations, potentially leading to improved forest management practices.
- Genetic engineering and synthetic biology could enhance the ability of microbes to perform specific nutrient recycling tasks. For instance, engineered microbes might be used to optimize the decomposition of organic matter or improve nitrogen fixation.
- As climate change alters forest ecosystems, microbes will also adapt for nutrient recycling processes to new environmental conditions. Research into how microbes respond to changing temperatures, moisture levels, and other factors will be important for maintaining ecosystem health and productivity need to be explored.
- Microbial communities are essential for the restoration and conservation of degraded forests. Future efforts may focus on enhancing microbial activity to support reforestation of mining and other land degraded projects and improve soil health in impacted areas.
- Microbes could be integral to developing sustainable forest management practices. By understanding and leveraging microbial contributions to nutrient cycling, forest managers can improve soil fertility and ecosystem resilience, reducing the need for chemical fertilizers and other interventions.
- Incorporating microbial dynamics into ecosystem models can provide more accurate predictions of nutrient fluxes and forest productivity. This integration helps in designing better management strategies and assessing the long-term impacts of different practices.

References

Alexander M (1999) Introduction to soil microbiology, 2nd edn. Wiley

Allison SD, Treseder KK (2008) Warming and drying suppress microbial activity and carbon cycling in boreal forest soils. Global Change Bio 14(12):2898–2909

Baldrian P (2017) Forest microbiome: diversity, dynamics, and functions. In: Soil microbiology, ecology, and biochemistry, 3rd edn. CRC Press, pp 103–124

Bardgett RD, van der Putten WH (2014) Below ground biodiversity and ecosystem functioning. Nature 515(7528):505–511

Bardgett RD, Freeman C, Ostle NJ (2008) Microbial contributions to climate change through carbon cycle feedbacks. ISME J 2(8):805–814

Berendsen RL, Pieterse CM, Bakker PA (2012) The rhizosphere microbiome and plant health. Trends Plant Sci 17(8):478–486

Berg B, McClaugherty C (2008) Plant litter: decomposition, humus formation, carbon sequestration. Springer

Birch HF (1958) The effect of soil drying on humus decomposition and nitrogen availability. Plant Soil 10(1):9–31

Bond-Lamberty B, Thomson A (2010) Temperature-associated changes in the global soil organic carbon pool. Nature 464(7289):799–802

Bosatta E, Berendse F (1984) The effect of nutrient availability on the decomposition of organic matter in soils. Oikos 42(3):266–272

Brown GG, Doube BM (2004) Functional interactions between earthworms, microorganisms, organic matter, and plants. Earthworm Ecol 2:213–239

Brussaard L (1997) Biodiversity and ecosystem functioning in soil. Appl Soil Ecol 5(1):1–6

Chapin FS, Matson PA, Mooney HA (2011) Principles of terrestrial ecosystem ecology. Springer

Clemmensen KE, Bahr A, Ovaskainen O, Dahlberg A, Ekblad A, Wallander H et al (2013) Roots and associated fungi drive long-term carbon sequestration in boreal forest. Science 339(6127):1615–1618

Cleveland CC, Liptzin D (2007) C: N: P stoichiometry in soil: is there a "Redfield ratio" for the microbial biomass? Biogeochemistry 85:235–252

Courty PE, Franc A, Garbaye J (2010) Temporal and functional pattern of secreted enzyme activities in an ectomycorrhizal community. Soil Biol Biochem 42(11):2022–2025

Crowther TW, Todd-Brown KEO, Rowe CW, Wieder WR, Carey JC, Machmuller MB et al (2016) Quantifying global soil carbon losses in response to warming. Nature 540(7631):104–108

David MB, Johnson DW, Kahl JS (1982) The role of nitrogen in the development of the structure and function of terrestrial ecosystems. Ecol Monogr 52(4):317–340

Davidson EA, Janssens IA (2006) Temperature sensitivity of soil carbon decomposition and feedbacks to climate change. Nature 440(7081):165–173

Drever JI (2005) Surface and ground water, weathering, and soils: treatise on geochemistry, vol 5. Elsevier

Driscoll CT, Lawrence GB, Bulger JB, Schaefer DA, Roy KM (2001) Acidic deposition in the Northeastern United States: sources and inputs, ecosystem effects, and management strategies. Bioscience 51(3):180–198

Fierer N, Jackson RB (2006) The diversity and biogeography of soil bacterial communities. Proc Natl Acad Sci 103(3):626–631

Firestone MK, Davidson EA (1989) Microbiological basis of NO and N_2O production and consumption in soil. Plant Soil 115(1):91–104

Fontaine S, Mariotti A, Abbadie L (2003) The priming effect of organic matter: a question of microbial competition? Soil Biol Biochem 35(6):837–843

Fowler D, Sutton MA, Smith R, Hill T (2013) The role of nitrogen in the biogeochemical cycles of ecosystems. J Environ Manag 130:175–182

Frossard E, Brossard M, Abadie J, Taboada M (2000) Soil phosphorus availability and plant uptake. Soil Sci Soc Am J 64(4):1372–1383

Galloway JN, Aber JD, Erisman JW, Seitzinger SP, Howarth RW, Cowling EB, Cosby BJ (2003) The nitrogen cascade. Bioscience 53(4):341–356

Giller KE (2001) Nitrogen fixation in tropical cropping systems. CABI Publishing

Gosz JR, Likens GE, Bormann FH (1972) Nutrient release from decomposing litter in a deciduous forest. Ecology 53(3):542–551

Gupta P, Samant K, Sahu A (2012) Isolation of cellulose-degrading bacteria and determination of their cellulolytic potential. Int J Microbiol 2012(1):578925

Hansen RA, Coleman DC (1998) Litter complexity and composition are determinants of the diversity and composition of the decomposer community. Oecologia 114(4):574–582

Harman GE, Howell CR, Viterbo A, Chet I, Lorito M (2004) Trichoderma species–opportunistic, avirulent plant symbionts. Nat Rev Microbiol 2(1):43–56
Hart SC, DeLuca TH, Newman GS, MacKenzie MD, Boyle SI (2005) Post-fire vegetative dynamics as drivers of microbial community structure and function in forest soils. For Ecol Manag 220(1–3):166–184
Hartmann M, Howes CG, VanInsberghe D, Yu H, Bachar D, Christen R et al (2012) Significant and persistent impact of timber harvesting on soil microbial communities in northern coniferous forests. ISME J 6(12):2199–2218
Hattenschwiler S, Tiunov AV, Scheu S (2005) Biodiversity and litter decomposition in terrestrial ecosystems. Annu Rev Ecol Evol Syst 36:191–218
Holland EA, Firestone MK, Vitousek PM (1999) The effects of soil moisture and temperature on microbial activity and nitrogen availability in tropical forests. Glob Chang Biol 5(5):389–398
Johnson DW, Matson PA (2009) Nitrogen and phosphorus availability and their influence on soil carbon dynamics. Annu Rev Environ Resour 34:35–58
Jones DL, Nguyen C, Finlay RD (2009) Carbon flow in the rhizosphere: carbon trading at the soil–root interface. Plant Soil 321(1–2):5–33
Keenan RJ, Reams GA, Achard F, de Freitas JV, Grainger A, Lindquist E (2015) Dynamics of global forest area: results from the FAO Global Forest Resources Assessment 2015. For Ecol Manag 352:9–20
Kögel-Knabner I (2002) The macromolecular organic composition of plant and microbial residues as inputs to soil organic matter. Soil Biol Biochem 34(2):139–162
Koprivova A, Kopriva S (2014) Sulfate uptake and assimilation in plants. Front Plant Sci 5:594
Lal R (2005) Soil degradation and the carbon cycle. Crit Rev Plant Sci 24(2):145–157
Landeweert R, Hoffland E, Finlay RD, Kuyper TW, van Breemen N (2001) Linking plants to rock, ectomycorrhizal fungi mobilize nutrients from minerals. Trends Ecol Evol 16:248e253
Likens GE, Bormann FH, Johnson NM (2002) Biogeochemistry of a forested ecosystem, 2nd edn. Springer
Liu J, Zhang L, Liu X (2020) Impacts of nitrogen deposition on ecosystem carbon dynamics and climate change. Nat Commun 11:4694
Malhi Y, Baldocchi DD, Jarvis PG (1999) The carbon balance of tropical, temperate, and boreal forests. Plant Cell Environ 22(6):715–740
Manzoni S, Schimel JP, Porporato A (2012) Responses of soil microbial communities to water stress: results from a meta-analysis. Ecology 93(4):930–938
Marschner P (2012) Marschner's mineral nutrition of higher plants, 3rd edn. Academic Press
Mohn WW, Stewart GR (2000) Microbial populations and enzymes involved in the degradation of polychlorinated biphenyls. Microbiol Mol Biol Rev 64(3):500–531
Norby RJ, Zak DR, Pregitzer KS, King JS, Holmes WE (2010) Atmospheric CO_2 enrichment of a northern hardwood forest: the role of nitrogen in the response of carbon and nitrogen cycles. Glob Chang Biol 16(2):378–396
Ojima DS, Parton WJ, Schimel DS, Owensby CE (1993) Simulated impacts of transient climate change on carbon storage and nitrogen cycling in grasslands. Ecol Appl 3(3):373–389
Paul EA, Clark FE (1996) Soil microbiology and biochemistry: an introduction. CRC Press
Perry DA, Oren R, Hart SC (2008) Forest ecosystems. JHU Press
Phillips RP, Finzi AC, Bernhardt ES (2011) Enhanced root exudation induces microbial feedbacks to N cycling in a pine forest under long-term CO_2 fumigation. Ecol Lett 14(2):187–194
Pieterse CMJ, Van der Does D, Zamioudis C, Léon-Reyes A, Van Wees SCM (2014) Induced systemic resistance by beneficial microbes. Annu Rev Phytopathol 52:347–375
Pilon-Smits EAH (2005) Phytoremediation. In: Hillel D (ed) Encyclopedia of soils in the environment. Elsevier, pp 307–315
Prescott CE (2010) Litter decomposition and nutrient cycling in forest ecosystems. Springer
Prosser JI (2002) Molecular and functional diversity in soil micro-organisms. Plant Soil 244:9–17
Quiza L, St-Arnaud M, Yergeau E (2015) Harnessing phytomicrobiome signaling for rhizosphere microbiome engineering. Front Plant Sci 6:507

Rodriguez H, Fraga R (1999) Phosphate solubilizing bacteria and their role in plant growth promotion. Biol Fertil Soils 29(3):203–210
Rousk J, Brookes PC, Bååth E (2010) The microbial PLFA composition as affected by pH in an arable soil. Soil Biol Biochem 42(3):516–520
Schimel JP (2001) Biogeochemical cycles: the feedbacks between climate and ecosystem processes. Glob Chang Biol 7(8):701–715
Schimel JP, Bennett J (2004) Nitrogen mineralization: challenges and opportunities. Ecol Appl 14(3):800–811
Schimel J, Balser TC, Wallenstein M (2007) Microbial stress-response physiology and its implications for ecosystem function. Ecology 88(6):1386–1394
Schindler DW, Turner MA, Muir DC (1986) Effects of nitrogen fertilization on nutrient cycling in a boreal forest. Ecology 67(4):1163–1175
Schlesinger WH, Andrews JA (2000) Soil respiration and the global carbon cycle. Biogeochemistry 48:7–20
Schlesinger WH, Bernhardt ES (2013) Biogeochemistry: an analysis of global change, 3rd edn. Academic
Schmidt MWI, Torn MS, Abiven S, Dittmar T, Guggenberger G, Lichter J et al (2011) Persistence of soil organic matter as an ecosystem property. Nature 478(7367):49–56
Sebastiana M, Martins J, Figueiredo A, Monteiro F, Sardans J, Peñuelas J et al (2017) Oak protein profile alterations upon root colonization by an ectomycorrhizal fungus. Mycorrhiza 27:109–128
Six J, Elliott ET, Paustian K, Coleman DC (2002) Soil organic matter, biota, and aggregation in temperate and tropical soils. Agronomie 22:755–775
Smith SE, Read DJ (2008) Mycorrhizal symbiosis, 3rd edn. Academic
Stevenson FJ (1994) Humus chemistry: genesis, composition, reactions, 2nd edn. Wiley
Swift MJ, Heal OW, Anderson JM (1979) Decomposition in terrestrial ecosystems. University of California Press
Swift MJ, Heal OW, Anderson JM (1998) Decomposition in terrestrial ecosystems. University of California Press
Tisdall JM, Oades JM (1982) Organic matter and water-stable aggregates in soils. J Soil Sci 33(2):141–163
Treseder KK (2008) Nitrogen additions and microbial biomass: a meta-analysis of ecosystem studies. Ecol Lett 11(10):1111–1120
Turner BL, Lambers H, Cramer MD (2014) The role of phosphorus in plant nutrition. Plant Soil 374(1–2):77–92
Vessey JK (2003) Plant growth promoting rhizobacteria as biofertilizers. Plant Soil 255(2):571–586
Vitousek PM, Howarth RW, Waring RH (2002) The role of nitrogen in the biogeochemical cycling of ecosystems. Glob Chang Biol 8(2):211–227
Vitousek PM, Pardo LH, Harpole WS (2010) Terrestrial nitrogen deposition: its effects on plant, soil, and ecosystem processes. Ecol Appl 20(1):6–15
Wang X, Wang X, Xie Y, Li Y (2017) Extreme weather events result in soil erosion, nutrient leaching, and changes in soil structure, all of which impact microbial communities. J Environ Manag 203:102–111
Wang Y, Xiao B, Wang W, Revillini D, Delgado-Baquerizo M (2023) Biocrust adaptations to microhabitat alter bacterial communities in a semiarid ecosystem. Plant Soil 492(1):413–427
Young IM, Crawford JW (2004) Interactions and self-organization in the soil-microbe complex. Science 304(5677):1634–1637

Guggulsterone Biosynthesis: Milestones, Prospects and Challenges

6

Shobha Mehra, Ashok Kumar Parmar, and Tarun Kant

Abstract

The plants produce different kinds of bioactive compounds for their primary and secondary growth and metabolism known as metabolites. These metabolites includes flavonoids or polyphenolic compounds, terpenoids and nitrogen containing alkaloids or sulphur compounds having potential to fight against pathogens, to interact with nearby plants and to attract pollinators. They are also important source of human nutrition and health and are used in Indian Ayurveda since ages. *Commiphora wightii* commonly known as 'Guggul' having medicinal values due to the presence of an oleo-gum-resin known as guggulsterone is found in desert regions. Over the time, *C. wightii* population in their natural habitat have been severely reduced due to the extraction of gum guggulu by local people. The poor seed germination, slow-growing nature, climate change and over-exploitation have brought the plant under critically endangered species recorded by the International Union for Conservation of Nature. The present chapter reviews the biosynthesis of guggulsterone, discovery, prospects and associated challenges with respect to plant tissue culture technology for sustainable production of guggulsterone.

Keywords

Guggul · Guggulsterone · Tissue culture · Bioreactors · Secondary metabolites

S. Mehra · A. K. Parmar · T. Kant (✉)
ICFRE-Arid Forest Research Institute, Jodhpur, Rajasthan, India
e-mail: tarunkant@icfre.org

R. K. Kalia, R. Pathak (eds.), *Tree Biology and Biotechnology*,
https://doi.org/10.1007/978-981-96-0002-1_6

6.1 Introduction

Plants synthesize different types of bioactive compounds for their primary and secondary growth and metabolism. Compounds that are essential for plant's basic functioning are called as primary metabolites, whereas compounds that are being synthesized against any kind of biotic and abiotic stress for the survival of plant is called as secondary metabolites. These metabolites are grouped into three major categories, that is, flavonoids or polyphenolic compounds, terpenoids and nitrogen containing alkaloids or sulphur compounds. These metabolites are useful for plants to fight against pathogens, to interact with nearby plants and to attract pollinators (Pichersky and Gershenzon 2002). They are also important in case of human nutrition and health as used in Indian Ayurveda since ages. Human uses these secondary metabolites from different medicinal plants for medicines, flavouring, pigments and dyes. Medicinal plants are the rich repertoire of these bioactive compounds; hence, these are used in pharmaceuticals and nutraceuticals.

Commiphora wightii (Arn.) Bhandari, belonging to Burseraceae family and commonly known as 'Guggul' having medicinal values, is found in desert regions having enormous therapeutic efficacy due to presence of an oleo-gum-resin known as guggulsterone. As a result, it is regarded as a '*Divyaushidhi*' in Ayurveda. This resin is a mixture of many bioactive compounds which are used to treat many disorders and diseases. The key signature bioactive compound obtained from guggul is guggulsterone (E and Z isomers) that has many applications in human medicines like hypocholesteramic, anti-obesity, anti-cancer properties, etc. Several ayurvedic companies sell their formulations made with guggul gum under different trade names. Additionally, it is employed in the incense, perfumery, fumigation, and silk dyeing industries. Over the time, *C. wightii* population in their natural habitat have been severely reduced due to the extraction of gum guggulu by local people through tapping methods wherein the trunk of the plant is tapped to collect the gum-resin, leading to the death of the plant. Along with this issue, other drawbacks are poor seed germination, slow-growing nature of plant, climate change and over-exploitation, which bring the plant under critically endangered species recorded by the International Union for Conservation of Nature (Ved et al. 2015).

6.2 Gum Guggulu

The trunk of the plant exudes oleo-gum-resin upon a sharp incision, which is a homogenous combination of resins (61%) and gum (29.3%), along with 6.1% water, 0.6% volatile oil, and 3.2% foreign matter. Monoterpenoids, diterpenoids, triterpenoids, flavonoids, steroids, carbohydrates, and lignan are among the principal ingredients. Myrcene, dimyrcene, and polymyrcene are found in the oil portion (Bhati 1950). Series of guggulsterol from I to VI, E-guggulsterone and Z-guggulsterone are the various steroidal ingredients that have been reported in guggul plant (Patil et al. 1972; Bajaj and Dev 1982), along with guggulsterol-Y and guggulsterone-M (Hanus et al. 2005). This oleo-gum-resin is synthesized in specialized tissues called resin canal or balsam canal which are present in the phloem of the larger veins (Sabnis 1920). The shape of gum resin ducts is round to oval and distributed irregularly in the phloem tissue. These ducts are prominently present in the bark of the plant in an irregular arrangement that forms a network (Nair 1995; Bhatt et al. 1989).

6.2.1 Guggulsterone

The main constituents of gum guggulu are guggulsterone E and Z having great medicinal values. These are the phytosterols that exist as two stereoisomers, which are interconvertible in nature. The structures of both the isoforms are shown in Fig. 6.1.

Different studies have been conducted for their activity and usages against various medical problems. In a systemic review and meta-analysis by Gupta et al. (2023), anti-cancer activity of guggulsterone was reported against various cancers disorders, viz., pancreatic cancer, hepatocellular carcinoma, oesophageal adenocarcinoma, prostate cancer, colon cancer, breast cancer, leukaemia, etc. Other use of these compounds has been described in the treatment of hyperlipidemia (Verma and Bordia 1988), cardioprotective agent (Wang et al. 2012) and weight loss agent (Kimura et al. 2001) and help to control dengue virus replication (Chen et al. 2021). Kunnumakkara et al. (2018) reviewed the potential of guggul for avoidance of chronic diseases and documented that it is effectively against various chronic diseases, viz., arthritis, Alzheimer's disease, asthma, cancer, dermatitis, diabetes, gingivitis, IBD, infectious diseases, intestinal metaplasia, otitis media, psoriasis, pancreatitis, respiratory diseases, etc. Moreover, due to its efficiency of targeting multiple signalling pathways, it also exhibits anti-inflammatory, anti-oxidant, cardioprotective, hepatoprotective, hypolipidemia, neuroprotective, and thyroid triggering effects. Use of guggul in various in vitro, in vivo and clinical studies is shown in Fig. 6.2.

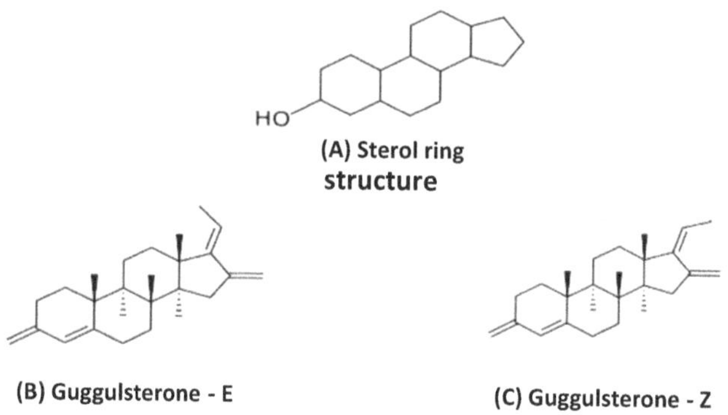

Fig. 6.1 (a) Structure of sterol backbone, (b) Guggulsterone E, (c) Guggulsterone Z

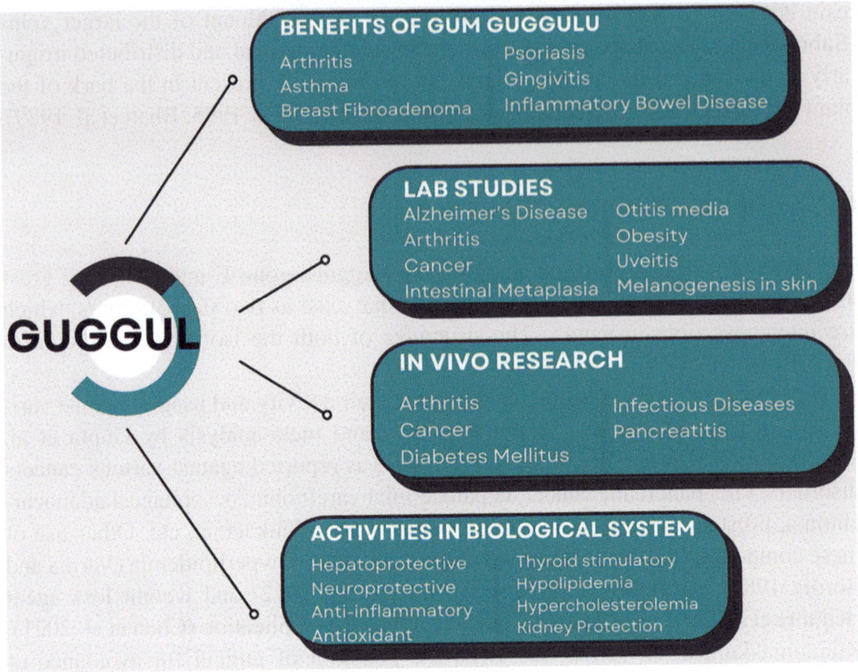

Fig. 6.2 Benefits of gum guggulu and different studies for use in treatment of diseases

6.2.2 Biosynthesis

The phytosterol biosynthesis pathway has been studied in *Arabidopsis thaliana* showing that this is a multistage pathway having many enzymes and proteins to carry out the reactions (Benveniste 2004, Schaeffer et al. 2001). Guggulsterone is the phytosterol present in *C. wightii* and their biosynthetic pathway is very crucial to understand. In a recent review, metabolic pathway of phytosterol has been discussed (Mehra and Kant 2021) and revealed that phytosterol metabolic pathway is having similarity with the synthesis of cholesterol in animals. Sterols are the structural components of biological membranes and have functional roles in signalling, growth and development. These are found in all eukaryotes as isopentenyl pyrophosphate derivatives.

After glycolysis, phytosterols are synthesized via the mevalonate pathway starting with the synthesis of Acetyl Co-A which is further converted to mevalonic acid by the enzyme HMGR (3-hydroxy-3-methyl glutaryl coenzyme A reductase). The FPP (Farnesyl Diphosphate) is further synthesized by mevalonic acid, and the formation of Squalene after many intermediate compound formation is the first enzymatic step towards the synthesis of phytosterols. Conversion of squalene to 2,3-oxidosqualene by squalene epoxidase enzyme is called as the cyclization step and also first diversifying step towards phytosterol synthesis. In higher plants, campesterol and sitosterol are being synthesized by the formation of cycloartenol catalysed by cycloartenol synthase, and this is the first committed step towards phytosterol biosynthesis. Sterol-C-methyltransferase (SMT) is the enzyme that catalyses the transfer of ethyl or methyl group at C24 position of cycloartenol heading the pathway towards final synthesis of phytosterols (Mehra and Kant 2021). The diagrammatic representation is shown in Fig. 6.3.

6.3 Plant Tissue Culture Technology for Sustainable Production

In the era of plant tissue culture technique, the increasing demand for natural products with medicinal properties has been the main catalyst for research and development. *C. wightii* arises as the centre of attraction in the exploration of biotechnological tools for sustainable production as the scientific interest deepens on unlocking the full potential of natural remedies. The optimization of plant cell cultures is a crucial task requiring a thorough exploration of physical, chemical and biological factors. Plant cell cultures act as a versatile platform for the development of bioactive compounds, secondary metabolites and pharmaceuticals making them valuable targets for process enhancement.

Production of large amount of plants using nodal segments can be done through micropropagation technique, which paved the path to conserve this endangered species. Micropropagation is affected by many factors like quality of explants, time of collection, genotype, and physiological conditions of the tissue. Micropropagation of guggul plant was carried out by the seedling explants (Yusuf et al. 1999; Kant et al. 2010), axillary shoot propagation obtained from nodal segments (Barve and Mehta 1993; Soni 2010), and shoot tips, nodes, internodes, and leaves (Singh et al. 2010).

It is possible to generate a significant amount of *C. wightii* biomass without harming plant's natural population through callus culture and cell suspension culture, the two biotechnological techniques for producing cell biomass under in vitro conditions. Plant cell suspension culture has become a promising means of generating secondary metabolites because of its consistency, simplicity and ease of chemical extraction.

Prajapati et al. (2010) reported callus culture of *C. wightii*, utilizing various growth regulators with different explants, viz., leaves, nodal segments, immature zygotic embryos, seedlings, etc. Maximum callusing had been reported on MS media enriched by 2 mg/L NAA for all four explants. Maximum biomass of 8.43%, which was higher than the rest of the explants, was produced by immature zygotic embryos (Mehra and Kant 2021). According to this study, immature zygotic embryos were found to be most appropriate explant to develop biomass under in vitro environments for guggulsterone production.

Immature zygotic embryos were used to produce embryogenic callus of Guggul as an explant by reciprocal transfer on to a media comprising of 2,4,5-trichlorophenoxy acetic acid (0.1 mg/L) plus kinetin (0.1 mg/L) without any growth regulators. The development of embryogenic callus was maximum on an

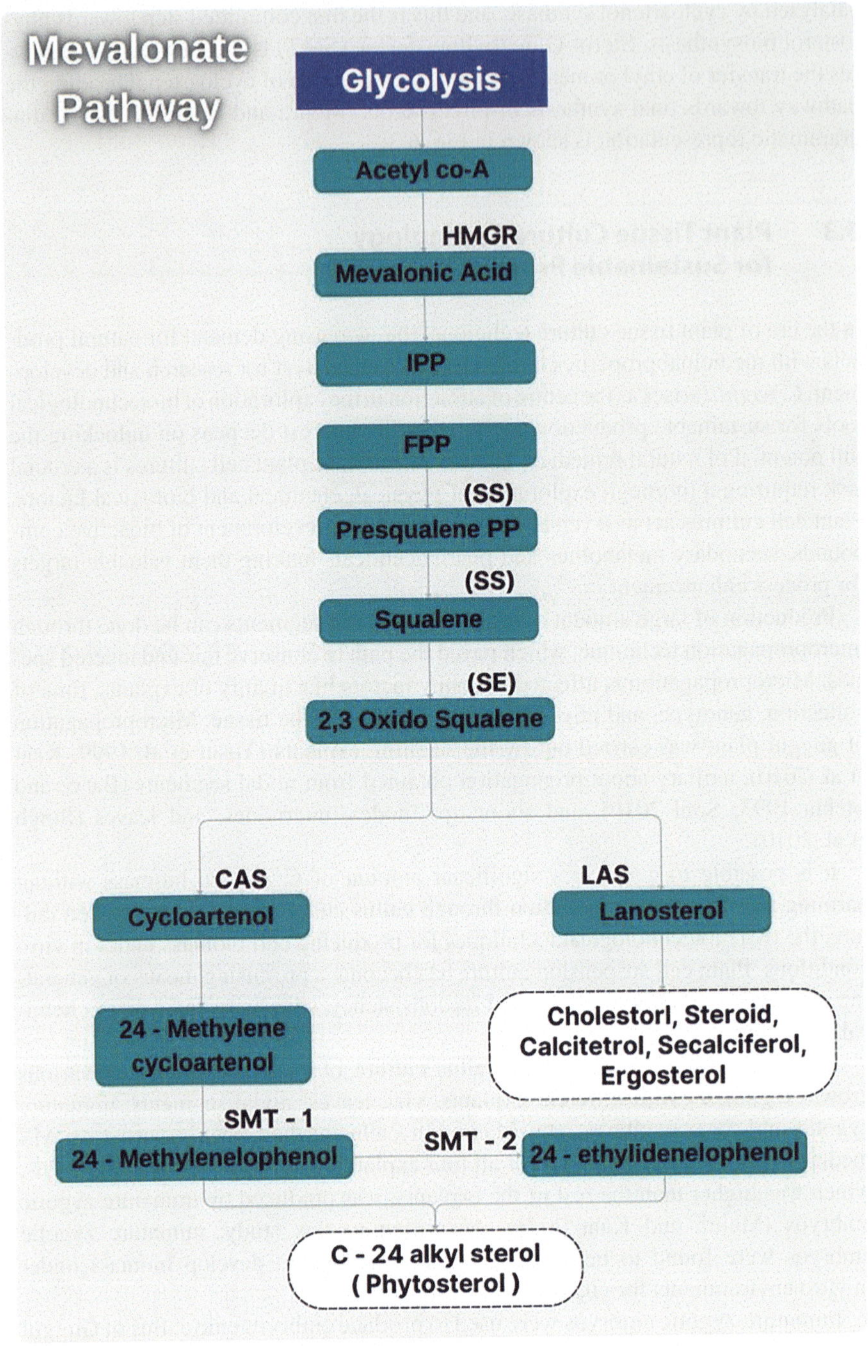

Fig. 6.3 Mevalonate pathway for phytosterol biosynthesis

altered MS media enriched with BA (0.25 mg/L) and IBA (0.1 mg/L) (Kumar et al. 2003). An efficient and refined protocol was developed by Parmar and Kant (2014) for somatic embryogenesis for regeneration of *C. wightii*. Immature fruits were subjected to culture grown on Gamborg's B5 media having 2.26 µM 2,4-D to acquire embryogenic callus. Further, a modified MS medium supplemented with 1.11 µM BAP plus 0.57 µM IBA enriched with 0.5% activated charcoal or hormone-free medium was used for a sub-cultured embryogenic callus. Somatic embryos were obtained then germinated, and emblings were transferred to field for hardening and acclimatization.

Suspension culture studies and growth pattern analysis were done by Mehra and Kant (2020), and the study reported a curve having different phases of growth of culture. The growth of cultures was higher in medium containing hormone as compared to hormone-free medium. The hormone-free medium also gave good cell biomass that can be further used because in hormone-free medium cells convert into embryogenic state. Cell suspension growth kinetics of guggul cultures revealed that the optimum time of cell growth was between 21 and 27 days (Exponential Phase).

6.4 Production of Guggulsterone

Guggulsterone production by different biotechnological methods can be achieved under in vitro conditions, and many studies have been done on this plant. Suri and Ramawat (1995) stated that cyto-differentiation is required to get a high yield of guggulsterone in callus and cell suspension cultures. Resin canal formation was observed in somatic embryos at the torpedo and cotyledonary phase by Kumar et al. (2004) during somatic embryogenesis in the callus cultures. They also reported that non-embryogenic calli produced a lower amount of guggulsterone as compared to the somatic embryos (Kumar et al. 2006). Plant tissue culture technologies provide a systematic approach in producing secondary metabolites under controlled conditions with further manipulations of growth conditions and growth parameters (Ramawat and Mathur 2006).

Gugggulsterone production under culture conditions was achieved by many researchers. Callus culture was initiated using leaf explants and zygotic embryos for good biomass and optimum gain of the secondary metabolites on modified Murashige and Skoog medium. Maximum production was achieved at 35th day of culture that was ~15 µg per g of dried sample (Mathur et al. 2007).

Cell suspension culture is another promising tool of biotechnology for production of bioactive compounds. A total of ~5 µg/g of guggulsterone production was achieved via cell suspension culture of *C. wightii* inoculated on altered MS medium having 2,4-Dichlorophenoxyacetic acid (0.5 mg/L) plus Kinetin (0.25 mg/L). Stirred tank bioreactor (2 L) was used to grow the cells, and 5.5 g/L growth was recorded along with production of 36 µg/L of guggulsterone. The optimum inoculum size was 10% (v/v) for best growth and guggulsterone production (Mathur and Ramawat 2007). Other studies focused on guggulsterone production by the use of different elicitors, growth retardants and plant gums.

6.5 Applications of Bioreactors in Secondary Metabolite Production

Bioreactors are closed vessel for the culture of cells in a controlled environment under aseptic conditions. There are three main categories (batch, fed-batch, and continuous culture) of bioreactors according to the medium supply and conditions required for the optimum production of the compounds. Plant bioreactors are used for the production of therapeutic products, nutritional products, vaccine antigens, biodegradable plastics, and industrial products. Utilization of bioreactors in clonal propagation (Gurel 2009) and cultivation of hairy roots for secondary metabolite production (Ziv 2000) are some of the examples of bioreactor applications. Dewir et al. (2007) utilized bubble column bioreactors in *Spathiphyllum cannifolium* to produce flowers, while Sreedhar et al. (2008) utilized this technique in *Stevia rebaudiana* for the production of biomass.

The production of anthraquinones from cultivation of *Frangula alnus* was done using airlift bioreactors (Sajc et al. 1995). A 1000-L airlift bioreactor was used by Wu et al. (2007) for the production of adventitious roots of *Echinacea purpurea*. Another type of bioreactor was developed by Shiau (2009) for fungus, plant cells, tissue, organs, hairy roots, and plantlets. The airlift bioreactors having several arrangements were constructed for utilization in the plant cell culture-based processes. The first attempt for cultivating plant cells in bioreactor was made in the 1970s. Since the plant cells are extremely sensitive to high shear, only airlift bioreactors were find to be more acceptable (Georgiev et al. 2009). The first industrial process was introduced in an airlift bioreactor of 750 L capacity for the cell culture of *Lithospermum erythrorhizon* (Georgiev et al. 2013).

The classical submerged bioreactor systems, for example, stirred tank and bubble column, have several challenges and constraints during their use in hyperhydricity (vitrification) of cultured plantlets; hence, a novel bioreactor, that is, Temporary Immersion System, was engineered and was specially engaged to carry out plant micro-propagation activities (Georgiev et al. 2014) and also in the field of biopharmaceuticals (Georgiev 2015) to create a controlled environment having aseptic conditions.

6.6 Conclusion

Commiphora wightii regarded as a '*Divyaushidhi*' in Ayurveda is found under adverse climates of desert regions. The plant commonly known as 'Guggul' contains an oleo-gum-resin known as guggulsterone that have immense medicinal values. This resin is a mixture of many bioactive compounds which are used to treat many disorders and diseases in human beings. Additionally, it is employed in the incense, perfumery, fumigation, and silk dyeing industries. Over the time, *C. wightii* population in their natural habitat has been rigorously affected due to the extraction of gum guggulu by local people through tapping methods leading to the death of the plant. Due to the poor seed germination, slow-growing nature of plant, climate change and over-exploitation, the plant species has become critically endangered. Hence, in vitro propagation of the plant species is an urgent need for its conservation and mass-production for exploitation of oleo-gum-resin.

References

Bajaj AG, Dev S (1982) Chemistry of Ayurvedic crude drugs-V: Guggulu (resin from *Commiphora mukul*)-5: some new steroidal components, and stereochemistry of guggulsterol-I at C-20 and C-22. Tetrahedron 38(19):2949–2954

Barve DM, Mehta AR (1993) Clonal propagation of mature elite trees of *Commiphora wightii*. Plant Cell Tissue Organ Cult 35:237–244

Benveniste P (2004) Biosynthesis and accumulation of sterols. Annu Rev Plant Biol 55:429–457

Bhati A (1950) Essential oil from the resin of *Commiphora mukul*, Hook. Ex Stocks. J Indian Chem Soc 27:436–440

Bhatt JR, Nair MNB, Mohanram HY (1989) Enhancement of oleo gum resin production in *Commiphora wightii* by improved tapping technique. Curr Sci 58(7):349–357

Chen WC, Wei CK, Hossen M, Hsu YC, Lee JC (2021) (E)-Guggulsterone inhibits dengue virus replication by upregulating antiviral interferon responses through the induction of heme oxygenase-1 expression. Viruses 13:712

Dewir YH, Chakrabarty D, Ali MB, Singh N, Hahn E, Paek K (2007) Influence of GA_3, sucrose and soil medium/bioreactor culture on in vitro flowering of *Spathiphyllum* and association of glutathione metabolism. Plant Cell Tissue Organ Cult 90:225–235

Georgiev V (2015) Mass propagation of plant cells—an emerging technology platform for sustainable production of biopharmaceuticals. Biochem Pharmacol 4:e180

Georgiev MI, Weber J, Maciuk A (2009) Bioprocessing of plant cell cultures for mass production of targeted compounds. Appl Microbiol Biotechnol 83:809–823

Georgiev MI, Eibl R, Zhong JJ (2013) Hosting the plant cells in vitro: recent trends in bioreactors. Appl Microbiol Biotechnol 97:3787–3800

Georgiev V, Schumann A, Pavlov A, Bley T (2014) Temporary immersion systems in plant biotechnology. Eng Life Sci 14:607–621

Gupta M, Singh D, Rastogi S, Siddique HR, Al-Dayan N, Ahmad A, Sikander M, Sarwat M (2023) Anti-cancer activity of guggulsterone by modulating apoptotic markers: a systematic review and meta-analysis. Front Pharmacol 14:1155163. Frontiers Media S.A.

Gurel A (2009) Propagation possibilities of commercially important plants through tissue culture techniques. Int J Nat Eng Sci 3(2):07–09

Hanus LO, Rezankab T, Dembitskya VM, Moussaieffa A (2005) Myrrh- Commiphora chemistry. Biomed Pap 149(1):3–28

Kant T, Prajapati S, Parmar AK (2010) Efficient micropropagation from cotyledonary node cultures of *Commiphora wightii* (Arn.) Bhandari, an endangered medicinally important desert plant. J Plant Dev 17:37–48

Kimura I, Yoshiwaka M, Kobayashi S, Sugihara Y, Suzuki M, Oominami H, Murakami T, Matsuda H, Doiphode VV (2001) New triterpenes, myrrhanol a and myrrhanone a, from guggul gum resins, and their potent anti-inflammatory effect on adjuvant induced air pouch granuloma of mice. Bioorg Med Chem Lett 11:985–989

Kumar S, Suri SS, Sonie KC, Ramawat KG (2003) Establishment of embryonic cultures and somatic embryogenesis in callus culture of guggul-*Commiphora wightii* (Arn.) Bhandari. Indian J Exp Biol 4:69–77

Kumar S, Suri SS, Sonie KC, Ramawat KG (2004) Development of resin canals during somatic embryogenesis in callus culture of *Commiphora wightii*. Indian J Biotechnol 3:267–270

Kumar S, Mathur M, Jain AK, Ramawat KG (2006) Somatic embryo proliferation in *Commiphora wightii* and evidence for guggulsterone production in culture. Indian J Biotechnol 5:217–222

Kunnumakkara AB, Banik K, Bordoloi D, Harsha C, Sailo BL, Padmavathi G, Roy NK, Gupta SC, Aggarwal BB (2018) Googling the Guggul (*Commiphora* and *Boswellia*) for prevention of chronic diseases. Front Pharmacol 9:686

Mathur M, Ramawat KG (2007) Guggulsterone production in cell suspension cultures of the guggul tree, *Commiphora wightii*, grown in shake flasks and bioreactors. Biotechnol Lett 29:979–982

Mathur M, Jain AK, Dass S, Ramawat KG (2007) Optimization of guggulsterone production in callus cultures of *Commiphora wightii* (Arnott.) Bhandari. Indian J Biotechnol 6:525–531

Mehra S, Kant T (2020) Pharmaceutically active cell biomass growth pattern under cell suspension culture of *Commiphora wightii*—a critically endangered medicinal plant. J Plant Dev 27:71–81

Mehra S, Kant T (2021) Gum Guggulu—Oleogum resin from *Commiphora wightii* (Arnot.) Bhandari: source, medicinal importance and biosynthesis. LS Int J Life Sci 10:214–224

Nair MNB (1995) Some notes on gum and resin ducts and cavities in angiosperms. In: Iqbal M (ed) The cambial derivatives, vol 9. Gebru der Borntraeger, Berlin, pp 317–340

Parmar AK, Kant T (2014) Efficient somatic embryogenesis and molecular marker based analysis as effective tools for conservation of the red-listed plant *Commiphora wightii*. J BioSci Biotech 2:169–182

Patil VD, Nayak UR, Dev S (1972) Chemistry of Ayurvedic crude drugs-I: Guggulu (resin from *Commiphora mukul*)-1: steroidal constituents. Tetrahedron 28(8):2341–2352

Pichersky E, Gershenzon J (2002) The formation and function of plant volatiles: perfumes for pollinator attraction and defense. Curr Opin Plant Biol 5:237–243

Prajapati S, Parmar AK, Kant T (2010) Growth optimization of *Commiphora wightii* callus – a potential source of *in vitro* guggalsterone production. J Phytological Res 23(1):35–40

Ramawat KG, Mathur M (2006) Factors affecting production of secondary metabolites. In: Ramawat KG, Merillon JM (eds) Biotechnology: secondary metabolites. Science Publishers, Inc., Enfield, pp 59–102

Sabnis TS (1920) The physiological anatomy of the plants of the Indian desert. J Indian Bot 1:183–205

Sajc I, Vunjak-Novakovic G, Grubisic D, Kovaccvic N, Vukovic D, Bugarski B (1995) Production of anthraquinones by immobilized *Frangula alnus* Mill. plant cells in four-phase air-lift bioreactor. Appl Microbiol Biotechnol 43:416–423

Schaeffer A, Bronner R, Benveniste P, Schaller H (2001) Plant sterolC24methyl transferases: different profiles of tobacco transformed with SMT1 or SMT2. Plant J 25:605–615

Shiau YJ (2009) Bioreactor for growing fungus, plant cell, tissue, organ, hairy roots and plantlet. US Patent 7531350

Singh N, Garg A, Yadav K, Kumari S (2010) Influence of growth regulators on the explants of *Commiphora mukul* (Hook. Ex Stocks) Engl. under *in-vitro* conditions. Researcher 2:41–48

Soni V (2010) Efficacy of *in-vitro* tissue culture versus stem cuttings for propagation of *Commiphora wightii* in Rajasthan, India. Conserv Evid 7:91–93

Sreedhar RV, Venkatachalam L, Thimmaraju R, Bhagyalakshmi N, Narayan MS, Ravishankar GA (2008) Direct organogenesis from leaf explants of *Stevia rebaudiana* and cultivation in bioreactor. Biol Plant 52(2):355–360

Suri SS, Ramawat KG (1995) In-vitro hormonal regulation of laticifers differentiation in *Calotropis procera*. Ann Bot 75:477–480

Ved D, Saha D, Ravikumar K, Haridasan K (2015) *Commiphora wightii*, the IUCN red list of threatened species. 2015:e.T31231A50131117

Verma SK, Bordia A (1988) Effect of *Commiphora mukul* (Gum Guggulu) in patients of hyperlipidemia with special reference to HDL-cholesterol. Indian J Med Res 87:356–360

Wang WC, Uen YH, Chang ML (2012) Protective effect of guggulsterone against cardiomyocyte injury induced by doxorubicin *in vitro*. BMC Complement Altern Med 12:138

Wu HC, du Toit ES, Reinhardt CF (2007) A protocol for direct somatic embryogenesis of *Protea cynaroides* L. using zygotic embryos and cotyledon tissues. Plant Cell Tissue Org Cult 89:217–224

Yusuf A, Rathore TS, Shekhawat NS (1999) Micropropagation of *Commiphora wightii* (Arn.) Bhandari—a threatened medicinal plant of semi-arid region. Indian J Plant Genet Res 12:371–375

Ziv M (2000) Bioreactor technology for plant micro-propagation. In: Janick J (ed) Horticultural reviews, vol 24., ISBN 0-471-33374-3. Wiley

Revolutionizing Tree Improvement Through Biotechnology

7

M. N. Ashwath, K. Bindyalaxmi, Sourav Ranjan Mohapatra, R. Shifin, and K. Shiran

Abstract

Tree improvement is aimed at excelling genetic makeup and productivity of trees to meet the evolving ecological and economic demands. Though conventional tree improvement techniques have served better, however, their intrinsic limitations impede further progress. Biotechnology tools offer a revolutionary path forward in tree improvement, addressing limitations of conventional methods and opening doors to accelerate breeding, precise trait selection, and various conservational efforts. This chapter explores the paradigm shift occurring in tree improvement by harnessing the biotechnological tools including molecular markers, genetic transformation, tissue culture, genomic sequencing, and marker-assisted selection (MAS). Each tool is discussed in depth, elucidating its applications and contributions to the realm of tree improvement. Molecular markers, such as SSRs and SNPs, have revolutionized tree improvement by enabling pre-

M. N. Ashwath
College of Forestry, Thrissur, Kerala Agricultural University, Thrissur, Kerala, India

College of Agriculture, Gangavathi, University of Agricultural Sciences, Raichur, Karnataka, India

K. Bindyalaxmi · R. Shifin
College of Forestry, Thrissur, Kerala Agricultural University, Thrissur, Kerala, India

S. R. Mohapatra
College of Forestry, Thrissur, Kerala Agricultural University, Thrissur, Kerala, India

College of Forestry, Bhubaneswar, Odisha University of Agriculture and Technology, Bhubaneswar, Odisha, India

K. Shiran (✉)
College of Forestry, Thrissur, Kerala Agricultural University, Thrissur, Kerala, India

ICAR-CAZRI-Regional Research Station, Jaisalmer, Rajasthan, India
e-mail: Shiran.k@icar.gov.in

R. K. Kalia, R. Pathak (eds.), *Tree Biology and Biotechnology*, https://doi.org/10.1007/978-981-96-0002-1_7

cise genetic mapping and diversity analysis. Genetic transformation involves introduction of desirable traits such as disease resistance and enhanced wood quality, along with micropropagation gained more importance in the commercial sector. Tree genome sequencing has provided invaluable insights into tree genetics and breeding, facilitating identification of beneficial genes and genomic regions which facilitated MAS. Despite ethical, regulatory, and cost-related challenges, potential biotechnological tools in tree improvement remain undeniable. Biotechnological advances continue to unfold and uphold the promise of revolutionizing tree cultivation and management.

Keywords

Forest biotechnology · Molecular markers · Genome sequencing · Tissue culture

7.1 Introduction

Biotechnology plays a crucial role in forestry and tree improvement by offering innovative techniques for breeding and conservation of agrobiodiversity. Biotechnology supports the management of sustainable genetic variation in the context of forest tree improvement by generating well-adapted genotypes of required quality for operational planting. Biotechnological tools considered as a solution for meeting the growing demands of forest and forest products open the door to understanding a wider range of composite biological issues associated with various tree species. Initial population establishment, species testing, breeding, gene conservation, and population enhancement through selection and recombination cycles are major concerns in forest biotechnology (Fang and Han 2019). Incorporating biotechnology into forestry lowers expenses, expands the wood supply, and tackles environmental issues such as sustainable forest management and habitat restoration (Martínez-Gómez 2019).

In the present era, the primary emphasis in tree-improvement efforts is directed towards the management of genetic resources encompassing selection of superior clones from existing forests, safeguarding genetic diversity, engaging in controlled propagation methods, and employing traditional breeding techniques to enhance desired characteristics (Panda et al. 2020). While numerous agricultural biotechnology firms are actively involved in the development of biotechnologically mediated crops, vegetables, and ornamental plants, it has become a standard practice in agriculture; however, application of these technologies to forest-tree species has been notably slower. Nevertheless, with the progression of plant-transformation methodologies, gene identification, and characterization techniques, it is imperative that the focus of molecular breeding and genetic engineering in the twenty-first century shifts towards forest trees. The field of forestry can harness the power of biotechnology to achieve efficient and sustainable tree improvement results by carefully balancing research objectives and integrating biotechnology with sound tree improvement programs.

Revolutionary technologies that combine classical morphology-based selection techniques with molecular marker technologies such as gene pyramiding, QTL documentation, MAS, and creation of transgenic crops characterize plant breeding practices. Molecular markers play a crucial role in understanding genetic relatedness among the individuals in the breeding population or germplasm. The emergence of high-throughput sequencing technologies has led to the development of numerous single-nucleotide polymorphism (SNP) markers across various germplasm or individuals (Brown et al. 2004; Ashwath et al. 2023). Marker-assisted gene pyramiding facilitates the creation of biotic- and abiotic-resistant varieties by consolidating multiple genes from different parental sources into a specific genotype. Furthermore, innovative methods such as restriction-site-associated DNA sequencing (RAD-seq), genotyping-by-sequencing, and multiplexed shotgun genotyping have unwrapped novel opportunities for plant breeding research.

7.2 Biotechnology Tools in Tree Improvement

In the dominion of tree improvement, three key areas hold promise for biotechnology applications; Genetic markers for identifying trees harbouring desirable genes, innovative tissue culture methods for generating enhanced planting stock, and genetic engineering for developing trees with novel traits and gaining insights into cell biology and function.

7.2.1 Molecular Markers

Molecular markers are an alternative way of understanding the variation in measurable features such as morphological appearance, vigour, and quality traits. These traits are governed by the cumulative effects of various molecules like proteins and DNA. These markers are employed in differentiating individuals, which are unaffected by environmental or the development stage of trees. Understanding genetic diversity and composition is becoming one of the prime objectives in any breeding program. This reduces inbreeding and promotes gene pool mixing, enhancing population resilience and adaptability (Frankham et al. 2011). Regulation of drift and having a powerful evolutionary force, artificial movements of real individuals in every generation of breeding are needed (Whitlock and McCauley 1999). The successful management and conservation of species are based on the extant knowledge of genetic variation with its natural distribution as well as introduced populations. The molecular diversity of forest tree species was explored by different researchers using different kinds of approaches, such as Allozymes (Kertadikara and Prat 1995), Sequence Characterized Amplified Region (SCAR) (Isoda et al. 2000), Random Amplified Polymorphic DNA (RAPD) (Watanabe et al. 2004), Amplified Fragment Length Polymorphism (AFLP) (Shrestha et al. 2005), Simple Sequence Repeat (SSR) (Huang et al. 2015), ISSR (Chaudhari et al. 2018), and SNPs (Win et al. 2015).

Isozymes have played a significant role in genetic variability research within forestry, serving to evaluate the genetic diversity within natural forest populations and to gauge potential reductions in diversity due to domestication practices (Chaisurisri and El-Kassaby 1994; Doligez and Joly 1997; Ritland et al. 2005). Nonetheless, the utility of these marker assays is hampered by their susceptibility to plant phenological stages and their restricted availability, thus precluding a comprehensive genome-wide examination of variability. An isozyme analysis of nine teak populations using 14 enzyme systems produced 20 putative loci with 18 polymorphic centers (Kertadikara and Prat 1995). Similarly, AFLP genetic diversity among different teak populations suggested 57% intra-population and 43% inter-population variation, and cluster analysis clearly distinguished the Indian teak population from Thailand and Indonesia (Shrestha et al. 2005).

The molecular diversity of 10 teak populations from South Gujarat was assessed using ISSR markers (Chaudhari et al. 2018). A total of 12 markers showed highest polymorphism (80.53%) with 91 loci. Based on Jaccard and Nei coefficient analysis, populations were divided into three major groups. Lyngdoh et al. (2013) examined the genetic variability of teak populations selected from Karnataka, using ISSR markers. In each location, an unimproved population, Seed Production Area (SPA), and Clonal Seed Orchards (CSO) were chosen to study the diversity differentiation with change in the level of improvement. Considering polymorphic loci and the Shannon index, the unimproved stands showed maximum diversity followed by SPA and then CSO. The natural populations of *Tectona grandis* from South India, North India, Thailand, and Laos were used to know genetic variability by employing 15 SSR markers (Fofana et al. 2009). Cluster analysis has divided populations into four main centers of genetic variability. Indian populations were divided into two clusters, and Thailand and Laos were categorized into individual groups representing the centre of origin of the species.

The genetic diversity among the 52 teak provenances assembled in a provenance trial in Ghana was assessed with SNPs using Genotyping by Sequencing and 23,182 SNPs were identified (Wanders et al. 2021). Evanno's delta K method classified all the provenances into two main clusters. 94% of SNPs were polymorphic in Asia–Pacific cluster, whereas, 74% polymorphism was found in South America–Africa cluster. Additionally, the South America-Africa cluster had only 1364 private alleles, whereas the Asia-Pacific cluster had 6201 private alleles. The plant genomes of 10 teak genotypes were sequenced using Illumina Technology for the identification of SNPs (Thongthawee and Volkaert 2014). The chromosomal genomes of the genotypes had 643,647 nucleotide changes, or 1 SNP per 500 bp, according to SNP detection. Three distinct chloroplast haplotypes were obtained that varied from one another by one SNP and one incidence of insertion/deletion.

7.2.2 Tissue Culture and Micropropagation

The in vitro propagation and plant tissue culture (PTC) have gained a significant advantage over the conventional propagation technique by rapid production of identical clonal materials. Its practical benefits encompass the generation and preservation of disease-free planting materials, conservation of germplasm, and augmentation of conventional plant improvement methodologies like genetic engineering. This method revolutionized plant production through enabling the mass propagation of superior plants, generating virus-free materials, and accelerating the introduction of new cultivars for commercial cultivation. The primary methods of PTC involve axillary budding, which essentially entails shrinking down propagation through cuttings, inducing the growth of new buds on non-meristematic tissue, thus prompting shoot formation. PTC is mainly influenced by the explant and media used for proliferation.

The effectiveness of most tissue culture endeavours hinges on the careful choice of explants. Various types of explants have been employed for PTC across the diverse tree crops such as tissue apex and auxiliary buds in *Quercus suber* (El Kbiach et al. 2004), leaves of *Ficus lyrate* (Abdolinejad and Salmi 2023), lateral and apical buds of *Quercus brantii* (Salari et al. 2018), internodes of *Quercus robur* (Gatti and Sgarbi 2015), Inflorescence of *Liquidambar styraciflua* (Merkle et al. 1998), zygotic embryos of *Fraxinus excelsior* (Nawrot-Chorabik et al. 2023), *Eucalyptus camaldulensis* (Prakash and Gurumurthi 2010), matured seeds of *Azadirachta indica* (Murthy and Saxena 1996), immature endosperm of *Acacia nilotica* (Garg et al. 1996), immature cotyledons of *Dalbergia latifolia* (Shirin et al. 2020), *Pterocarpus santalinus* (Chakraborty et al. 2023), etc. Additionally, the developmental stage and genetic composition of the explants play pivotal roles in the success of PTC. Genetic variability among explants resulted in differing responses to tissue induction.

The choice of base medium varies depending on the species and the specific explant used (Chirva et al. 2023). In the early 1960s, Murashige and Skoog made a ground-breaking impact on plant tissue culture by creating a standardized culture medium that has become the choice of culturing (Murashige and Skoog 1962). Some well-established base media in forest trees other than MS are, B5 (Gamborg et al. 1968), Y3 (Eeuwens 1976), SH medium (Schenk and Hildebrandt 1972), and WPM (Woody Plant Medium) (Lloyd and McCown 1980). Utilization of plant growth regulators in plant base media influence the tissue proliferation, morphogenesis, and rooting of the cells or explants. 2,4-dichlorophenoxyacetic acid (2,4-D) has been successfully employed as the primary auxin, either independently or in combination with other plant growth regulators, in the culture of *Pinus sylvestris* (Chirva et al. 2023). The callus or somatic embryo production was reported to be affected by different media compositions with varied concentrations. In *Dalbergia sissoo*, half-strength MS media enhanced with 0.68 mmol per litre L-glutamine improved somatic embryo maturation from 55–66% (Singh and Chand 2003). Whereas, the addition of Glutathione with ABA improved the embryo maturation in *Podocarpus lambertii* (Fraga et al. 2016). Akram and Aftab (2016) utilized *T. grandis* callus from softwood shoots for somatic embryo development on MS + TDZ (12 μmol per litre) + BA (2 μmol per litre), supplemented with 5 mmol per litre ascorbic acid. Using staminate inflorescence explants of *Liquidambar styraciflua* on MS media complemented with 3 mg per litre 2,4-D and 5 g per litre activated charcoal under dark conditions produced embryos for plant production (Merkle et al. 1998). Jouini et al. (2023) observed maximum embryo production of *Abies nebrodensis* occurred in SH media, in the presence of 4.27 μM ABA, 8% PEG-4000, and 4% maltose. Similarly, maximum proliferation of *Pinus massoniana* was observed on LP media (Von Arnold and Eriksson 1977) supplemented with maltose, activated carbon, inositol (1.0 g per litre), ABA (2 mg per litre), PEG8000, and Phytagel (3.5 g per litre) (Xia et al. 2021). The addition of TDZ in MS media maximized the proliferation of *Santalum album* (Ravishankar Rai and McComb 2002) and *Azadirachta* (Gairi and Rashid 2004) for further regeneration and rooting. In-vitro rooting of the shoots or somatic embryos is influenced by the germination media, time, light period, etc. (Corredoira et al. 2017). Merkle et al. (1998) reported rooting media without activated charcoal under 150 mmol per m^2 per second light intensity maintained under 25 °C for 2–12 weeks showing 90% rooting in *L. styraciflua*. In *D. sissoo*, rooting improved after transferring somatic embryos to MS media with 0.26 μmol per litre NAA + 2.22 μmol per litre BAP, and 10% sucrose (Singh and Chand 2003). Maximum root proliferation of teak was achieved in MS media supplemented with 8 mol per litre TDZ + 2 mol per litre BA with 5 mmol per litre ascorbic acid (Akram and Aftab 2016). For *F. mandshurica*, the highest rooting was found on 1/3 MS medium added with 0.01 mg per litre NAA, compared to 1 mg per litre IBA and IAA (Liu et al. 2021). Addition of 2,4-D and 2-Isopentenyladenine in rooting MS culture doubled the rooting potential of *Dendrocalamus asper* (Ornellas et al. 2022).

The acclimatization of plantlets is a pivotal stage for successfully transitioning them to soil or other growing substrates. The process of hardening makes rooted plants expose and adapt to slowly changing light intensity, air movement, and temperature with controlled conditions under specialized chambers (Maruyama and Hosoi 2019). Proper ventilation with ambient conditions is essential to minimize excessive humidity, reduce the risk of bacterial or fungal infections, and enhance gas exchange during acclimatization (Salma et al. 2019). Understanding of acclimatization process with different hardening media compositions is always a matter of research. During the hardening of teak plantlets, a maximum of 56.6% survival was found in media having coarse sand, soil and peat moss in the ratio of 1:1:1 (Akram and Aftab 2016). Similarly, in Shisham, Singh and Chand (2003) reported hardening in a media composition of peat moss and compost in the same ratio. Pine tree bark was found best hardening media blended with peat and perlite in 1:1:1 for Maritime pine achieving 50% survival rate (Miguel et al. 2004). Mixing of peat and perlite at different ratios found as best media in Scot's pine (1:1, Harju et al. 2022), Aleppo pine (3:1, Pereira et al. 2021), and Korean pine (2:3, Peng et al. 2021). The addition of vermiculite with peat soil and perlite improved the survival rate by up to 91% in Manchurian ash (Liu et al. 2021).

Although micropropagation improved the speedy multiplication of superior genotypes, challenges persist in its application for numerous tree species. Callus proliferation and regeneration from embryogenic cultures remain particularly challenging in many woody species. However, the effective utilization of PTC in tree breeding holds promising potential for forestry advancements.

7.2.3 Genomic Sequencing

Genome sequencing in forest trees will enable researchers to dive into the evolutionary history of woody trees, the structure of the genome and its functional biology (Plomion et al. 2016). On the contrary, genomic studies in trees are difficult due to their long vegetative growth period, delayed reproductive phase and the requirement of large areas for trials, which increases the time and cost. The large genome sizes of trees, especially the coniferous species with ~20 Gbps, along with the long life span and inability to control the mating designs, also hinder the sequencing of genomes, though high-throughput DNA sequencing methods are available, and only 200 out of the tens of thousands of known trees are sequenced, considering the economic importance and scientific worth (Cao et al. 2022). However, due to its importance in conservation and breeding programs, genome sequencing is accelerating among research programs. Though less than annual and horticultural crops, projects on genome sequencing in forest species are on the rise; the first complete sequencing in forest trees and the third in plant genome is done in *Populus trichocarpa* (Plomion et al. 2016). The properties of the genus, maximum genetic variations, and contrasting effective population size of the species made *Populus* a novel model system (Lin et al. 2018). After the first forest tree genomic sequencing, eucalyptus, oak, and other coniferous species have been taken up for genomic studies (Holliday et al. 2017). Further, teak being the most important tropical species, Zhao et al. (2019) reported a teak genome sequence with genomic size of 338 Mbp having 1474 contigs, and 936 scaffolds. Yasodha et al. (2018) assembled a teak draft genome of 317 Mb and 36,172 protein-coded genes. Compared to the model crops such as poplar and eucalyptus, teak has fewer publicly available gene sequences.

The use of genome sequences in phylogenetic studies maximizes the knowledge of the phylogenies estimates and the resolution of taxa that are closely related (Hall and Nisbet 2023). It also helps in understanding the epidemiological relationships between the outbreak of disease and ecology in forest trees (Besser et al. 2019). Studies on taxon sampling have grouped *Eucalyptus* and other members in the malvids family and one of the two clades of eurosids (fabids and malvids). Myburg et al. (2014) conducted a genome-wide analysis of 17 sequenced plant genomes to evaluate the phylogenetic position of *Eucalyptus* and found *Eucalyptus* to be in a sister taxon to eurosids and confirmed the malvids grouping of *Populus* and *Jatropha* instead of fabids. To understand the functional, comparative, and evolutionary genetics of *Populus*, Lin et al. (2018) undertook a genome sequencing in two species of *Populus* i.e., *P. tremula* and *P. tremuloides* and concluded that the genetic variation patterns and molecular evolution of the genus were a factor of expression level, coexpression network connectivity and governing variability. Patil et al. (2021) carried out the whole genome assembly in *Mesua ferrea* and compared it with the Malpighiales genomes to understand the evolution and influence of ecological and climatic events on forests and confirmed the presence of significant hindrance in all the tropical forest trees throughout Mid-Pleistocene glaciation. Using the high-quality reference genome of *Eucalyptus urophylla* × *Eucalyptus grandis*, Shen et al. (2023) were able to identify genes associated with starch and sucrose metabolism, flavonoid biosynthesis and plant-pathogen interactions.

In the future, releasing genomic sequences will enable researchers to understand tree's phylogenetic and evolutionary history and enhance the breeding cycles for productivity and wood quality for commercially economic species. Studying the diversities in functional genes will provide insight into the drift, selection of species, and mechanisms controlling the speciation and divergence adaptation. It will prove to be powerful information on how the species would respond to environmental changes (Myburg et al. 2014).

7.2.4 Genetic Transformation and Modified Trees

Unlike field crops, trees possess immense genetic potential that can be captured through new varieties through transgenic techniques (Burley 2001; Peña and Séguin 2001). With over 200 field trials involving at least 15 forest species, the development of genetically modified trees has accelerated in recent years. The most noticeable advantages of genetic transformation are the production potential of trees with novel traits such as reporter gene expression, resistance to pests, diseases, and herbicides, altered wood properties, altered flowering and fertility, and altered growth rate and stature. Fillatti et al. (1987) developed herbicide-resistant poplar against glyphosate, which was the first successful transformation in trees. The first field trial was established in 1988, in Belgium for herbicide-tolerant poplar. The State Forestry Administration of China allowed commercial cultivation of GM poplar trees in 2002. Utilization of transgenic trees is not widely accepted due to political, economic, and social issues. Genetically modified trees have been associated with various limitations including transformation, gene and or promoter stabilities along with organ-specific expression, genetic development, and propagation. Genetic transformation of trees has several advantages such as modified growth rate, modified flowering and fertility, resistance towards insect, pests and herbicides, etc. A schematic diagram of the tree transformation process is depicted in Fig. 7.1.

With a few reports on particle bombardment-mediated transformation, the processes for genetically transforming forest trees are primarily limited to the use of Agrobacterium and don't differ much from those for other plant species (Harfouche et al. 2011). Eighty per cent of transgenic plants created to date have been produced using Agrobacterium, making it the most commonly used transformation tool (Valenzuela et al. 2006). The agrobacteria can infect a broad range of dicotyledonous species and cause tumoric diseases in the infected host plants. *A. tumefaciens* was utilized to develop the first transgenic tree about 30 years ago (Fillatti et al. 1987). *A. rhizogenes* was used in European larch to create the first transgenic conifer plants.

Nowadays, the most preferred technique for the introduction of a gene into the cell organelles is biolistic transformation. Target cells are bombarded using small DNA-coated tungsten or gold microprojectiles, which are augmented initially by compressed gases such as carbon dioxide, nitrogen, or helium. Protoplasts are

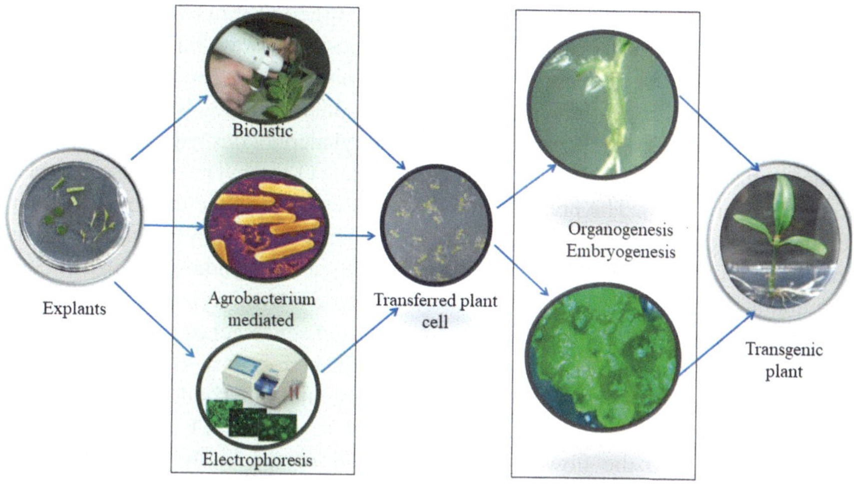

Fig. 7.1 Schematic diagram of the tree transformation process

necessary for protoplast transformation. The protoplast is developed by the enzymatic breakdown of plant cell walls and is typically grown in liquid suspension cultures. Plant cells are normally isolated from the leaf mesophyll. Protoplasts can be transformed by electroporation, microinjection, direct DNA uptake, or polyethylene glycol pre-treatment. Moreover, somatic embryogenesis or organogenesis are the two processes by which plants can be regenerated. The Btt and Btk genes isolated from *Bacillus thuringensis* proteins were introduced in *E. globulus* and *E. nitens* to control the eucalypt leaf beetle (*Chrysophtharta binaculata*) and autumn gum moth (*Mnesampela privata*) (Harfouche et al. 2011). The Bt protein was inserted into white spruce (*Picea glauca*) to control the spruce budworms (Peña and Séguin 2001). When the transgenic trees were scrutinised for spruce budworm toxicity, several of the trees showed total resistance. Nevertheless, resistance can be induced by introducing genes related to the synthesis of antifungal or antibacterial proteins (Séguin 1999). Zhang et al. (2005) utilized cowpea trypsin inhibitor gene (CpTI) to transform hybrid triploid poplars that showed resistance against insects such as *Stilpnotia candida*, *Lymantria dispar*, and *Malacosoma disstria*. Resistance to a pathogenic fungus, that is, *Cytospora chrysosperma* was observed by Chinese white poplar (*P. tomentosa*) that expressed the chitinase gene from *Beauveria bassiana* (Bbchit1). Likewise, Harfouche et al. (2011) observed substantially higher tolerance against drought, freezing, and salt stress due to amplification of pepper ERF/AP2 transcription factor, CaPF1, in eastern white pine. The biosynthesis of polyamines was linked to the elevated tolerance. Furthermore, in multiple lines of *E. globules*, amplification of the *Arthrobacter globiformis choline oxidase* (codA) gene led to increased tolerance against NaCl (Yu et al. 2009).

The paper industry would experience a major reduction in costs and pollution if plant material with lower lignin contents were produced (Pilate et al. 2002). Many studies have examined the biochemical pathways in lignin synthesis, and the corresponding genes of the enzyme have been identified (Tzfira et al. 1998; Merkle and Dean 2000). Down-regulating lignin biosynthesis pathways has been one of the methods for achieving forest trees with lower lignin content. The genes associated with genetic transformation for reducing lignin content includes 4-coumarate coenzyme A ligase (Pt4CL1) (Hu et al. 1999), cynnamyl alcohol deshydrogenase (CAD) (Baucher et al. 1996), and caffeate/5-hydroxyferulate O-methyl transferase (COMT) (Lapierre et al. 1999). The presence of lignin content or its structure may be modified by altering the expression of the genes. To confirm the claim that these characteristics are controlled by unique genes for monolignol biosynthesis, which encode 4-coumarate-CoA ligase (4CL) and conifer aldehyde 5-hydroxylase (CAld5H), the researchers transferred antisense genes for 4CL and sense genes for CAld5H into aspen (*Populus tremuloides*) using Agrobacterium. In antisense 4CL plants, lignin was reduced up to 40% with a 14% cellulose augmentation (Voelker et al. 2010).

In a study, the introduction of *BARNASE* ligated ribonuclease gene in silver birch (*Betula pendula*) to the flower-specific promoter of the birch gene *BpMADS1* (Lemmetyinen et al. 2004) resulted in to construction of 81 kanamycin-resistant lines. More significantly, no male inflorescences formed in 7 lines or they aborted at an early stage. Transgene containment and mass-controlled pollination of *Eucalyptus* are further facilitated by genetically engineered male sterility (Harfouche et al. 2011). Strong amino acid and DNA homology exist between the genes FLEI and FLE2 and the floral meristem genes FLORICAULA from Antirrhinum and LEAFY from Arabidopsis (Harfouche et al. 2011).

Due to the fact that the final products are not intended for human consumption, there should be less worry about transgenic trees than there would be with food crop plants. A lot of attention is being paid to the possible environmental impacts of genetically modified trees, particularly the threats posed by transgenes spreading over the native populations. Protoplastid transformation and reproductive sterility are two other measures that have been suggested to limit this possible transgene flow (Eriksson et al., 2000).

7.2.5 Marker-Assisted Selection (MAS)

MAS is a selection index in which the genotypes are selected using molecular markers to identify individuals carrying the genetic region responsible for the desired phenotype (Lande and Thompson 1990; Lema 2018). It comprises DNA-based approaches viz., QTL, associative mapping, linkage disequilibrium (LD), and genomic selection (Calazans et al. 2022). It helps accurately to predict genetic values of individual species (Ahmar et al. 2021). Therefore, it is defined as the application of molecular markers to predict the genetic worth and screen the genetics by harnessing the LD (Muranty et al. 2014). Simple traits influenced by a few QTLs were successfully used in MAS (Lebedev et al. 2020). The larger size of trees, long life span, and high outbreeding potential make the conventional phenotypic selection less accurate and require a significant amount of time and money. These limitations make MAS a profitable venture in tree breeding and improvement programs,

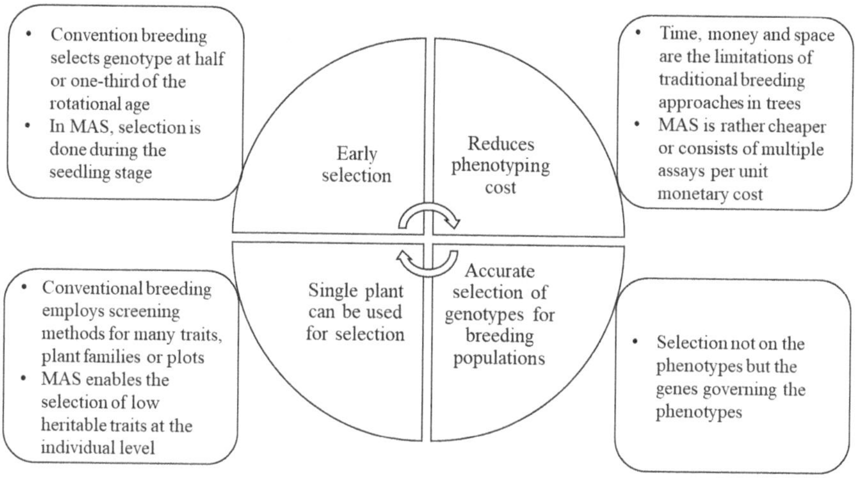

Fig. 7.2 Flowchart representing the benefits of MAS

and its benefits are depicted in Fig. 7.2. Identifying markers or gene regions for MAS is the prime stage in the breeding, hence, most of the studies reported in findings belong to the QTLs and markers for further genomic selection.

In eucalyptus breeding programs, the benefits of using MAS in forest breeding is the capability to identify the elite clones accurately with the help of markers such as RAPD, AFLP, etc. (Grattapaglia 2007). Disease and pest-resistant individuals were identified using ISSR and SSR markers in eucalyptus (Calazans et al. 2022). Zhou et al. (2020) used 839 microsatellite markers to identify the genomic loci influencing the wood and growth properties in *Eucalyptus cloeziana* and reported that 13 microsatellite markers were closely linked to height, diameter over bark at breast height, basic density and modulus of elasticity. These 13 microsatellite markers can be further exploited to identify and select individuals with superior wood and growth parameters.

In *Populus*, 51 SNPs exhibiting biosynthesis of cellulose and lignin were identified and the markers were found to amplify in 25 species of *Populus*, indicating such primers can be used in various *Populus* species or hybrid for selection of elite genotypes for superior wood properties (Köbölkuti et al. 2019). To know the markers highly associated with wood traits that can accurately identify seedling or clonal progenies by omitting the long progeny or clonal test, Kaya et al. (2023) used 384 SNP markers to find the loci responsible for wood traits in *Populus nigra* and reported that 11 SNP markers were closely associated with cellulose, 5 with lignin and 6 with glucose. These markers would enable in the early selection of progenies or clones for establishing superior poplar clones. In teak, Vaishnav et al. (2018) used 276 AFLP loci of markers to study the association with wood density in 46 genotypes and reported that three markers E-AAC/M-CAT were strongly associated with wood density, which can be used for recurrent selection in teak improvement and breeding program. Bano et al. (2024) identified seven SNPs linked with wood density traits in teak using lignin synthesis CAB1 gene and MYB1 and MYB2 transcription factors. However, only one locus from the exon of the MYB1 transcription factor was established to be highly associated with wood density showing 12% phenotypic variation, which is a promising molecular probe for marker-assisted breeding and picking for wood density traits in teak.

Using SNP genotyping, Liu et al. (2020) were able to identify blister rust-resistant wild parents and the progenies with 100% accuracy in *P. monticola*. Moriguchi et al. (2020) reported that with the help of MAS, 650 conifer trees were identified the presence of the male sterility gene, indicating that MAS reduces the time of breeding efforts. The resistance of Norway spruce (*Picea abies*) to root and butt rot caused by *Heterobasidion parviporum* was studied (Mukrimin et al. 2018) using SNPs and 10 SNPs loci located in eight genes were found to be closely linked to *H. parviporum* resistance suggesting that the identified loci may be the potential markers for screening of disease-susceptible and resistant trees. The success of MAS in tree breeding and improvement programs depends upon the choice of markers, availability of genomic information, knowledge of advanced tools such as QTL, SNPs, etc., and the need for trained manpower. MAS becomes an excellent tool for breeding and improving forest tree species if the requirements are fulfilled.

7.2.6 Other Advantages of Biotechnology

7.2.6.1 Cryopreservation and In Vitro Storage

The erratic changes in climatic conditions, loss and destruction of habitat, and the prevailing interspecific and intraspecific competitions in the ecosystem are the major reasons for the erosions of genetic variability. Cryopreservation or genetic conservation is the solution for such erosion, and it can be performed either in natural or artificial environment methods of germplasm conservation. However, the conservation in its natural environment is not a wise method due to the unavoidable change in climate accelerating biodiversity loss, while ex situ conservation, such as clonal orchards, seed banks, cryopreservation, etc., plays a pivotal role in the conservation and preservation of elite or critically endangered and threatened germplasm and re-introduction in its natural habitat when the environment becomes suitable for its establishment (Ruta et al. 2020).

The ex situ conservation technique of cryopreservation refers to the storage of samples in liquid nitrogen at a temperature of −196 °C. It is based on the principle of dehydration, eliminating freezable water from the tissues to avoid intracellular ice crystal formation and to arrest the biochemical and metabolic processes preventing death and destruction of cell organelles (Benelli 2021). The seed bank is the common method of genetic diversity conservation in forest tree species by freezing seeds. However, one of the disadvantages is the inability of nearly a third of all global forest species seed entities to withstand dry and cold storage (Wyse et al. 2018). The seed may also not represent the desired genotype (Wang et al. 2021); thus, a new technique for conserving forest species using cryopreservation of in vitro culture was developed during the twentieth century (Ballesteros et al. 2024).

This technique uses explants developed from shoot tips, embryogenic culture, leaf or stem-bearing adventitious buds, buds of rhizome, winter dormant buds, roots, seed embryonic axes or pollen; it is a low-cost, long-term conservation strategy (Ochatt et al. 2021). The first successful use of cortical cells of mulberry in cryo-preservation and its survival was demonstrated by Sakai (1956).

During in vitro culture, the long subculturing leads to a decline or loss of somatic embryonic potentials and genetic instability, the technique of cryo-biotechnology enables the retention of somatic embryonic potentials without hindering the survival rate, regeneration rate and proliferation coefficient of callus (Yu et al. 2022; Edesi et al. 2020). Pollen cryopreservation, a technique to tackle the phenological problems of asynchronous flowering and scarce pollen production, is an easy gene conservation method in tree breeding (Dinato et al. 2020). In forestry, clonal banks are usually used as the most common form of ex situ gene conservation method. However, climate change altering the pattern of rainfall, temperature and wind in general poses an unprecedented threat to such resources (Potter et al. 2017). As a result, winter buds cryopreservation proves to be an effective way of clonal conservation to overcome such risks (Endoh et al. 2023).

7.2.6.2 Genetic Fingerprinting

Advances in modern biotechnological tools help to monitor the changes at the gene level and detect the genetic diversity and genetic characterization of plants and the genetic relationship (Lanka et al. 2023). The correct identification of cultivars using such advanced tools is a must, considering the importance of accurate selection in any breeding technique. Genetic fingerprinting, developed by Alex Jeffreys, is one such tool that helps to identify the unique genetic pattern in the genome of the organism and is known by different terms such as DNA fingerprinting, DNA typing, and DNA profiling (Aydin et al. 2023). Both PCR-based and non-PCR-based markers can be used for genetic fingerprinting. It proves to be a reliable technique to differentiate accessions, correct the mislabelled cultivars, and identify the geographical origin. During the use of planting material where morphological differentiation becomes a vigorous task such as rootstock, it is an excellent tool for the identification (Wünsch and Hormaza 2002). Using SSR and ISSR markers, a successful selection of elite trees of *Quercus suber* having a high polymorphic pattern (94%) was reported and confirmed the importance of such markers in genetic fingerprinting (López-Aljorna et al. 2007). The technique is also noteworthy for the identification and germplasm repositories of trees having various synonyms like the European Cooperative Program for Plant Genetic Resources (ECPGR) and the US *Pyrus* Genetic Resources (USPGR) developed the *Pyrus* fingerprinting using SSR markers for pedigree and population structure studies to maintain the genetic resources (Zurn et al. 2020). Timber identification with the visual method using the macro and micro-anatomical features requires rigorous training and experience; the technique of genetic fingerprinting enables accurate identification at the species level using minimum wood materials (Domínguez-Delmás 2020).

7.3 Challenges and Considerations

The benefits and limitations of using biotechnological tools in tree improvement have no clear-cut demarcation, and it is a debatable topic. The complex nature of trees and the environmental interaction also makes it a confusing area; creation of disease resistance genetically modified trees may be advantageous from the consumer's point of maximizing the products; however, environmentally, it might not be a sustainable form of practice as such trees might become invasive and later classified as a weed (James 1997). Some ethical concerns are the insertion of foreign genes into the organism genome and are in disagreement with the traditional method of breeding; the monopoly in the distribution of improved materials by the profit-driven companies may isolate the local people and environmental interest at large. The monoculture forest plantation developed by the companies using such materials not only has a detrimental effect on the environment with accelerated water and nutrient requirements but also socially marginalised indigenous people, especially women, reducing the income generated from the natural forest (Boyd 2010).

Regardless of the drawbacks, the benefits of incorporating biotechnology into conventional forest tree breeding do not fall behind the limitations. The environmental impact of genetically modified trees is controversial and difficult to assess due to their long life span. Zuo et al. (2018) have proven that a 5-year-old genetically modified poplar had no effect on the environment or on the structure and diversity of microbial community. With the growing population and the increasing mortality of trees as a consequence of adverse climate stressors, the area available for and the area of forest declines annually. As a result, biotechnological techniques in forestry are applied to increase productivity in limited land areas through wood modifications by lignin reduction, increase resistance to disease and pests, combat the changing environmental conditions and shorten the rotation cycle. Such manipulations not only increase productivity but are also low-cost while helping in the reduction of pressure on the wood and wood products from natural forests, thus conserving biodiversity (Sedjo 2001). The use of genetically improved elite seedlings/planting materials has also proven to be a profitable investment for harvesting greater diameter and healthy trees at optimum economic rotational age as compared to unimproved materials (Chang et al. 2019). Considering the enhancing requirement for year round wood production and craze of carbon sequestration, and if carbon pricing becomes real in the near future, the forest plantations in bare land using the improved materials will become a profitable venture as well (Ahtikoski et al. 2020). The techniques of omics studies enhance the conservation strategies by improving seed sourcing through population genomics, generating novel genomes for the restoration of harsh environments with the help of genome editing and surveillance of the results through multi-omics technique (Breed et al. 2019). Some of the notable breakthroughs of transgenics in the forest are the popularisation of genetically modified poplar trees which can withstand beetles in China in 2003, the development of bacterial wilt-resistant *Eucalyptus urophylla* and the modification of *Eucalyptus* tree genetics for cold tolerance (Panda et al. 2020).

7.4 Future Prospects

The breeding techniques based on the phenotypic selections have considerably low transformational efficiencies and fluctuating effects on qualitative and quantitative traits of forest trees. The complex procedures for traditional breeding, coupled with the longer breeding period and adverse effects of changing environmental conditions, shift the conventional tree breeding paradigm towards the new transgenic biotechnological tools. The Agrobacterium-mediated method of gene transformation is the most common method used in forest tree gene editing, but it is specific to only a few species and tissues. The development of third-generation gene editing methods such as CRISPR/Cas sequences enabling the addition and deletion of genes at a specific site becomes a precise, reliable, and economical method for efficient tree breeding and diversification of woody traits (Min et al. 2022). This technique also helps in improving tree growth and its development and increases yield and fruit quality while amplifying the resistance of trees against the biotic and abiotic stress (Martín-Valmaseda et al. 2023). Using the modern techniques of multi-omics, the transformation efficiency of trees for ecosystem revival, bioenergy production, and sustainable productivity is enhanced to a greater extent (Yin et al. 2021).

7.5 Conclusion

The growing demands for wood and its products can be fulfilled by fusing the biotechnological tools into tree improvement programs. Such initiatives will elevate the resilience of trees to the detrimental effects and extremities of climate change and increase productivity with limited land availability within a short duration. The multi-omics techniques will open the gate to domestication of transgenic trees worldwide for their significant reduction in rotation age with increased production. Such enterprise will encourage more private and governmental participation and initiatives toward forestry for sustainability and a greener economy. Considering the environmental and economic role forest plays, such crucial icebreaking research and exploration in the domain of forest must be encouraged.

References

Abdolinejad R, Salmi MS (2023) Indirect regeneration in *Ficus lyrata* Warb. and metabolite profiles influenced by nitric oxide and plant growth regulators. BMC Plant Biol 23:325

Ahmar S, Ballesta P, Ali M, Mora-Poblete F (2021) Achievements and challenges of genomics-assisted breeding in forest trees: from marker-assisted selection to genome editing. Int J Mol Sci 22(19):10583

Ahtikoski A, Ahtikoski R, Haapanen M, Hynynen J, Kärkkäinen K (2020) Economic performance of genetically improved reforestation material in joint production of timber and carbon sequestration: a case study from Finland. Forests 11(8):847

Akram M, Aftab F (2016) Establishment of embryogenic cultures and efficient plant regeneration system from explants of forced softwood shoots of teak (*Tectona grandis* L.). Hortic Plant J 2:293–300

Ashwath MN, Lavale SA, Santhoshkumar AV, Mohapatra SR, Bhardwaj A, Dash U et al (2023) Genome-wide association studies: an intuitive solution for SNP identification and gene mapping in trees. Funct Integr Genomics 23(4):297

Aydin A, Kocak MZ, Kulak M (2023) DNA fingerprinting of crop plants. In: Husen A, Ahmad A (eds) Genomics. Academic, Transcriptomics, proteomics and metabolomics of crop plants, pp 229–247

Ballesteros D, Martínez MT, Sánchez-Romero C, Montalbán IA, Sales E, Moncaleán P, Arrillaga I, Corredoira E (2024) Current status of the cryopreservation of embryogenic material of woody species. Front Plant Sci 14:1337152

Bano N, Mohammad N, Ansari MI, Ansari SA (2024) Discovery of SNPs in lignin biosynthesis genes (CAD1, MYB1 and MYB2) and their association with wood density in teak. Mol Biol Rep 51:169

Baucher M, Chabbert B, Pilate G, Van Doorsselaere J, Tollier MT et al (1996) Red xylem and higher lignin extractability by down-regulating a cinnamyl alcohol dehydrogenase in poplar. Plant Physiol 112(4):1479–1490

Benelli C (2021) Plant cryopreservation: a look at the present and the future. Plan Theory 10(12):2744

Besser JM, Carleton HA, Trees E, Stroika SG, Hise K et al (2019) Interpretation of whole-genome sequencing for enteric disease surveillance and outbreak investigation. Foodborne Pathog Dis 16(7):504–512

Boyd E (2010) Societal choice for climate change futures: trees, biotechnology, and clean development. Bioscience 60(9):742–750

Breed MF, Harrison PA, Blyth C, Byrne M, Gaget V, Gellie NJ et al (2019) The potential of genomics for restoring ecosystems and biodiversity. Nat Rev Genet 20(10):615–628

Brown GR, Gill GP, Kuntz RJ, Langley CH, Neale DB (2004) Nucleotide diversity and linkage disequilibrium in loblolly pine. Proc Natl Acad Sci 101(42):15255–15260

Burley J (2001) Genetics in sustainable forestry: the challenges for forest genetics and tree breeding in the new millennium. Can J For Res 31(4):561–565

Calazans CC, Nunes VV, Souza JL, Santana NA, de Almeida FE, Silva-Mann R (2022) Marker-assisted selection and protein identification in Eucalyptus. Res Soc Dev 11(6):e45311629227

Cao HX, Vu GTH, Gailing O (2022) From genome sequencing to CRISPR-based genome editing for climate-resilient forest trees. Int J Mol Sci 23(2):966

Chaisurisri K, El-Kassaby YA (1994) Genetic diversity in a seed production population vs. natural populations of Sitka spruce. Biodivers Conserv 3:512–523

Chakraborty T, Chaitanya KV, Akhtar N (2023) Somatic embryogenesis and plantlet regeneration in red sandalwood (*Pterocarpus santalinus*). Plant Cell Tissue Organ Cult 153:547–558

Chang WY, Wang S, Gaston C, Cool J, An H, Thomas BR (2019) Economic evaluations of tree improvement for planted forests: a systematic review. BioProducts Bus 4(1):1–14

Chaudhari C, Jha SK, Dhaka RK, Parekh V, Sankanur MS, Prajapat P, Thakur S (2018) Assessment of ISSR based molecular variability in teak of South Gujarat. J Pharma Phytochem 7(6):562–568

Chirva OV, Ignatenko RV, Ershova MA (2023) *Pinus sylvestris* L. mature seeds megagamethophytes cultured in vitro. Influence of the genotype and the growth regulators content in nutrient medium on the initiation of callus formation. Plant Cell Tissue Organ Cult 152:299–308

Corredoira E, Cano V, Bárány I, Solís MT, Rodríguez H, Vieitez AM et al (2017) Initiation of leaf somatic embryogenesis involves high pectin esterification, auxin accumulation and DNA demethylation in *Quercus alba*. J Plant Physiol 213:42–54

Dinato NB, Imaculada Santos IR, Zanotto Vigna BB, de Paula AF, Fávero AP (2020) Pollen cryopreservation for plant breeding and genetic resources conservation. CryoLetters 41(3):115–127

Doligez A, Joly HI (1997) Genetic diversity and spatial structure within a natural stand of a tropical forest tree species, *Carapa procera* (Meliaceae). French Guiana Hered 79(1):72–82

Domínguez-Delmás M (2020) Seeing the forest for the trees: new approaches and challenges for dendroarchaeology in the 21st century. Dendrochronologia 62:125731

Edesi J, Tolonen J, Ruotsalainen AL, Aspi J, Häggman H (2020) Cryopreservation enables long-term conservation of critically endangered species *Rubus humulifolius*. Biodivers Conserv 29(1):303–314

Eeuwens CJ (1976) Mineral requirements for growth and callus initiation of tissue explants excised from mature coconut palms (*Cocos nucifera*) and cultured *in vitro*. Physiol Plant 36:23–28

El Kbiach MB, Lamarti A, Abdali A, Badoc A (2004) Micropropagation of cork oak (*Quercus suber* L.) by culture of apices from embryonic axes. Acta Bot Gall 151:401–413

Endoh K, Hanaoka S, Matsushita M, Ubukata M, Yamada H (2023) Ex situ conservation of birch trees by cryopreservation of dormant buds adapted to subzero temperatures by extracellular freezing. New For 54(3):515–523

Eriksson ME, Israelsson M, Olsson O, Moritz T (2000) Increased gibberellin biosynthesis in transgenic trees promotes growth, biomass production and xylem fiber length. Nat Biotechnol 18(7):784–788

Fang Y, Han T (2019) Advances in crop rapid breeding technology. Crops 2:1–7

Fillatti JJ, Sellmer J, McCown B, Haissig B, Comai L (1987) Agrobacterium mediated transformation and regeneration of Populus. Mol Gen Genet 206:192–199

Fofana IJ, Ofori D, Poitel M, Verhaegen D (2009) Diversity and genetic structure of teak (*Tectona grandis* L) in its natural range using DNA microsatellite markers. New For 37(2):175–195

Fraga HP, do Nascimento Vieira L, Puttkammer CC, Dos Santos HP, de Andrade Garighan J, Guerra MP (2016) Glutathione and abscisic acid supplementation influences somatic embryo maturation and hormone endogenous levels during somatic embryogenesis in *Podocarpus lambertii* Klotzsch ex Endl. Plant Sci 253:98–106

Frankham R, Ballou JD, Eldridge MDB, Lacy RC, Ralls K, Dudash MR et al (2011) Predicting the probability of outbreeding depression. Conserv Biol 25:465–475

Gairi A, Rashid A (2004) Direct differentiation of somatic embryos on different regions of intact seedlings of *Azadirachta* in response to thidiazuron. J Plant Physiol 161:1073–1077

Gamborg OL, Miller RA, Ojima K (1968) Nutrient requirements of suspension cultures of soybean root cells. Exp Cell Res 50:151–158

Garg L, Bhandari NN, Rani V, Bhojwani SS (1996) Somatic embryogenesis and regeneration of triploid plants in endosperm cultures of *Acacia nilotica*. Plant Cell Rep 15:855–858

Gatti E, Sgarbi E (2015) Micropropagation of *Quercus robur*: explant sources and cultural conditions affect *in vitro* responses differently. Acta Hortic 1083:303–310

Grattapaglia D (2007) Marker-assisted selection in eucalyptus. In: Guimaraes E, Ruane J, Sonnino A, Scherf B, Dargie JD (eds) Marker-assisted selection, current status and future perspectives in crops, livestock, forestry and fish. Food and Agriculture Organization of the United Nations, Rome, pp 251–281

Hall BG, Nisbet J (2023) Building phylogenetic trees from genome sequences with kSNP4. Mol Biol Evol 40(11):235

Harfouche A, Meilan R, Altman A (2011) Tree genetic engineering and applications to sustainable forestry and biomass production. Trends Biotechnol 29(1):9–17

Harju A, Heiska S, Julkunen-Tiitto R, Venäläinen M, Aronen T (2022) Somatic embryogenesis of *Pinus sylvestris* L. from parent genotypes with high- and low stilbene content in their heartwood. Forests 13:557

Holliday JA, Aitken SN, Cooke JE, Fady B, González-Martínez SC, Heuertz M, Jaramillo-Correa JP, Lexer C, Staton M, Whetten RW, Plomion C (2017) Advances in ecological genomics in forest trees and applications to genetic resources conservation and breeding. Mol Ecol 26(3):706–717

Hu WJ, Harding SA, Lung J, Popko JL, Ralph J, Stokke DD, Tsai CJ, Chiang VL (1999) Repression of lignin biosynthesis promotes cellulose accumulation and growth in transgenic trees. Nat Biotechnol 17(8):808–812

Huang GH, Liang KN, Zhou ZZ, Xu JM, Ma HM (2015) Genetic variation and origin of teak (*Tectona grandis* L.f.) native and introduced provenances. Silvae Genetica 64:33–46

Isoda K, Watanabe A, Widyatmoko A, Rimbawanto A, Shirachi S (2000) The simple and reliable management of teak (*Tectona grandis*) clones with SCAR (Sequence Characterized Amplified Region) marker. In: Proceeding of third regional seminar on teak. Potentials and opportunities in marketing and trade of plantation teak: challenge for the new millennium. Faculty of Forestry, Gadjah Mada University, Yogyakarta, pp 265–269

James RR (1997) Utilizing a social ethic toward the environment in assessing genetically engineered insect-resistance in trees. Agric Hum Values 14:237–249

Jouini N, Yahyaoui E, Tarraf W, İzgü T, Benelli C, Lambardi M, Germanà MA (2023) Somatic embryogenesis in *Abies nebrodensis*, an endangered Sicilian fir. Plant Cell Tissue Organ Cult 152:393–404

Kaya Z, Taşkıran B, Uluğ A, Güvendiren AD, Wegrzyn JL, Neale DB (2023) Association genetics of wood traits in European black poplar (*Populus nigra* L.). Tree Genet Genom 19(4):38

Kertadikara AWS, Prat D (1995) Isozyme variation among teak (*Tectona grandis* L.f) provenances. Theor App Genet 90:803–810

Köbölkuti ZA, Cseke K, Benke A, Báder M, Borovics A, Németh R (2019) Allelic variation in candidate genes associated with wood properties of cultivated poplars (Populus). Biologia Futura 70(4):286–294

Lande R, Thompson R (1990) Efficiency of marker-assisted selection in the improvement of quantitative traits. Genetics 124(3):743–756

Lanka CL, Ram M, Krishna SM (2023) DNA fingerprinting of crops and its significance in crop improvement. Int J Plant Soil Sci 35(16):232–242

Lapierre C, Pollet B, Petit-Conil M, Toval G, Romero J, Pilate G et al (1999) Structural alterations of lignins in transgenic poplars with depressed cinnamyl alcohol dehydrogenase or caffeic acid O-methyltransferase activity have an opposite impact on the efficiency of industrial kraft pulping. Plant Physiol 119(1):153–164

Lebedev VG, Lebedeva TN, Chernodubov AI, Shestibratov KA (2020) Genomic selection for forest tree improvement: methods, achievements and perspectives. Forests 11(11):1190

Lema M (2018) Marker-assisted selection in comparison to conventional plant breeding. Agric Res Technol 14:555914

Lemmetyinen J, Keinonen K, Sopanen T (2004) Prevention of the flowering of a tree, silver birch. Mol Breed 13:243–249

Lin YC, Wang J, Delhomme N, Schiffthaler B, Sundström G, Zuccolo A et al (2018) Functional and evolutionary genomic inferences in Populus through genome and population sequencing of American and European aspen. Proc Natl Acad Sci 115(46):e10970–e10978

Liu JJ, Williams H, Zamany A, Li XR, Gellner S, Sniezko RA (2020) Development and application of marker-assisted selection (MAS) tools for breeding of western white pine (*Pinus monticola* Douglas ex D. Don) resistance to blister rust (*Cronartium ribicola* JC Fisch.) in British Columbia. Can J Plant Pathol 42(2):250–259

Liu Y, Wei C, Wang H, Ma X, Shen H, Yang L (2021) Indirect somatic embryogenesis and regeneration of *Fraxinus mandshurica* plants via callus tissue. J For Res 32:1613–1625

Lloyd G, McCown BH (1980) Commercially-feasible micropropagation of mountain laurel, *Kalmia latifolia*, by use of shoot-tip culture. Comb Proc-Int Plant Propagator's Soc 30:421–427

López-Aljorna A, Bueno MÁ, Aguinagalde I, Martín JP (2007) Fingerprinting and genetic variability in cork oak (*Quercus suber* L.) elite trees using ISSR and SSR markers. Ann For Sci 64(7):773–779

Lyngdoh N, Joshi G, Ravikanth G, Vasudeva R, Shanker R (2013) Changes in genetic diversity parameters in unimproved and improved populations of teak (*Tectona grandis* Lf) in Karnataka state, Indian. J Genet 92(1):141–145

Martínez-Gómez P (2019) Editorial for special issue "plant genetics and molecular breeding". Int J Mol Sci 20:2659

Martín-Valmaseda M, Devin SR, Ortuño-Hernández G, Pérez-Caselles C, Mahdavi SME, Bujdoso G, Salazar JA, Martínez-Gómez P, Alburquerque N (2023) CRISPR/Cas as a genome-editing technique in fruit tree breeding. Int J Mol Sci 24(23):16656

Maruyama TE, Hosoi Y (2019) Progress in somatic embryogenesis of Japanese pines. Front Plant Sci 10:31

Merkle SA, Dean JF (2000) Forest tree biotechnology. Curr Opin Biotechnol 11(3):298–302

Merkle SA, Neu KA, Battle PJ, Bailey RL (1998) Somatic embryogenesis and plantlet regeneration from immature and mature tissues of sweet gum (*Liquidambar styraciflua*). Plant Sci 132:169–178

Miguel C, Gonçalves S, Tereso S, Marum L, Maroco J, Margarida Oliveira M (2004) Somatic embryogenesis from 20 open-pollinated families of Portuguese plus trees of maritime pine. Plant Cell Tissue Organ Cult 76:121–130

Min T, Hwarari D, Li DA, Movahedi A, Yang L (2022) CRISPR-based genome editing and its applications in woody plants. Int J Mol Sci 23(17):10175

Moriguchi Y, Ueno S, Hasegawa Y, Tadama T, Watanabe M, Saito R, Konno Y (2020) Marker-assisted selection of trees with male sterility 1 in *Cryptomeria japonica* D. Don. Forests 11(7):734

Mukrimin M, Kovalchuk A, Neves LG, Jaber EH, Haapanen M et al (2018) Genome-wide exon-capture approach identifies genetic variants of Norway spruce genes associated with susceptibility to *Heterobasidion parviporum* infection. Front Plant Sci 9:364454

Muranty H, Jorge V, Bastien C, Lepoittevin C, Bouffier L, Sanchez L (2014) Potential for marker-assisted selection for forest tree breeding: lessons from 20 years of MAS in crops. Tree Genet Genomes 10(6):1491–1510

Murashige T, Skoog F (1962) A revised medium for rapid growth and bio assays with tobacco tissue cultures. Physiol Plant 15:473–497

Murthy BNS, Saxena PK (1996) Somatic embryogenesis and plant regeneration of neem (*Azadirachta indica* A. Juss.). Plant Cell Rep 17:469–475

Myburg AA, Grattapaglia D, Tuskan GA, Hellsten U, Hayes RD, Grimwood J et al (2014) The genome of Eucalyptus grandis. Nature 510(7505):356–362

Nawrot-Chorabik K, Pluciński B, Latowski D (2023) Indirect auxiliary organogenesis of *Fraxinus excelsior* L. as a tool for ash dieback control. New For 55:323–344

Ochatt S, Lambardi M, Panis B, Pathirana R, Revilla MA, Wang QC (2021) Cryopreservation and *in vitro* banking: a cool subject–preface from the editors. Plant Cell Tiss Organ Cult 144:1–5

Ornellas TS, Fritsche Y, Cardona-Medina E, Guerra MP (2022) Somatic embryogenesis from young spikelets of the giant bamboo *Dendrocalamus asper* (Schult f.) backer ex Heyne. Plant Cell Tissue Organ Cult 149:635–644

Panda AK, Mishra R, Bisht SS, Kumar A, Kerketta A (2020) Technological advances in commercial forestry. Proc Natl Acad Sci India Section B Biol Sci 90:753–760

Patil AB, Shinde SS, Raghavendra S, Satish BN, Kushalappa CG, Vijay N (2021) The genome sequence of *Mesua ferrea* and comparative demographic histories of forest trees. Gene 769:145214

Peña L, Séguin A (2001) Recent advances in the genetic transformation of trees. Trends Biotechnol 19(12):500–506

Peng C, Gao F, Wang H, Shen H, Yang L (2021) Optimization of maturation process for somatic embryo production and cryopreservation of embryogenic tissue in *Pinus koraiensis*. Plant Cell Tissue Organ Cult 144:185–194

Pereira C, Montalbán IA, Pedrosa A, Tavares J, Pestryakov A, Bogdanchikova N et al (2021) Regeneration of *Pinus halepensis* (Mill.) through organogenesis from apical shoot buds. Forests 12:363

Pilate G, Guiney E, Holt K, Petit-Conil M, Lapierre C, Leplé JC et al (2002) Field and pulping performances of transgenic trees with altered lignification. Nat Biotechnol 20(6):607–612

Plomion C, Bastien C, Bogeat-Triboulot MB, Bouffier L, Déjardin A, Duplessis S et al (2016) Forest tree genomics: 10 achievements from the past 10 years and future prospects. Ann For Sci 73:77–103

Potter KM, Jetton RM, Bower A, Jacobs DF, Man G, Hipkins VD et al (2017) Banking on the future: progress, challenges and opportunities for the genetic conservation of forest trees. New For 48:153–180

Prakash MG, Gurumurthi K (2010) Effects of type of explant and age, plant growth regulators and medium strength on somatic embryogenesis and plant regeneration in *Eucalyptus camaldulensis*. Plant Cell Tissue Organ Cult 100:13–20

Ravishankar Rai V, McComb J (2002) Direct somatic embryogenesis from mature embryos of sandalwood. Plant Cell Tissue Organ Cult 69:65–70

Ritland K, Meagher LD, Edwards DGW, El-Kassaby YA (2005) Isozyme variation and the conservation genetics of Garry oak. Botany 83(11):1478–1487

Ruta C, Lambardi M, Ozudogru EA (2020) Biobanking of vegetable genetic resources by in vitro conservation and cryopreservation. Biodivers Conserv 29:3495–3532

Sakai A (1956) Survival of plant tissue of super-low temperatures. Low Temp Sci Ser B 14:17–23

Salari R, TabandehSaravi A, Kamali Aliabad K (2018) Determination of the optimized culture medium and growth conditions for micropropagation of Persian oak (*Quercus brantii* L.). Iran J Genet Plant Breed 7:1–8

Salma U, Kundu S, Ali MN, Mandal N (2019) Somatic embryogenesis-mediated plant regeneration of *Eclipta alba* (L.) Hassk. and its conservation through synthetic seed technology. Acta Physiol Plant 41:103

Schenk RU, Hildebrandt AC (1972) Medium and techniques for induction and growth of monocotyledonous and dicotyledonous plant cell cultures. Can J Bot 50:199–204

Sedjo RA (2001) Biotechnology in forestry. Resources for the Future. RFF, Washington, pp 10–12

Séguin A (1999) Transgenic trees resistant to microbial pests. For Chron 75(2):303–304

Shen C, Li L, Ouyang L, Su M, Guo K (2023) *E. urophylla* × *E. grandis* high-quality genome and comparative genomics provide insights on evolution and diversification of eucalyptus. BMC Genomics 24(1):223

Shirin F, Bhadrawale D, Mishra JP, Sonkar MK, Maravi S (2020) Evaluation of biochemical changes during different stages of somatic embryogenesis in a vulnerable timber tree *Dalbergia latifolia* (Indian rosewood). In Vitro Cell Dev Biol-Plant 56:894–902

Shrestha MK, Volkaert H, Straeten DVD (2005) Assessment of genetic diversity in *Tectona grandis* using amplified fragment length polymorphism markers. Can J For Res 35:1017–1022

Singh AK, Chand S (2003) Somatic embryogenesis and plant regeneration from cotyledon explants of a timber-yielding leguminous tree, *Dalbergia sissoo* Roxb. J Plant Physiol 160:415–421

Thongthawee S, Volkaert H (2014) Analysis of teak (*Tectona grandis*) genome and its diversity. In: The 26th annual meeting of the Thai Society for Biotechnology and international conference, pp 508–514

Tzfira T, Zuker A, Altman A (1998) Forest-tree biotechnology: genetic transformation and its application to future forests. Trends Biotechnol 16(10):439–446

Vaishnav V, Wali SA, Tripathi SB, Negi MS, Ansari SA (2018) A preliminary investigation on AFLP marker-wood density trait association in teak (*Tectona grandis* L. f.). Ann For Res 61(1):49–63

Valenzuela S, Balocchi C, Rodríguez J (2006) Transgenic trees and forestry biosafety. Electron J Biotechnol 9(3):335–339

Voelker SL, Lachenbruch B, Meinzer FC, Jourdes M, Ki C, Patten AM et al (2010) Antisense down-regulation of 4CL expression alters lignification, tree growth, and saccharification potential of field-grown poplar. Plant Physiol 154(2):874–886

Von Arnold S, Eriksson T (1977) A revised medium for growth of pea mesophyll protoplasts. Physiol Plant 39(4):257–260

Wanders TH, Ofori JN, Amoako A, Postuma M, Wagemaker CA, Veenendaal E, Vergeer P (2021) Teak genetic diversity in Ghana shows a narrow base for further breeding and a need for improved international collaboration for provenance exchange. Genet Res 2(4):44–54

Wang MR, Lambardi M, Engelmann F, Pathirana R, Panis B, Volk GM, Wang QC (2021) Advances in cryopreservation of in vitro-derived propagules: technologies and explant sources. Plant Cell Tissue Org Cult 144:7–20

Watanabe A, Widyatmoko A, Rimbawanto A (2004) Discrimination of teak (*Tectona grandis*) plus trees using selected random amplified polymorphic DNA (RAPD) markers. Trop For Sci 16(1):17–24

Whitlock M, McCauley D (1999) Indirect measures of gene flow and migration FST6¼1/(4Nm+1). Heredity 82:117–125

Win TT, Hirao T, Watanabe A, Goto S (2015) Current genetic structure of teak (*Tectona grandis*) in Myanmar based on newly developed chloroplast single nucleotide polymorphism and nuclear simple sequence repeat markers. Trop Conserv Sci 8(1):235–256

Wünsch A, Hormaza JI (2002) Cultivar identification and genetic fingerprinting of temperate fruit tree species using DNA markers. Euphytica 125(1):59–67

Wyse SV, Dickie JB, Willis KJ (2018) Seed banking not an option for many threatened plants. Nat Plants 4(11):848–850

Xia XR, Yang F, Ke X, Chen YM, Ye JR, Zhu LH (2021) Somatic embryogenesis of masson pine (*Pinus massoniana*): initiation, maturation and genetic stability analysis at SSR loci. Plant Cell Tissue Organ Cult 145:667–677

Yasodha R, Vasudeva R, Balakrishnan S, Sakthi AR, Abel N, Binai N, Rajashekar B, Bachpai VKW, Pillai C, Dev SA (2018) Draft genome of a high value tropical timber tree, teak (*Tectona grandis* L.f): insights into SSR diversity, phylogeny and conservation. DNA Res 25(4):409–419

Yin Y, Wang C, Xiao D, Liang Y, Wang Y (2021) Advances and perspectives of transgenic technology and biotechnological application in forest trees. Front Plant Sci 12:786328

Yu X, Kikuchi A, Matsunaga E, Morishita Y, Nanto K, Sakurai N, Suzuki H, Shibata D, Shimada T, Watanabe KN (2009) Establishment of the evaluation system of salt tolerance on transgenic woody plants in the special netted-house. Plant Biotech 26(1):135–141

Yu X, Liu Y, Liu X, Tretyakova IN, Nosov AM, Shen H, Yang L (2022) Cryopreservation of *Fraxinus mandshurica* Rupr. by using the slow cooling method. Forests 13(5):773

Zhang Q, Zhang ZY, Lin SZ, Lin YZ (2005) Resistance of transgenic hybrid triploids in *Populus tomentosa* Carr. against 3 species of lepidopterans following two winter dormancies conferred by high level expression of cowpea trypsin inhibitor gene. Silvae Genética 54(1–6):108–116

Zhao D, Hamilton JP, Bhat WW, Johnson SR, Godden GT, Kinser TJ, Boachon B, Dudareva N, Soltis DE, Soltis PS, Hamberger B (2019) A chromosomal-scale genome assembly of *Tectona grandis* reveals the importance of tandem gene duplication and enables discovery of genes in natural product biosynthetic pathways. Giga Sci 8(3):005

Zhou C, Wang L, Weng Q, Li F, Li M, Chen J et al (2020) Association of microsatellite markers with growth and wood mechanical traits in *Eucalyptus cloeziana* F. Muell (Myrtaceae). Ind Crops Prod 154:112702

Zuo L, Yang R, Zhen Z, Liu J, Huang L, Yang M (2018) A 5-year field study showed no apparent effect of the Bt transgenic 741 poplar on the arthropod community and soil bacterial diversity. Sci Rep 8(1):1956

Zurn JD, Nyberg A, Montanari S, Postman J, Neale D, Bassil N (2020) A new SSR fingerprinting set and its comparison to existing SSR and SNP-based genotyping platforms to manage *Pyrus germplasm* resources. Tree Genet Genomes 16:1–10

8 Genomic Selection: An Innovative Approach for Tree Improvement

Desha Meena, Aditi Tailor, Drishti Kataria, and Suresh Kumar Meena

Abstract

Tree breeding aims to produce quality products like seeds and clones for different end uses. The major challenge in tree breeding is the lengthy gestation period of tree species for completing a full cycle of breeding and selection. Breeding forest tree species is much more complicated, longer, and costlier than agricultural crops. However, recent breakthroughs in the fields of molecular biology and genomics have transformed the traditional plant breeding methods, which relied heavily on visual assessment of phenotypes. The advent of an assortment of molecular markers has enabled genotype selection. Genomic selection (GS), a modern improvement strategy, involves analyzing the influence of quantitative trait loci (QTLs), making use of numerous molecular markers spread across the genome. This allows for the evaluation of an individual genomic estimated breeding value (GEBV). GS is particularly effective for predicting complex quantitative traits such as productivity and wood quality. Overall, GS holds great promise for substantially shortening the breeding cycle and enhancing genetic gains. In this chapter, we review the genomic selection in detail and discuss different methods and its perspectives in forest trees.

Keywords

Breeding cycle · Genomic selection · Molecular markers · Quantitative trait loci · Tree improvement

D. Meena (✉) · A. Tailor · D. Kataria · S. K. Meena
Genetics and Tree Improvement Division, ICFRE-Arid Forest Research Institute, Jodhpur, Rajasthan, India

R. K. Kalia, R. Pathak (eds.), *Tree Biology and Biotechnology*,
https://doi.org/10.1007/978-981-96-0002-1_8

8.1 Introduction

Tree breeding entails applying genetic and reproductive biology in conjunction with silvicultural practices to improve and manage forest trees for desired traits. The initial step in any tree improvement program is to identify the presence of variability within a population and to select genetically superior individuals. These programs focus on isolating and evaluating genetic variation for one or more targeted traits. The choice of suitable breeding strategy and mating design plays a key role in any breeding program. The breeding of forest tree species is more complicated in comparison to agricultural crops. Trees have a long juvenile phase, characterized by late flowering and seed production, as well as substantial physical size, requiring extensive areas and long-term monitoring for progeny testing. Assessing certain important traits in forest trees, such as tolerance to biotic and abiotic stresses, can be difficult in field conditions. Consequently, many forest tree breeding programs are still in their initial or second breeding cycles, with only the most advanced programs reaching their third or fourth cycles (Isik 2014). Traditional tree breeding methods, which rely solely on phenotypic assessment, have improved tree genotypes and reduced rotation times, but they have not significantly accelerated the breeding process.

Advanced breeding and propagation methods, like top-grafting and somatic embryogenesis, have been shown to shorten the breeding cycle of coniferous trees from 10 to 4 years and from 8 to 1 year, respectively (Harfouche et al. 2012), but no change in the duration of field testing could be recorded. The lengthy breeding cycles of forest trees hinder the timely development of valuable genotypes that meet societal needs and require large-scale plantations. Consequently, breeders have consistently sought new methods to enhance the efficiency of forest breeding and expedite the selection of valuable genotypes (Lebedev et al. 2020).

Since the 1980s, plant geneticists and breeders have been incorporating molecular marker technology into breeding programs. These recent advances in the field of molecular biology and genomics have revolutionized the traditional tree breeding methods. Exploiting these advanced technologies initially enabled the discovery of quantitative trait loci (QTL) (Kearsey and Farquhar 1998) and subsequently facilitated marker-assisted introgression and selection (Lorenz et al. 2011). While these methods have been effective for manipulating alleles characterized by large effects and with recognized associations to markers, they have not been as useful for quantitative traits, which still require extensive field testing (Lorenz et al. 2011).

Most QTL studies employ shared methodologies that focus on the phenotypic effects of QTLs and struggle with repeatability across environments. Marker-assisted selection (MAS), which emerged as a significant step forward in tree breeding in the 1990s (Coors 2008), proved valuable for identifying major genes and transferring them to desirable traits. MAS is efficient for handling oligogenes and QTLs with large effects but lesser so for minor QTLs. The inclusion of only a few alleles or genes is insufficient to improve complex quantitative traits, which are influenced by numerous minor-effect QTLs (Kearsey and Farquhar 1998). MAS-based breeding approach faces constraints on account of the use of low-to-medium density markers, which often fail to detect small-effect loci and focus on a few large effect loci, potentially overlooking significant genetic variation (Jannink et al. 2010; Ben-Ari and Lavi 2012). As a result, MAS has struggled with improving traits with complex inheritance, such as grain yield and abiotic stress tolerance. To address these limitations, genomic selection (GS) has emerged as a promising approach for enhancing complex traits in breeding programs. Unlike traditional MAS, GS uses markers to measure effects more comprehensively. The main advantage of GS is its ability to capture several small genetic factors and concurrently improve multiple traits. GS selects genotypes based on genomic estimated breeding values (GEBVs), offering significant potential to boost selection efficiency. It surpasses pedigree selection and MAS, especially in enhancing complex traits like yield by reducing the required number of breeding cycles. GS has been proven effective in rectifying the deficiency of MAS by utilizing genome-wide marker data, regardless of the significance of their associations with the traits. This chapter provides a detailed review of this upcoming valuable tree breeding approach, i.e., genomic selection (GS), its applications, and various methods therein.

8.2 Genomic Selection

Genomic selection (GS), also referred to as genome-wide selection, is a novel approach that employs genetic markers across the entire genome to ensure that all QTLs are in linkage disequilibrium (LD) with at least one marker. This method is anticipated to be significantly more efficient for forest tree improvement compared to traditional breeding techniques.

With the advancement witnessed in the next-generation sequencing technologies, the use of a large number of markers, including thousands of single nucleotide polymorphisms (SNPs), to estimate genomic breeding values through genomic prediction and GS has become standard practice in both animal and plant breeding (Misztal et al. 2020; Sandhu et al. 2022). GS has proven to be a well-established strategy in animal breeding; however, its application in plant breeding is still in a nascent stage. Fundamentally, GS represents a specialized form of MAS applied on a genome-wide scale. Unlike association analysis (mapping), which aims to identify the impact of specific genes, GS concentrates on forecasting phenotypes based on multi-locus genotypes. It utilizes random effect models to make approximations about the effects of all markers without testing individual variants for significance. The capturing QTLs that account for small fractions of genetic variation, might not meet conventional significance thresholds (Heffner et al. 2011).

In GS, preliminary information such as phenotype-marker linkage, QTL localization, and their effects on phenotypes are not required. A key feature of GS is the utilization of high-density markers, where each trait locus is likely to be in the LD with at least one marker locus across the entire target population. This approach eliminates the need for gene mapping and the exploration for individual linked QTL marker loci. Additionally, GS helps to speed up breeding cycles, reducing the time and cost associated with annual genetic gains. The efficacy and precision of genomic

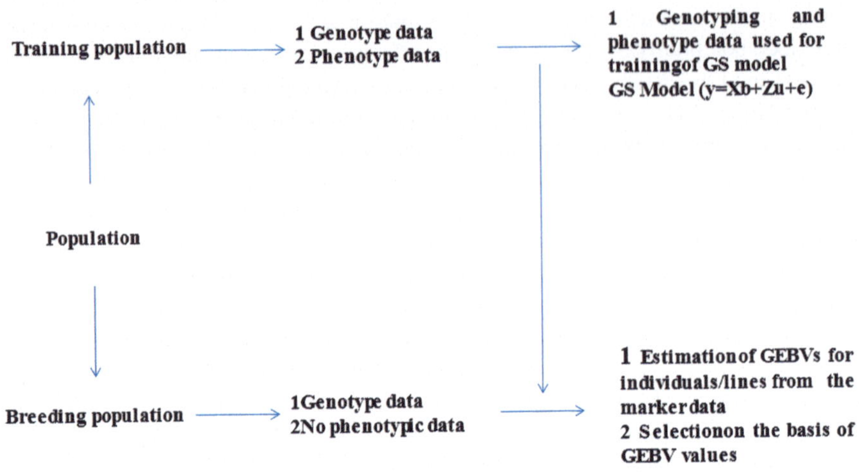

Fig. 8.1 Schematic representation of genomic selection strategy

predictions have been evaluated using various models (Grattapaglia and Resende 2011; Pérez-Enciso et al. 2017).

8.2.1 Steps of Genomic Selection

It is difficult to predict locus identification and effect estimation for low heritability quantitative traits. However, molecular marker technology has made it possible for identifying QTLs, marker-assisted introgression, and selecting large-effect alleles, but it is less suited for quantitative traits exhibiting low heritability, which still require extensive field testing. GS can be understood through the following three steps as reported by Jannink et al. (2010). A schematic representation of genomic selection strategy is presented in Fig. 8.1.

8.2.1.1 Prediction Model Training and Validation

This step involves creating a training population from selected lines in a breeding population. The training population comprises germplasm with both phenotypic data and genome-wide marker data. These data are used to make estimations about marker effects and validate the results. A second group, designated as selection candidates (breeding materials), is genotyped to compute their Genomic Estimated Breeding Values (GEBVs). These GEBVs are subsequently employed for the construction of a statistical model that correlates variations in detected marker loci with individual phenotypes (Gidamo 2022).

8.2.1.2 Prediction of Breeding Value

The GEBV is computed by integrating marker effect estimates with marker data originating from a single cross. For each individual, the GEBV is derived as the aggregate of all marker effects incorporated in the model. These GEBVs are then utilized to guide for selection decisions (Gidamo 2022).

8.2.1.3 Selection Based on Predicted GEBV

In the context of GS, selection is based on the predicted GEBVs that single crosses are subjected to. The breeding value of a single cross is determined directly from genotype data, allowing for precise and efficient selection based on these estimates (Gidamo 2022).

8.2.2 Models Used for Prediction in GS

Various GS models have been developed to predict genotypic performance; however, selection of an appropriate statistical model is crucial for achieving higher predictive accuracy and overall success in GS. Commonly used models in GS-based investigations include RR-BLUP (Ridge Regression Best Linear Unbiased Prediction), G-BLUP (Genomic Best Linear Unbiased Prediction), and Bayesian models (Bayes A, Bayes B, Bayes C pie, LASSO). The RR-BLUP model operates under the assumption that all markers exhibit equal variance, each contributing a small but non-zero effect (Bernardo and Yu 2007). This model particularly performs well and enables higher prediction precision when the trait is influenced by numerous loci with small effects (Buckler et al. 2009). Conversely, the G-BLUP utilizes genome-wide markers to estimate both the genetic and phenotypic values of selection candidates (Wang et al. 2015). Model Bayes A is well-suited for traits controlled by a moderate number of genes, while Bayes B is effective for traits governed by QTLs with substantial effects and with high genetic variation (Munkvold et al. 2009). On the other hand, Bayes C pie, by allowing the parameter π to be derived from experimental data, facilitates the estimation of the shrinkage level. Bayesian models typically provide enhanced prediction accuracy by effectively capturing large-effect QTLs. Bayesian LASSO combines features offered by subset selection with the shrinkage effect produced by Bayesian regression, optimizing both feature selection and model accuracy.

8.2.3 Genomic Selection in Trees

GS has been successfully studied in various agricultural crops including wheat and barley (Schmidt et al. 2016). Prominent diseases like wheat rust, Fusarium head blight, and rice blast represent some of the most thoroughly studied examples wherein GS has been utilized (Ornella et al. 2012; Huang et al. 2019). Some reports also underscore the application of GS in hybrid breeding for maize and rice (Xu et al. 2018; Cui et al. 2020). The use of GS in tree breeding has garnered significant interest in recent years owing to its efficiency and accuracy, particularly for breeding of slow-growing coniferous tree. GS has been successful in reducing the breeding cycles for species such as Eucalyptus (Grattapaglia 2017) and loblolly pines

(Lorenzana and Bernardo 2009). Additionally, promoting early flowering in individual plants can further shorten the breeding cycle, achieving reductions up to 1–4 years in Eucalyptus breeding programs and 3 years in *Pinus taeda* by grafting (Grattapaglia et al. 2011). Given the constraints of limited cultivation of trees, extended conventional breeding cycles, and the considerable genetic variability inherent in nearly all traits, GS is likely to be more effective in forest breeding, offering a significant advantage over traditional methods (Isik 2014). GS has demonstrated efficacy across several key tree species including *Eucalyptus*, *Picea*, and *Pinus*, which are integral to global forestry plantation operations. Additionally, in recent years, the application of GS has expanded to lesser-known and infrequently studied forest species, i.e., Douglas-fir, Japanese cedar, and rubber trees. Historically, phenotypic assessments for tree breeding have predominantly concentrated on growth parameters and wood quality traits, with less emphasis on tree architecture and pulp yield, and even less attention to species-specific traits such as essential oil content and composition in *Eucalyptus polybractea* (Kainer et al. 2018) or rubber production in *Hevea brasiliensis* (Cros et al. 2019). Contemporary GS, however, has been increasingly investigated for resistance to various stresses, including diseases (Stocks et al. 2019; Westbrook et al. 2020), pests (Lenz et al. 2020; Beaulieu et al. 2020), and drought (Bouvet et al. 2020).

8.2.4 Factors Responsible for the Estimation Accuracy of GS Models

The accuracy of GS models is influenced by several critical factors. These include the extent and distribution of LD among individuals, the performance of the model, sample size and relatedness, marker density, the magnitude of gene effects, trait heritability, and the overall genetic design of the study.

8.2.4.1 Marker Density

The accuracy of GS is closely linked to marker density, which can be assessed by examining the rate of LD decay across the genome. Achieving high accuracy necessitates an appropriate marker density which, in turn, is governed by the effective population size and the number of QTLs. Ideally, the genome should be densely covered by an adequate number of markers to ensure that at least one marker is in LD from each gene region. Higher marker density and more extensive LD typically led to improved prediction accuracy (Liu et al. 2018). Furthermore, the type of used markers impacts marker density; bi-allelic markers such as SNPs generally require twice or thrice the density compared to multi-allelic markers like SSRs (Solberg et al. 2008; Jannink et al. 2010; Lozada et al. 2019).

8.2.4.2 Size and Composition of Training Population

Another factor greatly influencing the GS accuracy is the size of the training population. Studies revealed that larger training populations yield higher accuracy in GS predictions (VanRaden et al. 2009). Additionally, factors such as population structure, the age of the training population, the inclusion of older or related lines, and the number of generations represented can further enhance the accuracy (Meuwissen et al. 2001).

8.2.4.3 Number of QTLs

The number of QTLs and the heritability of the trait dictate the necessary marker density and training population size. For traits with low heritability, a large training population can improve the accuracy of predictions using models such as BLUP (Hayes et al. 2009).

8.2.4.4 Heritability

Prediction accuracy for traits with low heritability can be improved with a larger training population and extensive phenotypic data (Bernardo and Yu 2007; Hayes et al. 2009; Liu et al. 2018). For traits with high heritability, an average LD value (r^2) of 0.15 is adequate, while up surging r^2 to 0.2 can enhance the accuracy of GEBV estimates for lower heritability characters (Gidamo 2022).

8.2.4.5 Linkage Disequilibrium

Linkage disequilibrium represents the nonrandom association of alleles at different loci and is a critical determinant of marker density and GS accuracy. The rate of LD decay across the genome provides insight into the required marker density for effective GS.

8.2.4.6 Model Used

Literature shows that accuracy of various statistical models has negligible impact on GS. However, studies carried out on rice hybrid breeding resulted into variation in the prediction accuracies among available GS models (Huang et al. 2019; Cui et al. 2020; Islam et al. 2020).

8.2.5 Advantages of Genomic Selection

The advantages of GS are summarized as under:

- GS leverages marker effects derived through a training population and applies them directly to the breeding population, thus circumventing the need for traditional steps such as QTL discovery and mapping.
- Research based on simulation and empirical investigations has demonstrated that GS achieves greater gains per unit of time compared to traditional selection approaches. This can be exemplified by a simulation study in maize that revealed that GS outperforms Marker-Assisted Recurrent Selection (MARS), especially for traits with lower heritability (Gidamo 2022). More accurate predictions of breeding line performance can be derived through GS as compared to using pedigree data alone. Hence, GS is perceived as a highly powerful approach for enhancing breeding efficiency.

- The approach of selection index, a core component of GS, integrates data from several traits into a unified index, which then guides subsequent trait selections. This approach incorporates the genome-wide marker data with phenotypic data into a range index, either individually or across multiple traits. Several simulation inquiries suggest that the integrated selection index approach of GS significantly improves the selection efficiency, especially with respect to the traits exhibiting lower heritability.
- The implementation of GS can possibly reduce inbreeding rates and mitigate genetic variability loss, unlike selection methods based on phenotype-derived breeding values. This advantage is realized without diminishing selection gains, which is especially beneficial for species prone to severe inbreeding depression.
- One of the major advantages of GS is that it eliminates the need for phenotyping in each selection cycle within the breeding population. This reduction in selection cycle time is due to avoidance of interbreeding performance analysis, which is essential for phenotypic selection. As a result, GS can have special significance in perennial species with long gestation periods and is expected to accelerate the development and commercialization of improved genotypes. For instance, the employment of GS has been estimated to reduce the selection cycle time for oil palm (*Elaeis guineensis*) from 19 years to just 6 years (Gidamo 2022). Additionally, GS is anticipated to be superior to MARS and phenotypic selection in terms of selection gains per unit of time and cost, even with relatively small populations, with sizes as small as fifty. This reduction in the selection cycle is on account of GS excluding any analysis of the interbreeding performance of plants under selection, which is a factor indispensable to phenotypic selection. In this context, GS is expected to expedite the development and subsequent commercialization of improved genotypes in perennial species, achieving this in a shorter time spans compared to the longer durations typically required by traditional phenotypic selection.
- GS is anticipated to enable selection of parent lines for crossbreeding programs from among lines for which data regarding target environment-specific evaluation of performance has not been generated. The predicted GEBV values of these lines in relation to their target environment, thus, lay the foundation for such a selection, thereby facilitating the use of germplasm in breeding programs.
- It is well-established that a phenotype is an outcome of the genotype × environment (G × E) interaction. This renders phenotype estimation a complex and challenging task. GS can leverage accumulated data on marker genotypes and trait phenotypes from various research programs across different environments. This integration can be included in the GEBV approximations for different individuals or lines, allowing for prediction of GEBV even for traits that have not been previously evaluated.
- GS can also prove valuable for hybrid variety development by employing GEBV estimates for selecting parents in crossing programs. However, practical validation of these applications is still needed to confirm their effectiveness.

8.2.6 Limitations of Genomic Selection

The disadvantages and or limitations of GS are summarized as under:

- GS has limited popularity and yet to achieve widespread adoption among plant and tree breeders owing to the availability of limited practical evidences. Much of the debate around its benefits has been primarily statistical and based on simulations, which may not be readily comprehensible or valued by the breeders.
- There is a necessity for cautious approach in evaluating the value of GS, as its benefits have predominantly been evaluated only through simulation studies. This mandates the usage of GS in actual crop breeding programs as a crucial step for demonstrating its practical effectiveness.
- The marker effects are subjected to modifications by shifts in gene frequencies and epistatic interactions. These fluctuating marker effects, and consequently GEBV estimates, necessitate the regular updating of the GS model with each breeding cycle in order to optimize the gains.
- The simulation models employed for deciphering the accuracy of GEBV estimates often rely on additive genetic variance and overlook epistatic effects, potentially limiting their realism. While additive genetic models might maximize selection gains, this is primarily applicable to self-fertilizing species where both parent and variety lines are homozygous.
- A prerequisite for development of accurate GS models is a comprehensive understanding of the genetic architecture of quantitative traits, which for now is quite inadequate, thus posing a big hindrance in achieving GS models with maximum prediction accuracy.
- Another potential disadvantage of GS is the accelerated decline in selection response compared to pedigree selection. However, this issue can be mitigated by continually incorporating new markers for GEBV prediction and prioritizing favorable alleles with low frequencies by assigning higher weights, especially during the early phases of a GS program. GS certainly remains more effective than phenotypic selection provided facilities such as greenhouses are available. In the absence of such facilities, the practical utility and cost-effectiveness of GS becomes uncertain.
- GS involves genotyping a large number of marker loci in each generation of selection, which significantly increases the cost of breeding programs. Future advancements will likely focus on leveraging marker data rather than phenotypic data.
- Infrastructure and resource requirements for implementing GS are likely to be prohibitive for many medium-sized public breeding programs, especially in developing countries. Additionally, the planning and execution of GS impose significant demands and challenges, necessitating substantial adjustments to existing breeding strategies.

8.3 Conclusion

Traditional breeding approaches, which rely on phenotypic data, involve lengthy assessment periods for desired traits. The extended breeding cycles of forest trees impede the rapid development of new economically important genotypes to meet the growing demand. Conventional tree improvement programs struggle to evaluate critical traits of economic significance, such as growth rate and wood quality. However, genomic selection (GS) has emerged as a transformative approach, enhancing the efficiency of forest breeding. GS facilitates breeding without the need for extensive gene or QTL mapping and characterization, thereby reducing costs while acquiring functional data. By utilizing dense markers, increasing training population sizes, and adopting precise GEBV prediction models, GS can achieve substantial genetic gains per unit of time in tree breeding programs. The effectiveness of GS, however, will be contingent upon the specific species and traits of interest. Nonetheless, GS holds great promise for the future tree breeding efforts.

References

Beaulieu J, Nadeau S, Ding C, Celedon JM, Azaiez A, Carol Ritland C, Laverdière JP, Marie Deslauriers M, Adams G, Fullarton M, Bohlmann J, Lenz P, Bousquet J (2020) Genomic selection for resistance to spruce budworm in white spruce and relationships with growth and wood quality traits. Evol Appl 13(10):2704–2722

Ben-Ari G, Lavi U (2012) Marker-assisted selection in plant breeding. In: Plant biotechnology and agriculture. Elsevier, Amsterdam, pp 163–184

Bernardo R, Yu JM (2007) Prospects for genome wide selection for quantitative traits in maize. Crop Sci 47:1082–1090

Bouvet JM, Ekomonod CGM, Brendelf O, Laclaug JP, Bouillet JP, Epron D (2020) Selecting for water use efficiency, wood chemical traits and biomass with genomic selection in a *Eucalyptus* breeding program. For Ecol Manag 465:118092

Buckler ES, Holland JB, Bradbury PJ, Acharya CB, Brown PJ et al (2009) The genetic architecture of maize flowering time. Science 325:714–718

Coors JG (2008) Who are plant breeders, what do they do, and why? In: Kendall RL, Michael L (eds) Plant breeding: the Arnel R Hallauer international symposium, pp 51–60

Cros D, Mbo-Nkoulou L, Bell JM, Oum J, Masson A, Soumahoro M, Tran DM, Achour Z, Le Guen V, Clement-Demange A (2019) Within-family genomic selection in rubber tree (*Hevea brasiliensis*) increases genetic gain for rubber production. Ind Crop Prod 138:111464

Cui Y, Li R, Li G, ZhanG F, Zhu T, Zhang Q et al (2020) Hybrid breeding of rice via genomic selection. Plant Biotechnol J 18:57–67

Gidamo GH (2022) Genomic selection: a faster strategy for plant breeding. In: Case studies of breeding strategies in major plant species. https://doi.org/10.5772/intechopen.105398

Grattapaglia D (2017) Status and perspectives of genomic selection in forest tree breeding. In: Varshney RK, Roorkiwal L, Sorrells ME (eds) Genomic selection for crop improvement. Springer International Publishing AG, Cham, pp 199–249

Grattapaglia D, Resende MD (2011) Genomic selection in forest tree breeding. Tree Genet Genomes 7(2):241–255

Grattapaglia D, Resende MDV, Resende MR, Sansaloni CP, Petroli CD, Missiaggia AA, Takahashi EK, Zamprogno KC, Kilian A (2011) Genomic selection for growth traits in *Eucalyptus*: accuracy within and across breeding populations. BMC Proc 5(7):O16

Harfouche A, Meilan R, Kirst M, Morgante M, Boerjan W, Sabatti M, Mugnozza GS (2012) Accelerating the domestication of forest trees in a changing world. Trends Plant Sci 17:64–72

Hayes B, Bowman P, Chamberlain A, Goddard M (2009) Invited review: genomic selection in dairy cattle: Progress and challenges. J Dairy Sci 92:433–443

Heffner EL, Jannink JL, Sorrells ME (2011) Genomic selection accuracy using multifamily prediction models in a wheat breeding program. Plant Genome 4(1):65–75

Huang M, Balimponya EG, Mgonja EM, McHale LK, Luzi-Kihupi A, Wang GL et al (2019) Use of genomic selection in breeding rice (*Oryza sativa* L.) for resistance to rice blast (*Magnaporthe oryzae*). Mol Breed 39:114

Isik F (2014) Genomic selection in forest tree breeding: the concept and an outlook to the future. New For 45:379–401

Islam MS, Fang DD, Jenkins JN, Guo J, McCarty JC, Jones DC (2020) Evaluation of genomic selection methods for predicting fiber quality traits in upland cotton. Mol Genet Genomics 295:67–79

Jannink JL, Lorenz AJ, Iwata H (2010) Genomic selection in plant breeding: from theory to practice. Brief Funct Genom 9:166–177

Kainer D, Stone EA, Padovan A, Foley WJ, Külheim C (2018) Accuracy of genomic prediction for foliar terpene traits in *eucalyptus polybractea*. G3 Genes Genomes Genet 8:2573

Kearsey MJ, Farquhar AGL (1998) QTL analysis in plants-where are we now? Heredity 80:137–142

Lebedev VG, Lebedeva TN, Chernodubov AI, Shestibratov KA (2020) Genomic selection for forest tree improvement: methods, achievements and perspectives. Forests 11:1190

Lenz PRN, Nadeau S, Mottet MJ, Perron M, Isabel N, Beaulieu J, Bousquet J (2020) Multi-trait genomic selection for weevil resistance, growth, and wood quality in Norway spruce. Evol Appl 13:76–94

Liu X, Wanga H, Wanga H, Guoa Z, Xua X, Liua J et al (2018) Factors affecting genomic selection revealed by empirical evidence in maize. Crop J 6:341–352

Lorenz AJ, Chao S, Asoro FG, Heffner EL, Hayashi T, IwataH SKP, Sorrells ME, Jannink JL (2011) Genomic selection in plant breeding: knowledge and prospects. Adv Agron 110:7–123

Lorenzana RE, Bernardo R (2009) Accuracy of genotypic value predictions for marker-based selection in biparental plant populations. Theor Appl Genet 120(1):151–161

Lozada DN, Mason RE, Sarinelli JM, Brown-Guedira G (2019) Accuracy of genomic selection for grain yield and agronomic traits in soft red winter wheat. BMC Genet 20:82

Meuwissen THE, Hayes BJ, Goddard ME (2001) Prediction of total genetic values using genome-wide dense marker maps. Genetics 157:1819–1829

Misztal I, Lourenco D, Legarra A (2020) Current status of genomic evaluation. J Anim Sci 98(4):14

Munkvold JD, Tanaka J, Benscher D, Sorrells ME (2009) Mapping quantitative trait loci for pre-harvest sprouting resistance in white wheat. Theor Appl Genet 119:1223–1235

Ornella L, Singh S, Perez P, Burgueño J, Singh R, Tapia E et al (2012) Genomic prediction of genetic values for resistance to wheat rusts. The plant. Genome 5:136e148

Pérez-Enciso M, Forneris N, de los Campos G, Legarra A (2017) Evaluating sequence-based genomic prediction with an efficient new simulator. Genetics 205(2):939–953

Sandhu KS, Shiv A, Kaur G, Meena MR, Raja AK, Vengavasi K, Mall AK, Kumar S, Singh PK, Singh J et al (2022) Integrated approach in genomic selection to accelerate genetic gain in sugarcane. Plan Theory 11:2139

Schmidt M, Kollers S, Maasberg-Prelle A, Großer J, Schinkel B, Tomerius A et al (2016) Prediction of malting quality traits in barley based on genome wide marker data to assess the potential of genomic selection. Theor Appl Genet 129(2):203–213

Solberg TR, Sonesson AK, Woolliams JA, Meuwissen TH (2008) Genomic selection using different marker types and densities. J Anim Sci 86:2447–2454

Stocks JJ, Metheringham CL, Plumb WJ, Lee SJ, Kelly LJ, Nichols RA, Buggs RJA (2019) Genomic basis of European ash tree resistance to ash dieback fungus. Nat Ecol Evol 3:1686–1696

VanRaden PM, Van Tassell CP, Wiggans GR, Sonstegard TS, Schnabel RD, Taylor JF et al (2009) Reliability of genomic predictions for north American Holstein bulls (invited review). J Dairy Sci 92:16–24

Wang X, Yang Z, Xu CA (2015) Comparison of genomic selection methods for breeding value prediction. Sci Bull 60:925–935

Westbrook JW, Zhang Q, Mandal MK, Jenkins EV, Barth LE et al (2020) Optimizing genomic selection for blight resistance in American chestnut backcross populations: a trade-o with American chestnut ancestry implies resistance is polygenic. Evol Appl 13:31–47

Xu Y, Wang X, Ding X, Zheng X, Yang Z, Xu C et al (2018) Genomic selection of agronomic traits in hybrid rice using an NCII population. Rice 11:32

9 Sacred Grove: A Reservoir of Plant Diversity and Conservation

Ruby Susana Kujur, Somnath Singh Munda, Alok Kumar, V. Rathna, and Ravinder R. Lall

Abstract

Sacred groves, traditional forest fragments preserved by indigenous and local communities, represent invaluable reservoirs of plant diversity. The groves are protected due to cultural, religious, or spiritual beliefs, which have inadvertently contributed to biodiversity conservation. Being a part of forests, these groves are revered as sacred places connected to the local community's ancestral and spiritual beliefs and have been conserved through generations due to their spiritual and cultural importance. The intrinsic value of these groves lies in their rich biodiversity, harboring diverse plant species, among which few are rare, endangered, or endemic. They are significant reservoirs of plant diversity and a treasure trove of species, which includes rare, endemic, and medicinal plants. They exhibit unique biological diversity and ecological traits according to geographical location, resulting in varied species richness with a wide range of species, such as trees, shrubs, herbs, climbers, epiphytes, parasites, and wild varieties of cultivated plant species. Although there is no comprehensive nationwide study, yet ~13,270 existing records of sacred groves have been chronicled thus far. These areas often remain undisturbed by human activity, allowing for preservation of natural ecosystems and continuation of ecological processes. By analyzing various sacred groves across different regions, the present chapter highlights the unique species compositions, endemic, and rare plants, and the ecological services provided by these ecosystems.

Keywords

Sacred groves · Biodiversity · Conservation · Ecological · Tribals

R. S. Kujur · S. S. Munda · A. Kumar · V. Rathna (✉) · R. R. Lall
ICFRE – Institute of Forest Productivity, Ranchi, Jharkhand, India
e-mail: rathnav@icfre.org

R. K. Kalia, R. Pathak (eds.), *Tree Biology and Biotechnology*,
https://doi.org/10.1007/978-981-96-0002-1_9

9.1 Introduction

In ancient Megalithic times, sacred groves emerged as significant forms of local participatory conservation due to the prevalence of shifting agriculture. These clusters of pristine forests were safeguarded and preserved by local communities and dedicated to deities or spirits of ancestors and have been conserved through generations due to their spiritual and cultural importance. Sacred groves have safeguarded diverse plant species, conserved soil and water, and served as hubs for socioeconomic gatherings, religious festivals, and other social interactions. These groves blend religious, spiritual, and cultural beliefs; associating deities; and ancestral spirits with conservation principles (Jeevan et al. 2023). These beliefs highlight the need to re-evaluate indigenous approaches to biodiversity conservation, as sacred groves are revered as divine dwellings for Gods, ancestral spirits, and other supernatural beings. Such beliefs have historically supported the preservation of biodiversity within these ecosystems. With over 100,000 sacred groves, India holds the highest concentration in the world (Sharma et al. 2014).

Sacred groves contribute significantly to biodiversity and ecological balance and are traditionally protected for religious or cultural reasons. They contribute substantially to the cultural heritage of local communities. The prime value of the groves lies in their diverse biodiversity, which includes a plethora of plant species, and some among them are rare, endangered, or endemic. These areas typically remain untouched by human activities, which allows natural ecosystems to thrive and ecological processes to continue in an undisturbed manner. In addition to their environmental significance, sacred groves serve as repositories of traditional knowledge concerning plant uses, including medicinal properties. Across various cultures worldwide, sacred groves play a significant role in biodiversity conservation by acting as reservoirs of plant diversity and promoting the coexistence of human communities with natural ecosystems.

Conservation efforts for sacred forest groves are vital for the preservation of biodiversity and supporting ecological balance. By protecting these natural sanctuaries, communities also uphold the spiritual and cultural traditions that nurture the nature to promote sustainable environmental practices. Sacred groves indicate unique intersection of cultural heritage and natural conservation, highlighting the relevance of interlinking indigenous ecological wisdom with modern conservation strategies. This chapter explores the role and importance of sacred groves in maintaining plant diversity and examines the ecological, cultural, and conservation significance of these areas. Additionally, it also discusses the threats to these groves from urbanization, deforestation, changing cultural wisdom and practices, and proposed strategies for their preservation. It is imperative to state that sacred groves can also be seen as secure preservation plots that are being demarcated for climate change studies by several organizations and institutions.

9.2 Sacred Groves at Global Level

Globally sacred groves have been recorded (Hughes and Chandran 1998) in many countries like Syria, Turkey, Nigeria, and Japan (Gadgil and Vartak 1976). In Sumatra, Senegal, and Ghana, rich vegetation was documented in sacred groves (MAB 1995). In Bangladesh, Hussain (1998) reported sacred sites linked with abundant vegetation. Sacred groves of Dubla Island (Sundarbans mangrove forest) of Bangladesh support diverse plant life and serve as a place of worship for Hindus (Islam et al. 1998). In rural communities of Afghanistan, creation and preservation of sacred groves have become integral to their historical and geographical traditions (Mohamed 1998).

9.3 Sacred Groves in India

Studies estimated about 100,000 sacred groves in India (Malhotra et al. 2007), but many of them are disappearing due to cultural changes and increasing pressure to exploit the natural resources (Chandrakanth et al. 2004; Ormsby and Bhagwat 2010). The terminology of sacred groves differs in various regions and languages across the country. The size of groves varies from less than one hectare to several hundred hectares (Ntiamoa-Baidu 1995; Malhotra et al. 2007). In some instances, these patches represent the last remaining natural forests outside of protected areas, making them vital for the safeguarding of flora and fauna. Many researchers reported that the preservation of sacred groves is closely connected to ancient world cultures (Malhotra et al. 2001). This tradition in India is exceptionally ancient, originating during the pre-agrarian hunting-gathering stage, prior to humans settlement to domesticate livestock or cultivate the land (Kosambi 1962).

Indian society encompasses diverse cultures, each with unique native conservation strategies. Sacred groves are widespread across India, particularly in tribal-majority regions, and are referred to by various native or tribal communities such as Sarna or Dev in Madhya Pradesh, Devrai or Deovani in Maharashtra, Sarna or Jaherthan in Bihar/Jharkhand, Orans in Rajasthan, Devaravana or Devarakadu in Karnataka, Sarpakavu and Kavu in Tamil Nadu and Kerala, and Dev van in Himachal Pradesh. In Meghalaya, they are referred to as Law Lyngdoh, or Law Kyntang, among others. Majorly, they are observed in Western Ghats. Although a comprehensive study encompassing all sacred groves in India has not been found, few reports suggest that the presence of sacred groves could range between 100,000 and 150,000 as mentioned in Table 9.1.

9.4 Sociocultural Practices

The positive impact of sacred groves on the socioeconomic and traditional life of many rural communities in Ghana is attributed to the joint initiatives of local people to protect them (Michaloud and Durry 1998). Groves often serve as sites for various

Table 9.1 State-wise documented sacred forest of India with their local term

State	Local term for Sacred Forests	No. of documented Sacred Forests
Andhra Pradesh	Pavithravana	728
Arunachal Pradesh	Gumpa Forests (Sacred Groves attached to Buddhist monasteries)	159
Assam	Sacred Grove	29
Bihar	Sarhuli Mander	43
Chhattisgarh	Matagudi	77
Goa	Deorai, Pann	93
Gujarat	Sabarkantha, Dahod, Banaskantha	42
Haryana	Gurudwara grove	57
Himachal Pradesh	Kul Deveta	329
Jammu & Kashmir	Bani	92
Jharkhand	Sarana/Jaherthan	29
Karnataka	Devara Vana, Devara Kadu, Huli Devarakadu, Nagavan, Bhatappavana, Jatakappanbana, Ghowdibana, Kan	1476
Kerala	Kavu, Sarp Kavu	1096
Madhya Pradesh	Sharana,Devkot, Matikot, Devsthali, Budhadev	170
Maharashtra	Pavithravan, Deovan, Deorai, Devgudi, Pen Gada/ Gonds, Devarahati	2820
Manipur	Gamkhap, Mauhak (sacred bamboo reserves)	166
Meghalaya	Ki Law Lyngdoh, Ki Law Kyntang, Ki Law Niam	105
Odisha	Jahera, Thakuramma	188
Puducherry	Kovil Kadu	108
Rajasthan	Vani, Malvan, Kenkri, Orans, Shamlatdeh, Devabani, Jogmaya	588
Sikkim	Pandam	16
Tamil Nadu	Swami Shola, Koil Kadu, Katttu Koil, Vanakkoil	1275
Telangana	Pavithravana	57
Uttarakhand	Deo Bhumi, Bugyal (sacred alpine meadows)	133
Uttar Pradesh	Dev van, Pavithravan	32
West Bengal	Gramthan, Santalburitan Shitalatan, Haritan Sabitritan, Jahera, Deo Tasara, Mawmund	562

Source: CPREEC ENVIS (2021)

religious rituals, ceremonies, and festivals (Fig. 9.1). These events include prayers, offerings, and sacrifices to deities believed to dwell in the sacred groves. Many groves are linked to local myths, legends, and folklore. These stories often explain the source of the grove and its sacred status, reinforcing the community's cultural identity and continuity, and are protected through taboos and cultural norms that prohibit activities like logging, hunting, and grazing. These practices help to conserve biodiversity and maintain ecological balance. The spiritual reverence for

Fig. 9.1 Ritual performed during tribal Sarhul festival

sacred groves cultivates a profound connection between people and nature, promoting a sense of responsibility for environmental stewardship and sustainability.

9.5 Biodiversity in Sacred Groves

Sacred groves are forest patches that are preserved through traditional practices and beliefs, which results in their conservation over long periods. The vegetation found in the groves has distinct ecological characteristics that harbor a plethora of trees, shrubs, herbs, climbers, epiphytes, parasites, and other cultivated plants of wild relatives that vary significantly depending on their geographical area and proximity to urban settlements (Fig. 9.2). Since some of the species are sensitive to changes in habitat, these groves serve as protected natural sanctuaries. The preservation of sacred groves is closely linked to the cultural practices and beliefs of local communities, making a strong connection between cultural heritage and biodiversity conservation. The protected groves have rich genetic reservoirs, preserving genetic diversity of several plant species. This diversity is crucial for the adaptation and survival of species under changing climatic conditions. Sacred groves can be labeled as biodiversity hotspots, supporting a high number of plant species within a relatively small area. The undisturbed nature of groves allows for the maintenance of complex ecological interactions, soil fertility, water quality, and climate regulation.

The different tribal communities of Jharkhand worship in sacred groves, namely, "Sarna," "Marang Buru," and "Jaherthan," by Oraon, Munda, and Santhal tribes, respectively. Documentation of 29 sacred groves has been done in Jharkhand. The Sarna, Marang Buru, and Jaherthan are clusters of trees worshipped by tribal communities on various ceremonies and occasions. It is believed that, in addition to

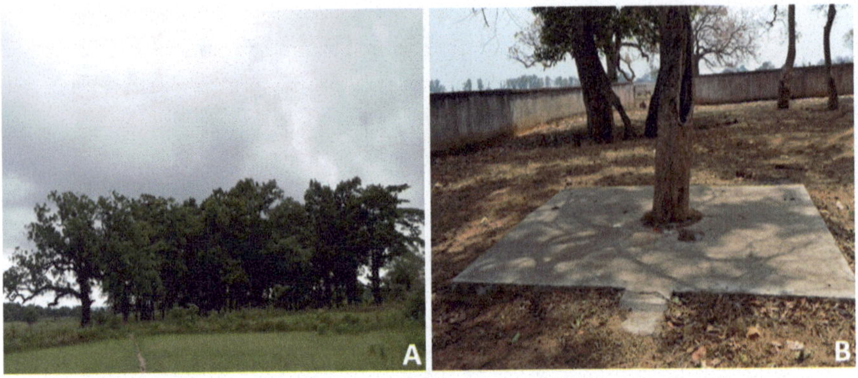

Fig. 9.2 Plant diversity in sacred groves. (**A**) Dense and diversified sacred grove, (**B**) Moderate dense and less diversified sacred grove

other tree species, a grove must have a minimum of five Sal (*Shorea robusta*) trees, called "Sarjom" by local tribes. Local plant species observed in the groves are Sal (*Shorea robusta*), Peepal (*Ficus religiosa*), Banyan (*Ficus benghalensis*), Indian black plum (*Syzygium cumini*), Indian plum (*Ziziphus mauritiana*), White marudah (*Terminalia arjuna*), Tulsi (*Ocimum tenuiflorum*), Indian gooseberry (*Phyllanthus embelica*), Neem (*Azadirachta indica*), Mango (*Mangifera indica*), Malabar nut (*Justicia adhatoda*), Thorn apple (*Datura stramonium*), and Champak (*Magnolia champaca*) (CPREEC ENVIS 2021).

9.5.1 Sacred Species

The worshiping of species by local communities or tribes at sacred groves, called sacred species, varies across the groves in Jharkhand. Along these species, *Shorea robusta* (Sal), *Ficus religiosa* (Peepal Tree), and *Ocimum tenuiflorum* (Tulsi or Holy Basil) are often worshiped in India and considered sacred in Hindu religion. These trees are associated with various Gods and deities. *Shorea robusta* (Fig. 9.3) and *Madhuca indica* (Mahua) are the major worshiping trees of tribal communities. *Adina cordifolia* (Karam), *Schleichera oleosa* (Kusum), *Bombax ceiba* (Semal), and *Ficus benghalensis* (Banyan) are also found to be worshiped in the sacred groves. The list of trees worshiped and conserved over the years in Jharkhand is given in Table 9.2.

9.6 Management Practices of Sacred Groves

The sacred groves in India have diverse ownership and management. Some groves are held under private or community ownership, distinct from formal government protected areas or parks. According to the global framework of Indigenous and

Fig. 9.3 Sacred species (*Shorea robusta*) worshiped at Bandhia village, Chatra district, Jharkhand

Table 9.2 List of sacred species revered in the sacred groves of Jharkhand

Local name of sacred species	Scientific name of sacred species	Major tribes of Jharkhand
Hesah	*Ficus religiosa*	Santhal, Munda, and Oraon
Baade	*Ficus benghalensis*	Santhal, Munda, and Oraon
Sarjom	*Shorea robusta*	Santhal, Munda, and Oraon
Madukam	*Madhuca indica*	Santhal, Munda
Tulsi	*Ocimum tenuiflorum*	Santhal, Munda, and Oraon
Adel	*Bombax ceiba*	Munda, Oraon
Baru	*Schleichera oleosa*	Oraon, Munda
Korom	*Adina cordifolia*	Oraon, Munda

Community Conserved Areas (ICCAs), in India they are primarily governed by local customary systems. They are managed by community groups rather than government agencies. Management and ownership of sacred groves vary widely across Indian states and don't have legal clarity (Chandrakanth et al. 2004). Chandrashekara and Sankar (1998) observed that management of sacred groves was often entrusted to individual families, groups of families, or statutory agencies responsible for temple management in Kerala. Chandrakanth et al. (2004) noted two management systems, either family-owned or community-managed, in Kodagu district in Karnataka. In Meghalaya, groves were managed by a community committee led by a priest (Tiwari et al. 1998).

In contrast, at the international level, the International Union for Conservation of Nature (IUCN) and UNESCO have formulated guidelines for managing sacred sites (Wild and McLeod 2008). Additionally, within the IUCN, expert group on cultural and spiritual values of protected areas (CSVPA) contributes to policies concerning sacred sites (http://www.csvpa.org/). Many UNESCO World Heritage Sites encompass sacred forests, mountains, and springs (Robson and Berkes 2010; Schaaf and Rossler 2010). Over the past few years, there has been growing cooperation and synergy among international organizations like IUCN, UNESCO, and Convention on Biological Diversity (CBD) in addressing problems faced at sacred sites. India's Forest Acts and Forest Rights Act are key legal frameworks for managing sacred groves at the national level. The primary categories of sacred groves are determined by their boundaries and the level of protection provided by local communities or tribes. It is crucial to emphasize that conservation efforts for sacred groves persist because communities recognize the consequence of climate change and understand that preserving such green spaces can significantly mitigate these effects. There is an increasing acknowledgment of the benefits provided by ecosystems of sacred groves and the biodiversity they harbor. This awareness emphasizes the importance of continuing to protect these groves, not only for their cultural relevance but also for their essential purpose in maintaining biodiversity and mitigating climate change impacts.

Furthermore, it has been noted that even among people having good educational background, the key reason for advocating the conservation of sacred groves is to sustain their religious beliefs. While education enhances awareness of biodiversity and conservation issues, it can also lead to skepticism or loss of faith in traditional norms. Therefore, initiatives aimed at increasing ecological awareness to protect and sustain these groves are crucial. It is a vital effort that concentrates on integrating cultural and religious values with scientific knowledge. This integration can foster sustainable conservation practices that both honor and safeguard the ecological and sacred value of these natural sites. By bridging traditional wisdom with modern conservation strategies, we can effectively promote the long-term protection and resilience of these invaluable cultural and environmental assets.

Different types of following boundaries represent distinct ways in which the sanctity of the groves is maintained and traditional practices are upheld and safeguarded:

9.6.1 Open Sacred Grove

These groves have well-defined boundaries that are strictly protected by the local community. The boundaries are often marked by natural features like rivers, hills, or specific tree lines. Access is restricted and strict rules govern activities within the grove. These rules are enforced by community leaders or religious authorities.

9.6.2 Concrete Boundary

Sacred groves are maintained and protected by the entire community which are bounded by concrete walls. Access may be restricted or open, depending on customs and rules of the community.

9.6.3 Bamboo Boundary

These groves are temporarily bound by bamboo for demarcation and necessary signage are displayed for information. Access is prohibited for the trespassers and are only permitted to the local community with various religious rules.

9.6.4 Stone Boundary

This is the traditional way of demarcating the sacred groves, which are bounded by stones. Protection of these groves is closely linked to the religious practices and rituals conducted by the local communities.

9.7 Degradation or Threats to the Sacred Groves

The world faces major environmental issues like biodiversity decline and climatic variability (Saha et al. 2009). Anthropogenic activities have accelerated extinction of species to about 1000 times the natural rate (Pimm et al. 2014). Traditions such as beliefs and taboos have played crucial roles in conserving sacred groves. However, these protective customs are gradually eroded due to various factors, currently placing these sacred groves in a precarious position. Previous reports highlighted that sacred groves are under threats from ecological, social, institutional, religious, and economic changes within the communities that have historically protected them. Additionally, approximately 80% of groves that were once rich in vegetation now exist solely as sacred sites due to human interventions such as development, urbanization, resource exploitation, and population growth posing significant threats (Khan et al. 2008) and extensive removal of fallen wood, litter, and twigs (Patil 2011). Forest fires were observed in some sacred groves of Srinagar, Jammu, and Kashmir (Kumar et al. 2011).

A study on the attitudes of residents in Meghalaya and Karnataka toward the conservation of sacred groves noted that both groups were aware of local taboos, their consequences, and associated rituals. In spite of knowledge about the consequences, these groves faced significant challenges, due to cultural changes and increased demands for natural resources. For instance, in Meghalaya, economic pressures led to a decline in rituals and a reduction in grove size for activities like coffee plantations, highlighting the conflict between conservation and resource extraction (Ormsby 2013). Habitat destruction, land degradation, and the displacement of local communities are significant threats to sacred natural resources (Kumar et al. 2022). Community-conserved sacred forests in the Garhwal Himalaya, play a crucial role in long-term greenhouse gas mitigation. However, these biodiversity-rich patches are being converted into farmland at an alarming rate. Tree species found in lower densities are vulnerable to extinction, highlighting the urgent need for enhanced protection measures (Pala et al. 2013).

Various regions and individual sacred groves face diverse threats, but some of the common challenges are: (a) Fading away of traditional belief systems, once central to sacred groves but now often dismissed as superstition, diminishing their cultural significance. (b) Rapid urbanization and development projects have resulted in the encroachment and shrinking of many sacred groves across India, (c) Conversion of ancient nature of worship in sacred groves into formal temple rituals has led to changes in traditional conservation practices, (d) Intrusion of invasive species like *Eupatorium odoratum, Lantana camara, Prosopis juliflora*, etc. is disrupting native ecosystems within the groves, (e) Increased livestock grazing and fuel wood collection are imposing pressure on sacred groves (CPREEC ENVIS 2021).

9.8 Conclusion

Previous studies proved that sociocultural ancestral beliefs and nature worship were the constructive tools to deter the exploitation of local sacred groves, acting as invisible barriers against economic exploitation. Developing urbanization pressures, like, industrialization, railways, roads etc. often erode traditional norms and social taboos. These growing demands for creation of new infrastructure have led to exploitation of many sacred groves. Moreover, the rituals connected with these groves and their protection are often subjected to scientific queries. Thus, it is essential to emphasize that sacred groves can additionally be considered as secured preservation plots that are being established for climate change studies by several organizations and institutions, where nothing can be introduced or removed.

9.9 Future Prospects

Numerous studies indicate that religious traditions and related beliefs are major factors promoting the safeguarding of sacred groves. Larger groves accommodate more number of species; however, the erosion of cultural beliefs has significantly contributed to the decline of sacred sites, presenting a new challenge for their future conservation. It is imperative that each sacred grove be documented with geo-tagged location data for accurate demarcation and inclusion in the Protected Area Network to guarantee proper conservation. Awareness campaigns highlighting the ecological services provided by sacred groves are crucial. Communities must understand that conserving these groves is essential for their own sustenance and for maintaining biodiversity. Sacred groves are not just unique forest patches representing relict vegetation of their respective areas but also preserve the traditions and beliefs of local communities. Therefore, their conservation is essential for the continuity of these cultural practices and for safeguarding the ecological integrity of these invaluable natural sites.

Acknowledgments The funding provided under the ICFRE project on "Documentation and assessment of plant diversity of sacred groves of major tribes in Jharkhand; IFP131/S and FM-31/2021-2024" is acknowledged.

References

Chandrakanth MG, Bhat MG, Accavva MS (2004) Socio-economic changes and sacred groves in South India: protecting a community-based resource management institution. Nat Res Forum 28:102–111

Chandrashekara UM, Sankar S (1998) Ecology and management of sacred groves in Kerala, India. For Ecol Manag 112(1–2):165–177

CPREEC ENVIS (2021) https://cpreec.org/wp-content/uploads/2021/09/Environmental-Information-System-ENVIS.pdf

Gadgil M, Vartak VD (1976) Sacred groves of Western Ghats of India. Ecol Bot 30:152–160

Jeevan K, Boony PB, Gopalkumar S (2023) Climate change perspectives and conservation of sacred groves: case of Sharngakavu, Kerala. Indian J Tradit Knowl 22(2):433–443

Hughes DJ, Chandran SMD (1998) Sacred grove around the earth: an overview. In: Ramakrishnan PS, Saxena KG, Chandrashekara UM (eds) Conserving the sacred for biodiversity management. UNESCO/IBH Publishing, Oxford/New Delhi, pp 69–86

Hussain ABME (1998) Scared sites in Bangladesh: country report. In: Ramakrishnan PS, Saxena KG, Chandrashekara UM (eds) Conserving the sacred for biodiversity management. UNESCO/IBH Publishing, Oxford/New Delhi, p 167

Islam AKMN, Islam MA, Hoque AE (1998) Species composition of sacred groves, their diversity and conservation in Bangladesh. In: Ramakrishnan PS, Saxena KG, Chandrashekara UM (eds) Conserving the sacred for biodiversity management. UNESCO/IBH Publishing, Oxford/New Delhi, pp 163–165

Khan ML, Khumbongmayum AD, Tripathi RS (2008) The sacred groves and their significance in conserving biodiversity: an overview. Int J Ecol Environ Sci 34(3):277–291

Kumar K, Manhas RK, Magotra R (2011) The Shankaracharya sacred grove of Srinagar, Kashmir, India. Curr Sci 101(3):262–263

Kumar R, Prajapati U, Koli VK (2022) Factors driving the tree species richness in sacred groves in Indian subcontinent: a review. Biodivers Conserv 31:2927–2943

Kosambi DD (1962) Myth and reality: studies in the formation of Indian culture. Popular Prakashan, Bombay

MAB (1995) Sacred places and vernacular conservation. Man and Biosphere Programme, UNESCO, Paris

Malhotra KC, Gokhale Y, Chatterjee S, Srivastava S (2007) Sacred groves in India. Aryan Books International, New Delhi

Malhotra KC, Gokhale Y, Chatterjee S, Srivastava S (2001) Cultural and ecological dimensions of sacred groves in India. INSA, New Delhi, pp 1–30

Michaloud G, Dury S (1998) Sacred trees, groves, landscapes and related cultural situations may contribute to conservation and management in Africa. In: Ramakrishnan PS, Saxena KG, Chandrashekhara UM (eds) Conserving the sacred: for biodiversity management. UNESCO/IBH Publication, Oxford/New Delhi, pp 25–38

Mohamed Z (1998) A note on sacred groves in Afghanistan. In: Ramakrishnan PS, Saxena KG, Chandrashekhara UM (eds) Conserving the sacred: for biodiversity management. UNESCO/IBH Publishing, Oxford/New Delhi, pp 151–152

Ntiamoa-Baidu Y (1995) Indigenous vs. introduced biodiversity conservation strategies: the case of protected area systems in Ghana. Biodiversity Support Program

Ormsby A (2013) Analysis of local attitudes toward the sacred groves of Meghalaya and Karnataka, India. Conserv Soc 11(2):187–197

Ormsby A, Bhagwat SA (2010) Sacred forests of India: a strong tradition of community-based natural resource management. Environ Conserv 37:320–326

Pala N, Negi A, Gokhale Y, Todaria N (2013) Tree regeneration status of sacred and protected landscapes in Garhwal Himalaya. J Sustain For 32:230–246

Patil VK (2011) Wildlife reserves v/s sacred groves: conservation assessment from peoples' perspective. Ashoka Trust for Research in Ecology and the Environment (ATREE), Bangalore, p 64

Pimm SJ, Clinton A, Robin B, Thomas G, Joppa Lucas R et al (2014) The biodiversity of species and their rates of extinction, distribution, and protection. Science 344:1246752

Wild R, McLeod C (2008) Sacred natural sites: guidelines for protected area managers. IUCN World Commission on Protected Areas. Task Force on Cultural and Spiritual Values of Protected Areas. UNESCO Programme on Man and the Biosphere, IUCN

Robson JP, Berkes F (2010) Sacred nature and community conserved areas. In: Pilgrim S, Pretty J (eds) Nature and culture: rebuilding lost connections. Earthscan, Washington DC, pp 197–216

Saha SK, Nair PKR, Nair VD, Kumar BM (2009) Soil carbon stock in relation to plant diversity of home gardens in Kerala, India. Agrofor Syst 76:53–65

Schaaf T, Rossler M (2010) Sacred natural sites, cultural landscapes and UNESCO's action. In: Verschuuren B, Wild R, McNeely JA, Oviedo G (eds) Sacred natural sites: conserving nature and culture. Earthscan, Washington DC, pp 161–169

Sharma R, Aggarwal N, Kumar S (2014) Ecological sustainability in India through the ages. Int Res J Environ Sci 3(1):70–73

Tiwari BK, Barik SK, Tripathi RS (1998) Biodiversity value, status, and strategies for conservation of sacred groves of Meghalaya, India. Ecosyst Health 4(1):20–32

Agroforestry Tree Species: *Acacia tortilis*, Biology, Importance, Agroforestry Production, and Biotechnology Application

10

Gladys Cheruto, Emmanuely Z. Nungula, Lukelysia Nyawira, Luciana R. Chappa, Ruth Kahuthia-Gathu, Riziki Mwadalu, Jerry C. Dlamini, Shivani Ranjan, Sumit Sow, Sagar Maitra, Jamal Nasar, and Harun I. Gitari

G. Cheruto (✉)
Environment and Natural Resource Management Programme, Kenya Agricultural and Livestock Research Organization, Nairobi, Kenya

E. Z. Nungula
Department of Environment and Sustainable Development, Mzumbe University, Morogoro, Tanzania

L. Nyawira
World Agroforestry, United Nations Avenue, Nairobi, Kenya

L. R. Chappa
Department of Economics, Swedish University of Agricultural Sciences, Uppsala, Sweden

R. Kahuthia-Gathu · H. I. Gitari
Department of Agricultural Science and Technology, School of Agriculture and Environmental Studies, Kenyatta University, Nairobi, Kenya

R. Mwadalu
Central Highlands Eco-Region Research Programme, Kenya Forestry Research Institute, Nairobi, Kenya

J. C. Dlamini
Unit for Environmental Sciences and Management, Faculty of Natural and Agricultural Sciences, North-West University, Potchefstroom, South Africa

S. Ranjan · S. Sow
Department of Agronomy, Dr. Rajendra Prasad Central Agricultural University, Pusa, Samastipur, Bihar, India

S. Maitra
Department of Agronomy and Agroforestry, Centurion University of Technology and Management, Sitapur, Odisha, India

J. Nasar
College of Agriculture, Institute of Rice Industry Technology Research, Guizhou University, Guiyang, Guizhou, People's Republic of China

R. K. Kalia, R. Pathak (eds.), *Tree Biology and Biotechnology*,
https://doi.org/10.1007/978-981-96-0002-1_10

Abstract

Trees, especially indigenous species play a key role in the improvement or maintenance of soil fertility, and the overall ecosystem functioning. Trees have long been utilized in various ways including in agricultural production to optimize land productivity. *Acacia tortilis* is one of the trees that play a critical role in arid and semi-arid environments and is widely distributed in Africa and the Middle East. *A. tortilis* provides numerous benefits in the form of products e.g., timber, fuel wood, fodder, food, and gum, services, for example, improvement of soil health, cultural benefits, and shade, and medicinal properties. It is highly valued among dwellers of arid and semi-arid regions due to its ability to tolerate extreme environmental conditions such as drought, salinity, drifting of sand dunes, and management practices like cutting and browsing. The biology, origin, distribution, ecology, morphology, and growth management of *A. tortilis* and its mutual associations with other organisms such as arbuscular mycorrhizal fungi and nitrogen-fixing bacteria, along with biotechnological studies carried out in the species, has been outlined in the chapter.

Keywords

Acacia · Utilization · Climate change · Perception · Management · Challenges

10.1 Introduction

Trees form an integral part of human development by providing products and services to fulfil humankind's needs and the well-being of the planet. Human welfare is influenced by the benefits provided by trees as well as tree diversity and density (Seth 2003). There has always been intentional integration of trees by farmers in agricultural production, a practice that has run through generations and is now well known as agroforestry practice (Nair 1993; Chappa et al. 2024). The importance of trees is well-established and widely recognized. Such benefits are grouped into products and services. Products of value to humankind include timber, fuel wood, edible fruits, seeds, vegetables, gums, resins, fodder, and medicinal products (Maitra et al. 2024). Services include improvement and maintenance of soil health, air and water purification, shade, shelter, climate regulation through avenues like carbon sequestration, cooling effect, and provision of habitat for animals, recreational amenities, windbreaks, etc. (Shackleton et al. 2015; Tiruneh 2017; Turner-Skoff and Cavender 2019; Nungula et al. 2024). Trees also have social and cultural values that invoke a sense of identity and pride, preserve religious heritage, and may contribute to economic growth through tourism and recreational activities (Seth 2003; Elmendorf 2008; Hobbs et al. 2014; Gitari et al. 2024). The trees have also been used in the rehabilitation of soils in areas with high water tables.

Acacia is one of the keystone species in arid and semi-arid environments. More than a thousand species of acacia have been identified. Acacia trees are tolerant to drought, and salinity, and have been utilized for the production of wood, gum, and fodder, to improve soil properties, hence it is highly popular in agroforestry (Mahdhi et al. 2020). *Acacia tortilis* provides numerous benefits in the form of products e.g., timber, fuel wood, fodder, food, and gum (Maundu and Tengnäs 2005; Orwa et al. 2009; Al Dhanhani et al. 2023; Taha et al. 2022), services e.g., improvement of soil health, cultural benefits, and shade (Noumi and Chaieb 2012; Hobbs et al. 2014; Tiruneh 2017; Tasisa and Nemomissa 2019), and medicinal properties (Orwa et al. 2009). It is highly valued among dwellers of arid and semi-arid regions (ASALs) because of its ability to tolerate extreme environmental conditions such as drought, salinity, drifting of sand dunes, and management practices like cutting and browsing (Yadav et al. 2013). Understanding the biology of a tree helps in site selection, the selection of the right species, establishment, the purpose for which it is planted, and better management, including improvement (Judice and Gordon 2020).

Plant biotechnology has found its application in forestry. Biotechnology tools such as genetic engineering, molecular markers, and genomics, among other tools, have been utilized in breeding trees (Saini et al. 2020; Ahuja 2021). Although there are no biotechnological studies reported on *A. tortilis*, such tools have been used in forestry and hence have potential applications in the breeding of *A. tortilis* to address various problems like seed pest infestation. Molecular markers, for example, can be used to explore genetic information in *A. tortilis* and use the information to breed trees with desired traits such as pest resistance and develop conservation plans for genetic resources. Thwe-Thwe-Win et al. (2015) used molecular markers to identify the need and developed conservation plans for teak in Myanmar.

10.2 Biology of *Acacia tortilis*

10.2.1 Origin, Ecology and Distribution

Acacia tortilis (Forssk) Hayne (common name: umbrella thorn) is a pod-bearing tree species belonging to the Fabaceae family, sub-family Mimosoideae. It is considered a keystone species in arid ecosystems in Africa and the Middle East (Andersen and Krzywinski 2007). The tree is native to dry Africa (Tropical, South, and North Africa) and the Middle East and grows in most parts of these regions. The tree has also been adopted in various countries across Tropical Africa, and some parts of the Middle East. It may grow in pure or mixed stands and may occur in dry bushland, bushed grasslands, riverine vegetation, wooded grassland, and arid scrub (Maundu and Tengnäs 2005). It can also be found scattered in homesteads, and croplands, planted on soil conservation structures such as contours (Kuria et al. 2017). The tree grows in a wide range of soils, is drought tolerant, and adapts well to seasonal rains (Kebbas et al. 2015). However, *A. tortilis* thrives well in alkaline soils, also in rocky, sandy loam, well-drained soils, and sand dunes, with a pH range of 7.9–8.3 (Orwa et al. 2009). The tree can tolerate a maximum of 50 ° C and a minimum of 0 ° C temperature and requires an average of 100–1000 mm of annual rainfall. It grows well between 23.4 °C and 31.3 °C mean annual temperature. It can grow at altitude 0–1000 m above sea level (Orwa et al. 2009).

10.2.2 Morphology and Growth

Acacia tortilis grows up to a height of 8–20 m depending on the growing environment hence can be found as a tree, a shrub, or a bush. It is evergreen and propagated by seed (hardly succeeds due to seed dormancy), wildings, and seedlings (Masamba 1994; Argaw et al. 1999; Maundu and Tengnäs 2005; Andersen and Krzywinski 2007). The tree has a narrow crown when young and flat or umbrella-like when mature (Fig. 10.1). The tree has dark grey bark, fissured along the trunk; white to cream and pleasantly scented flowers; pairs of small hooked and long white thorns; greenish yellow to yellow brown pods; greenish grey seeds; and bears both sexual reproductive organs. The roots are deep-rooted growing up to 35 m deep into the soil, making the tree exploit large soil volumes for water and nutrients and thus enabling it to withstand drought. In Kenya, for example, *A. tortilis* grows in agro-ecological zones IV to VII and flowers in February and fruits in September to October (Maundu and Tengnäs 2005; Orwa et al. 2009; Tiruneh 2017). In the United Arab Emirates (UAE), *A. tortilis* flowers from March to May (Al Dhanhani et al. 2023). In Nigeria and India, the tree produces flowers between May and June (Orwa et al. 2009). Seeds are mostly dispersed by invertebrates which are attracted due to strong scents and highly nutritious seeds (Rugemalila et al. 2017). The tree has high longevity hence most stands constitute old-aged trees (Andersen and Krzywinski 2007). *Acacia tortilis* may show disparity in growth across different regions, for instance, in Ethiopia, it exhibits slow growth, but in Kenya, it has fast growth (Kuria et al. 2017). The difference in growth rate across regions could be due to variation in environmental conditions; hence, this should be put into consideration when growing *Acacia tortilis*.

Fig. 10.1 *Acacia tortilis* trees

10.2.3 Mycorrhizal and Bacterial Interactions

Mycorrhizal fungi can play a key role in the growth and establishment of *A. tortilis* because of the benefits they offer to plants like enhanced nutrient uptake, tolerance to adverse environmental conditions like high salinity and drought, and resistance to diseases and pathogens. Mycorrhizal fungi can alleviate stress related to water deficit in *A. tortilis* since this tree mostly grows in arid and semi-arid environments where water deficiency is a common phenomenon. This is possible through increased water uptake by mycorrhizal hyphae, alteration of plant's physiological processes, and maintenance of the soil-root continuum which helps in plant recovery from drought (Smith and Read 2008; Al-Barakah and Mridha 2014). The tree has also been shown to respond to mycorrhizal inoculation with enhanced mycorrhizal infection and growth rate, implying that the growth and establishment of *A. tortilis* can be improved using mycorrhiza inoculum (Munro et al. 1999). A high diversity of Arbuscular mycorrhizal fungi (AMF) species colonizing the roots of *A. tortilis* compared to other Acacia species has been reported (Mahdhi et al. 2020).

A. tortilis also establishes a symbiotic relationship with nitrogen-fixing root bacteria (rhizobia) and nourishes the soil by fixing atmospheric nitrogen (Noumi and Chaieb 2012; Mahdhi et al. 2020; Mirriam et al. 2023). Inoculation of *A. tortilis* subspecies *raddiana* with strains of native bacteria was found to increase growth parameters (Hnini and Aurag 2024). Better performance in Acacia trees has been found when colonized by both AMF and rhizobia (Al-Barakah and Mridha 2014).

10.2.4 Management of *Acacia tortilis*

Tree management in agroforestry systems is conducted to maximize the positive interactions between the woody and non-woody species and/or animal production (Chavan et al. 2017; Kisaka et al. 2023; Alkharabsheh et al. 2023). Management of *A. tortilis* may begin from the seed to a mature tree. Seed dormancy is common among trees that grow in arid and semi-arid environments due to unfavourable environmental conditions for germination and establishment of seedlings. Seed dormancy may lead to low germination rates leading to low regeneration of trees. To overcome this challenge, seed management such as breaking of seed dormancy may be required to increase seed germination and regeneration of trees. Methods to break seed dormancy may include mechanical scarification, acid treatment, heating, and treating with boiling water; among these mechanical scarification and acid treatment are more effective. Mechanical scarification and sulphuric acid treatment of *A. tortilis* seeds led to a higher germination rate. Care should be taken into consideration when treating seeds with dry heating as some seeds may be damaged due to higher heating temperatures. Seeds of *A. tortilis* can be treated with dry heating up to 80 °C temperature, beyond which seeds get damaged. The thickness of the seed coat is an important determining factor in the selection of methods to manage seed dormancy (Teketay 1996). Under natural conditions, seed dormancy is broken by wild or man-set fire, ingestion by animals, soil organisms, abrasion, and/or acids in the soil (Teketay 1996). Soil, and water conservation measures should be incorporated in the early stages of *A. tortilis* for its successful establishment and growth. The tree tends to develop lateral roots, which may be problematic when planted along roads and paths, and grows in a shrub-like manner. Therefore, wide spacing should be taken into consideration when planting (Orwa et al. 2009).

Tree management practices for *A. tortilis* may include lopping and pollarding. Lopping, which is commonly practiced in ASALs, involves chopping off the leafy biomass for fodder (Tengnas 1994; Kuria et al. 2017). In addition, lopping is done to meet food, fuel wood, and other seasonal needs of the household (Chavan et al. 2017). Pollarding involves cutting off the top part of the tree including all the branches leaving only the bole. *A. tortilis*, particularly subspecies *raddiana*, can withstand pollarding without its growth being negatively affected (Andersen and Krzywinski 2007; Hobbs et al. 2014). However, browsing (i.e., pollarding, removal of canopy branches and barks) by large herbivores such as elephants led to a decline in *A. tortilis* density and low survival rates after browsing. Although poor coppicing ability in *A. tortilis* has been reported (MacGregor and O'Connor 2004), elsewhere it is reported to coppice well after being felled in the absence of interference (Orwa et al. 2009). Other management practices may include farmer-managed natural regeneration. Pastoralists, for example, in Southern Ethiopia, direct most preservation efforts to the regeneration of naturally grown *A. tortilis* due to its invaluable benefits especially to animal nutrition. Pastoralists also protect the trees from livestock damage and fire destruction especially those grown around the homesteads (Retta et al. 2022). The cultural and religious values attached to this Acacia species have contributed to its conservation in dryland areas (Hobbs et al. 2014).

10.3 *Acacia tortilis* in Agroforestry System

The agroforestry system involves the integration of trees and or shrubs into agriculture, either in crop cultivation livestock production, or both (Chappa et al. 2024; Kisaka et al. 2023). Agroforestry systems are grouped into three distinct categories namely, agrosilvopastoral, silvopastoral, and agrisilvicultural systems (Nair et al. 2021; Nungula et al. 2024). *A. tortilis* could be used either in silvopastoral agroforestry systems (i.e., combining trees/shrubs and livestock production) or in agrosilvopastoral agroforestry systems (i.e., combining trees/shrubs, livestock production, and crop production) as a forage tree or in agrisilvicultural system (i.e., combining trees/shrubs and crop production). The species dominates most parkland agroforestry systems in the Central Rift Valley of Ethiopia. Farmers in Oromia in Ethiopia deliberately keep *A. tortilis* on their farmlands, which could be due to the tree's desirable attribute of being drought-resistant, and the numerous benefits it offers. In Ethiopia, farmers have resorted to retaining *A. tortilis* on their farms to address low soil fertility due to the increasing cost of inorganic fertilizers which have proven costly to use to supplement soil nutrients for crop production (Tiruneh 2017). When used in silvopastoral systems, *A. tortilis* can improve the quality of forages. Forages planted under *A. tortilis* had higher quality than grasses in open fields (Ludwig et al.

2008; Tasisa and Nemomissa 2019). This is because acacia trees provide microhabitats and resources to companion plants (Hobbs et al. 2014). The tree has also been promoted as a multipurpose agroforestry tree in Botswana (Nakafeero et al. 2007).

10.3.1 Utilization of *Acacia tortilis*

Acacia trees provide services and products which are beneficial to both the environment and mankind. Acacia trees can be used in the restoration of degraded areas since they improve soil properties (Noumi and Chaieb 2012; Retta et al. 2022). *A. tortilis* fixes atmospheric nitrogen into the soil thus improving soil nitrogen content. It also improves other soil properties like moisture content, organic matter, organic carbon, electrical conductivity, and phosphorus content through litter addition (Noumi and Chaieb 2012; Tiruneh 2017; Tasisa and Nemomissa 2019). Other ecosystem services provided by the tree include shade, beautification, soil conservation, windbreak, and live fence. The tree also provides cultural benefits to rangeland dwellers (Hobbs et al. 2014) and products such as fuel wood, timber, edible pods, poles, fodder for livestock, fiber, tannin, dye, bee forage, and medicine (Maundu and Tengnäs 2005).

Traditionally, tree parts such as roots, leaves, barks, fruit, and gum are extracted and used to treat ailments such as asthma, diabetes, burns, malaria, and hepatitis. The tree could also be used in the medical field for treatment since phytochemical analysis of extracts from the tree has expressed biological properties like antifungal, anti-inflammatory, anti-bacterial, antioxidant, and anti-parasitic (Taha et al. 2022). *A. tortilis* is one of the most valued legume fodders for animals such as Arabian tahrs, goats, camels, and sheep. The pods, leaves, flowers, and twigs are used as fodder either by direct animal browsing or stall feeding to the animals. It is a preferred fodder because of the good amount of minerals and fiber the tree can offer to meet the animal's nutritional requirements (Al Dhanhani et al. 2023; Sow et al. 2024; Abdelmalik et al. 2024). Turkana and Maasai communities in Kenya use pods and immature seeds as part of their food (Orwa et al. 2009). Acacia trees, including *A. tortilis*, can be used in the enrichment of rangelands, which will be beneficial to pastoralists due to improved forage availability for their livestock.

10.3.2 Challenges Facing *Acacia tortilis*

Acacia tortilis is vulnerable to high infestation by arthropod pests that attack both seed and timber (Kahuthia-Gathu et al. 2018a, b). Beetles make up about 30% of all insects and play significant roles, such as natural enemies, herbivores, fungivores, predators, and acting as food sources for other organisms (Kahuthia-Gathu et al. 2021). In addition, they help in decomposing organic matter and hence cycling nutrients in forest ecosystems (Grove 2002). Nonetheless, some arthropod pests attack comparatively healthy Acacia trees, and untreated wood (Kahuthia-Gathu et al. 2018a; Thomas et al. 2002). Seed pests include beetles (*Bruchidius raddianae* and *Bruchidius aurivillii* (Coleoptera: Chrysomelidae)) which have been associated with low regeneration of *A. tortilis* trees (Noumi and Chaieb 2012). Pod infestation by these seed beetles was observed in Acacia trees in Tunisia proving a major threat to regeneration (Noumi and Chaieb 2012; Noumi et al. 2010).

The pests (adult and larvae beetles) bore the stems and branches of dead or stressed trees and damage the timber that may be seen in form of circular exit holes. Such feeding behaviour alters not only the tree's structural integrity but also affects its health, making it susceptible to environmental stresses. Treatment of timber with chemicals through a prophylactic method is recommended to protect it from pest attacks (Orwa et al. 2009). The use of biological control as a management option needs to be investigated since numerous families and species of predators and parasitoids have been documented from the various insect pests attacking Acacia trees. Stresses such as frequent dry spells, new economic and market opportunities, and sedentarization pose a threat to *A. tortilis*. For example, the Ma'aza community in Egypt in the late nineteenth century through their charcoal-making enterprise caused a decline in Acacia trees. Abandonment of Acacia trees because of migration of people to other areas and changes in their way of living have also led to mortality and low regeneration (Hobbs et al. 2014).

Acacia trees inhibit their regeneration through the allelopathic effects of considerable amounts of bioactive chemicals they produce (Hoque et al. 2003). Acacia seeds also possess seed dormancy attributes due to the hard seed coat which may require pre-sowing treatment such as scarification and acid treatment (Masamba 1994; Argaw et al. 1999). High variation in soil water availability could also lower the regeneration rates of acacia trees (Fig. 10.2). Therefore, old-aged trees would dominate Acacia plantations (Noumi and Chaieb 2012). Furthermore, the tree may show slow growth suggesting the need for careful management to reduce death, particularly in unfavourable environments where there is a high likelihood of mortality and poor recruitment of new trees (Andersen and Krzywinski 2007).

10.3.3 Perception of Pastoralists on *Acacia tortilis*

Pastoralists highly regard *A. tortilis* as a source of fodder for their livestock since they extract pods, leaves, and fruits during both dry and wet seasons when there is shortage and abundance of forage, respectively. Conserving *A. tortilis* is vital for pastoralists (Retta et al. 2022). The pastoral community also ascribes great cultural value to this tree as revealed by these statements: 'desert organisms depend on the trees' and 'there is no life in the desert in the absence of Acacia'. Pastoralists also consider *A. tortilis* to be drought-tolerant, being able to survive even in dry periods. Acacia trees are critical in the livelihoods of the pastoralists residing in the Red Sea Hills in Egypt and Sudan (Hobbs et al. 2014). The cultural values and traditional ecological knowledge possessed by pastoralists can be utilized to sustainably use and manage *A. tortilis* trees.

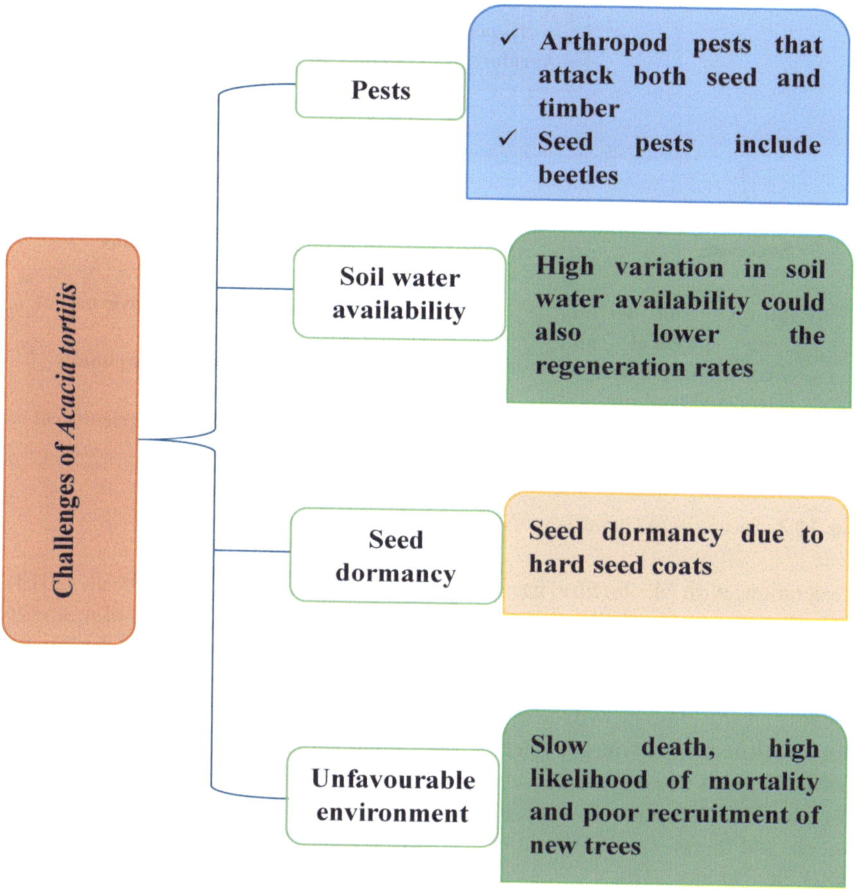

Fig. 10.2 Challenges facing *Acacia tortilis*

10.3.4 Success and Limitation of *Acacia tortilis* in Agroforestry Systems

A. tortilis has been reported to improve soil fertility including physical and chemical properties of the soil, and thus can be integrated into existing land uses for soil fertility improvement and maintenance (Tiruneh 2017). However, *A. tortilis* may be influenced negatively due to clearing of natural vegetation to pave the way for crop cultivation (Sulieman 2018). *A. tortilis* could also deteriorate because of overgrazing and the expansion of agricultural crop cultivation (Abdelmalik et al. 2024). Based on the analysis of allelopathic compounds (which have an inhibitory effect on companion crops), the presence of phenolic compounds and alkaloids found in the leaves of *A. tortilis* (Fig. 10.3) may have a negative influence on the growth of companion crops, hampering their productivity (Nakafeero et al. 2007). Poor growth in herbaceous crops such as cowpea, clusterbean, and mung bean growing in

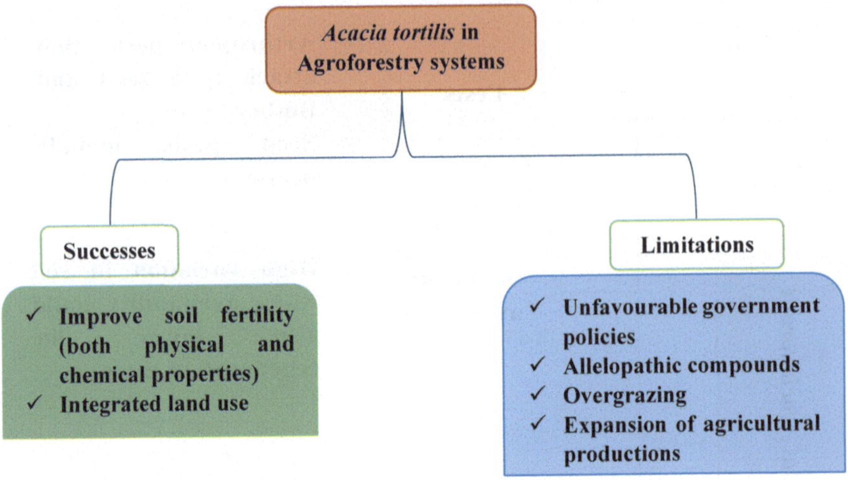

Fig. 10.3 Successes and limitations of *Acacia tortilis* in agroforestry systems

association with *A. tortilis* has been reported in India (Yadav et al. 2013). Nonetheless, improved growth in crops can be expected when the roots of *A. tortilis* are properly managed (Orwa et al. 2009). However, it may take long before the accumulation of these allelopathic compounds becomes toxic to companion crops because of the slow growth of *A. tortilis*. Furthermore, crops have been shown to respond differently to the presence of these allelopathic chemicals (Nakafeero et al. 2007). In an in situ experiment, *A. tortilis* has shown low tolerance to soil salinity compared to Egyptian balsam *Balanites aegyptiaca* and tamarind *Tamarindus indica*, and therefore, it may not be suited to reclaiming saline lands (Tomar et al. 2003; Seid et al. 2016).

10.3.5 Effect of Climate Change on *Acacia tortilis*

A. tortilis has been reported to be significantly affected by climatic factors, particularly rainfall, and temperature (Uni et al. 2023; Abdelmalik et al. 2024). Since the Acacia species is an important legume fodder for pastoralists, a reduction in production due to climate change could negatively affect animal production and subsequently the livelihoods of the pastoralist community. This may be adversely felt during drought where forage is scarce, and because it is also the most preferred forage tree by livestock in ASALs. Reduction in the amount of rainfall has been associated with the degradation of Acacia trees (Alredaisy and Zubair 2011; Gitari et al. 2024). It is also projected that the quality of Acacia seeds, and thereof Acacia succession, may be affected by the changing climate (Pramanik et al. 2017). For example, the seed bank in the soil of *A. tortilis*, and *A. mellifera* in the Butana region of Sudan showed an average low seed viability (Abdelmalik et al. 2024).

10.4 Tree Biotechnology

Biotechnology is essentially applied in forestry for the improvement of tree characteristics and management (Fialka and News 2019). Tree improvement involves a series of processes of selecting trees with desirable traits, testing, and breeding to produce superior offspring (Hadley et al. 2001). Biotechnology involves applying science and engineering to directly or indirectly use living organisms or their parts either in the natural or modified form. Several methods have been developed and used in plant biotechnology, which include gene and genome transfer, recombinant DNA, tissue culture, molecular markers, cell culture, and metabolite production, etc. Biotechnology may be applied in forestry in three ways; (i) devising new methods of vegetative reproduction (e.g., somatic embryogenesis) for the production of improved planting materials, (ii) utilization of genetic markers in the identification of trees with desirable genes, and (iii) using genetic engineering to generate information that can be utilized in the future to produce trees with new traits (Hanover 1973; Hadley et al. 2001).

Molecular methods, genomics, and genetic engineering have been used in forestry for effective forest management and reforestation. Molecular markers have been used to study genetic variation in tree species in natural and planted forests to generate information to utilize in the conservation of forest genetic resources. Molecular markers can also be useful in further differentiating tree species that are complex and difficult to characterize by using morphological characteristics. Genomics has been introduced in tree science for high throughput analysis. DNA sequencing was first applied in black cottonwood trees (*Populus trichocarpa*) (Tuskan et al. 2006; Saini et al. 2020). Genomics is purposely applied in forestry to improve the genetic makeup of trees and manage existing tree genetic resources. Genomics tools in existence have been resourceful in providing information on tree traits that are complex and proved difficult to predict. It is possible to monitor and evaluate changes in tree populations using genomic tools coupled with computational biology. Genetically modified trees such as poplars have been produced for commercial purposes. More advanced technologies in genetic engineering in crop science may have potential applications in forestry. Clonal propagation through somatic embryogenesis and organogenesis techniques has gained much progress in tree biotechnology. These techniques have allowed the transfer of genes in trees and paved the way for breeding trees of desired traits such as pest resistance, flowering control, and increased wood biomass production. Biotechnology has been applied in the establishment and management of several trees such as pine, teak, poplar, eucalyptus, Douglas fir, coast redwood, Norway spruce, oak, and willows.

However, tree biotechnology may have shortcomings that are technical, such as unintended unstable genetic issues, and stringent regulations on genetically modified trees (Chang et al. 2018; Ahuja 2021; Mohapatra et al. 2023). There are also concerns about biosafety issues even though genetically modified trees pose minimal threat to food and human safety (Saini et al. 2020). To counteract technical challenges, incorporating strategies such as clustered regularly interspaced short palindromic repeats (CRISPR), and conducting both greenhouse and field trials in monitoring the stability of the phenotypic and genetic characteristics of the clonal materials, have been proposed. While social hurdles could be addressed by reforms in regulatory systems that favour genetically modified products (Chang et al. 2018; Ahuja 2021).

10.4.1 *Acacia tortilis* and Biotechnology

The first focus of improvement in acacia breeding programmes was growth, survival, and tree form, but there is a call to incorporate improvement of wood properties and resistance to abiotic and biotic stresses. Genetic improvement in *A. auriculiformis*, *A. crassicarpa*, and a hybrid (*A. auriculiformis* and *A. mangium*) has been successfully conducted in Vietnam (Harwood et al. 2015). Wood density properties in *A. auriculiformis*, and stem form characteristics in *A. mangium* and *A. auriculiformi* has been shown to improve in bred progenies. Production and use of clones have been utilized for the establishment of Acacia hybrids such as *A. auriculiformis*, *A. crassicarpa* and *A. mangium* in Southeast Asia. The establishment of seed orchards of the improved progenies (*A. auriculiformis*, *A. crassicarpa*, and *A. mangium*) has been successful. Production of improved seed from the same orchards has been achieved and successfully used to establish hectares of the bred Acacia species (Harwood et al. 2015; Saini et al. 2020). There is no published data on the breeding of *A. tortilis* necessitating the need for tree biotechnology research to be conducted on the Acacia species considering its invaluable significance to pastoralists, coupled with the threats posed by climate change.

Biotechnological methods applied in forestry could also be used in *A. tortilis*. Molecular and genomic tools can be potentially applied to study the genetic diversity in Acacias established naturally and planted (Saini et al. 2020). The information generated provides important insights that may guide the conservation of genetic resources of acacia trees, including *A. tortilis*. Additionally, insights derived from biotechnology applications could be helpful in tree breeding programmes to produce trees of desired traits. The application of biotechnology in Acacia trees in breeding, production and utilization, and conservation is increasingly growing (Saini et al. 2020). Biotechnology offers the potential to provide a mechanism to breed superior trees of *A. tortilis*, which can cope with changing climatic conditions and continue to benefit pastoralists and sustain their livelihoods. Low seed viability in *A. tortilis*, which has constrained regeneration could be addressed by developing vegetative propagation methods via biotechnology to increase regeneration by mass seedling production. To develop breeds of *A. tortilis* with traits of interest, best provenances must be targeted to achieve superior breeds. In addition to choosing the best provenances, the factors viz., inbreeding from self-fertilization and proper collection of seed must also be considered (Harwood et al. 2015). In general, tree breeders can capitalize on molecular markers, genomics, and genetic engineering to explore genetic information on *A. tortilis* (Saini et al. 2020).

10.4.2 Prospects of *Acacia tortilis* for Biological Control

To combat the wood boring beetles, the tree relies on a complex network of natural enemies, which include predators and parasitoids. They play a crucial role in regulating beetle populations and maintaining the balance of the ecosystem. The predators feed on the pests, thus playing a critical role in reducing beetle populations before they can cause extensive damage to the tree. Parasitoids lay their eggs on or in the bodies of other insects (ectoparasitoids and endoparasitoids, respectively), eventually killing the host as the parasitoid larvae develop.

Recent research has underscored the importance of these natural enemies in maintaining the health of *A. tortilis* populations. Predators recovered from Acacia species belong to the families Elateridae, Laemopholoidae, Histeridae, Cleridae, Latridiidae, Anthocoridae, Ciidae, and Pseudoscorpionida while the parasitoids are from Braconidae, Eulophidae, Chalcididae, Ichneumonidae, and Eurytomidae and Eupelmidae (Kahuthia-Gathu et al. 2018a). Smith et al. (2023) demonstrated that parasitoid wasps from the family Braconidae significantly reduced populations of wood boring beetles on *A. tortilis* in the Serengeti region of Tanzania. The study found that areas with higher parasitoid activity showed lower levels of beetle infestation, highlighting the significant role of parasitoids in protecting these trees.

The interactions between *A. tortilis,* wood boring beetles, and their natural enemies such as predators and parasitoids form a complex ecological web essential for the survival of trees. By preying on or parasitizing beetle larvae, these natural enemies maintain the balance of *A. tortilis* ecosystems, ensuring the trees can thrive and continue to support diverse wildlife. Understanding and conserving these natural controls is crucial for the long-term health and sustainability of *A. tortilis* and similar tree species in their native habitats.

10.5 Conclusions

Acacia tortilis plays a critical role in arid and semi-arid environments and is widely distributed in Africa and the Middle East. It is highly appreciated among dwellers of arid and semi-arid regions due to its ability to tolerate extreme environmental conditions such as drought, salinity, and drifting of sand dunes and management practices such as cutting and browsing. The species provides numerous benefits in the form of products such as timber, fuel wood, fodder, food, and gum, and services such as improvement of soil health, cultural benefits, shade and medicinal properties. *A. tortilis* are integral parts of several agroforestry systems, such as silvopastoral, agrosilvopastoral, and agrisilvicultural agroforestry systems. The slow growth of species suggest the need for careful management to reduce death, particularly in unfavourable environments where there is a high likelihood of mortality and poor recruitment of new trees. Utilization of biotechnological approaches and natural control measures may be essential requirement for the long-term health and sustainability of *A. tortilis* in its native habitats.

Author Contribution The author's contribution to the chapter was as follows: study conception was contributed by GC, EZN, LN, LRC, RK, RM, JCD, SR, SS, SM, JN, and HIG. Specifically, the first section (introduction) was contributed by GC, EZN, LN, and LRC while RM, RK, JCD, and SR focused on the second section on *Acacia tortilis* in agroforestry system. Lastly, GC, SS, RK, SM, JN, and HIG summed up by compiling the third part on Tree biotechnology. EZ drew the flow chart figures. All authors reviewed and approved the final version of the chapter.

References

Abdelmalik AM, Babikir IEA, Ajloon FHI, ElHag FM, Ibrahim IA, Khatir AA (2024) Production of browse trees/shrubs under climate change conditions in the Butana Rangelands of Sudan. J Rangeland Sci 14(2):1–9

Ahuja MR (2021) Fate of forest tree biotechnology facing climate change. Silvae Genet 70(1):117–136

Al Dhanhani ASSJ, Mirza SB, Alhefeiti FRMO, Ridouane FL (2023) Investigating the nutritional potential and vegetation blueprint mapping of *Acacia tortilis* and *Acacia ehrenbergiana* from the origin of Fujairah, UAE, for Arabian Tahr as native fodder plants. Advancements Life Sci 10(1):99–103

Al-Barakah FN, Mridha MAU (2014) Status and need of research on arbuscular mycorrhizal fungi and Rhizobium for growth of Acacias. J Pure Appl Microbiol 8:129–140

Alkharabsheh HM, Mwadalu R, Mochoge B, Danga B, Raza MA, Seleiman MF, Khan N, Gitari H (2023) Revitalizing the biochemical soil properties of degraded Coastal soil using *Prosopis juliflora* biochar. Life 13:2098

Alredaisy SMAH, Zubair MAHA (2011) Deterioration of Acacia in western Butana Plain, Sudan. J Horticult Forest 3(3):78–87

Andersen GL, Krzywinski K (2007) Longevity and growth of *Acacia tortilis*; insights from 14 C content and anatomy of wood. BMC Ecol 7:1–14

Argaw M, Teketay D, Olsson M (1999) Soil seed flora, germination and regeneration pattern of woody species in an acacia woodland of the Rift Valley in Ethiopia. J Arid Environ 43(4):411–435

Chang S, Mahon EL, MacKay HA, Rottmann WH, Strauss SH, Pijut PM, Jones TJ (2018) Genetic engineering of trees: progress and new horizons. In Vitro Cell Dev Biol Plant 54:341–376

Chappa LR, Nungula EZ, Makwinja YH, Ranjan S, Sow S, Alnemari AM, Maitra S, Seleiman MF, Mwadalu R, Gitari HI (2024) Outlooks on major agroforestry systems. In: Raj A, Jhariya MK, Banerjee A, Jha RK, Singh KP (eds) Agroforestry. Wiley-Scrivener, pp 21–48

Chavan SB, Kumar N, Uthappa AR, Keerthika A, Handa AK, Sridhar KB, Singh M, Kumar D, Ram N (2017) Tree management practices in agroforestry. In: Forests, climate change and biodiversity. ICAR, New Delhi, pp 87–101

Elmendorf W (2008) The importance of trees and nature in community: a review of the relative literature. Arboricult Urban For 34(3):152

Fialka J, News EE (2019) Biotech could modify trees to protect against pests. Tree breeding and gene editing could help reverse the deteriorating health of U.S. forests. https://www.scientificamerican.com/article/biotech-could-modify-trees-to-protect-against-pests/

Gitari HI, Nungula EZ, Chappa LR, Raza MA, Ranjan S, Sow S, Alnemari AM, Seleiman MF, Mwadalu R, Maitra S (2024) Agroforestry for climate security. In: Raj A, Jhariya MK, Banerjee A, Jha RK, Singh KP (eds) Agroforestry. Wiley-Scrivener, pp 317–342

Grove SJ (2002) Saproxylic insect ecology and the sustainable management of forests. Annu Rev Ecol Syst 33:1–23

Hadley MJ, Fraser J, Tanz JS (2001) Biotechnology: potential applications in tree improvement. Forest Genetics Council of British Columbia

Hanover JW (1973) Applications of biotechnology in forest tree improvement. In: Proceedings, No. 19. Louisiana State University, Division of Continuing Education, p 59

Harwood CE, Hardiyanto EB, Yong WC (2015) Genetic improvement of tropical acacias: achievements and challenges. South For J For Sci 77(1):11–18

Hnini M, Aurag J (2024) Unlocking the diversity and plant-growth promoting potential of the bulk soil bacteria associated with the Saharan tree *Vachellia tortilis* subsp. *raddiana*. Soil Sci Soc Am J 88(4):1161–1185

Hobbs JJ, Krzywinski K, Andersen GL, Talib M, Pierce RH, Saadallah AEM (2014) Acacia trees on the cultural landscapes of the Red Sea Hills. Biodivers Conserv 23:2923–2943

Hoque ATMR, Ahmed R, Uddin M, Hossain M (2003) Allelopathic effect of different concentration of water extracts of *Acacia auriculiformis* leaf on some initial growth parameters of five common agricultural crops. Pak J Agron 2(2):92–100

Judice A, Gordon J (2020) Know your tree biology. UGA Warnell School of Forestry and Natural Resources, University of Georgia, p 11

Kahuthia-Gathu R, Thungu D, Wangu L, Kimani R (2018a) Wood-boring beetles associated with *Acacia xanthophloea* in Nairobi and Machakos Counties, Kenya. PLoS One 13:e0188773. https://doi.org/10.1371/journal.pone.0188773

Kahuthia-Gathu R, Kirubi DT, Gitonga D (2018b) Composition and abundance of wood-boring beetles of *Acacia xanthophloea* and their associated natural enemies in Thika, Kenya. J Asia-Pacific Biodivers 11:248–254

Kahuthia-Gathu R, Abonyo A, Sakalian V, Georgiev G (2021) New and interesting records of jewel beetles (Coleoptera: Buprestidae) from Kenya. Acta Zool Bulg 73(2):479–480

Kebbas S, Lutts S, Aid F (2015) Effect of drought stress on the photosynthesis of *Acacia tortilis* subsp. *raddiana* at the young seedling stage. Photosynthetica 53(2):288–298

Kisaka MO, Shisanya C, Cournac L, Manlay JR, Gitari H, Muriuki J (2023) Integrating no-tillage with agroforestry augments soil quality indicators in Kenya's dry-land agroecosystems. Soil Tillage Res 227:105586

Kuria A, Dawit S, Endale Y, Derero A, Iiyama M, Muthuri C et al (2017) Interactive suitable tree species selection and management tool for Ethiopia. [Database], World Agroforestry Centre (ICRAF). https://www.worldagroforestry.org/publication/interactive-suitable-tree-species-selection-and-management-tool-east-africa-ethiopia

Ludwig F, De Kroon H, Prins HH (2008) Impacts of savanna trees on forage quality for a large African herbivore. Oecologia 155:487–496

MacGregor SD, O'Connor TG (2004) Response of *Acacia tortilis* to utilization by elephants in a semi-arid African savanna. S Afr J Wildl Res 34(1):55–66

Mahdhi M, Tounekti T, Abada E, Al-Faifi Z, Khemira H (2020) Diversity of arbuscular mycorrhizal fungi associated with acacia trees in southwestern Saudi Arabia. J Basic Microbiol 60(4):322–330

Maitra S, Bhattacharya U, Pramanick B, Sagar L, Gaikward DJ, Pramanick B et al (2024) Agroforestry: a resource conserving technology for efficient utilization of agricultural inputs, leads to food and environmental security. In: Jatav HS, Rajput VD, Minkina T, Van Hullebusch ED, Dutta A (eds) Agroforestry to combat global challenges. Sustainable development and biodiversity, vol 36. Springer, Singapore. https://doi.org/10.1007/978-981-99-7282-1_2

Masamba C (1994) Presowing seed treatments on four African Acacia species: appropriate technology for use in forestry for rural development. For Ecol Manag 64(2–3):105–109

Maundu P, Tengnäs BO (2005) Useful trees and shrubs for Kenya. ICRAF Technical handbook series, 40.

Mirriam A, Mugwe J, Nasar J, Kisaka O, Ranjan S, Gitari H (2023) Role of phosphorus and inoculation with *Bradyrhizobium* in enhancing soybean production. Adv Agric 2023:3231623

Mohapatra PP, Seleiman MF, Mandal R, Pramanik K, Maity TK, Tarafdar J, Das SP, Raj AC, Jena C, Alenazi MM, Sow S, Ranjan S, Ahmad A, Gitari H, Khan N (2023) Efficiency of RAPD and SSR markers in assessing genetic diversity in summer onion (*Allium cepa* L.) genotypes. Notulae Botanicae Hortic Agrobotanici Cluj-Napoca 51:13369

Munro RC, Wilson J, Jefwa J, Mbuthia KW (1999) A low-cost method of mycorrhizal inoculation improves growth of *Acacia tortilis* seedlings in the nursery. For Ecol Manag 113(1):51–56

Nair PKR (1993) An introduction to agroforestry. Kluwer academic publishers, Dordrecht

Nair PKR, Kumar BM, Nair VD (2021) Classification of agroforestry systems. In: An introduction to agroforestry: four decades of scientific developments, pp 29–44

Nakafeero AL, Reed MS, Moleele NM (2007) Allelopathic potential of five agroforestry trees, Botswana. Afr J Ecol 45(4):590–593

Noumi Z, Chaieb M (2012) Dynamics of *Acacia tortilis* (Forssk.) Hayne subsp. *raddiana* (Savi) Brenan in arid zones of Tunisia. Acta Botanica Gallica 159(1):121–126

Noumi Z, Dhaou SO, Abdallah F, Touzard B, Chaieb M (2010) *Acacia tortilis* subsp. *raddiana* in the North African arid zone: the obstacles to natural regeneration. Acta Botanica Gallica 157(2):231–240

Nungula EZ, Chappa LR, Ranjan S, Sow S, Alnemari AM, Seleiman MF, Mwadalu R, Maitra S, Gitari HI (2024) Ecosystem services through agroforestry systems and its sustainability. In: Raj A, Jhariya MK, Banerjee A, Jha RK, Singh KP (eds) Agroforestry. Wiley-Scrivener, USA, pp 221–252

Orwa C, Mutua A, Kindt R, Jamnadass R, Simons A (2009) Agroforestree database: a tree reference and selection guide version 4.0. http://www.worldagroforestry.org/af/treedb/

Pramanik P, Maity A, Krishnan P, Mukherjee A (2017) Dynamics of soil seed bank under changing climate: a potential threat to natural vegetation. J Agric Phys 17(2):230–236

Retta KS, Kanu B, Hassen A (2022) Farmers perception, abundance and utilization practices of Acacia species and its pod as animal feed in Borana Zone, Mio District, Southern Ethiopia. J Rangeland Sci 12(4):396–409

Rugemalila DM, Morrison T, Anderson TM, Holdo RM (2017) Seed production, infestation, and viability in *Acacia tortilis* (synonym: *Vachellia tortilis*) and *Acacia robusta* (synonym: *Vachellia robusta*) across the Serengeti rainfall gradient. Plant Ecol 218:909–922

Saini I, Kumar T, Chauhan J, Kaushik P (2020) Acacia from a breeding and biotechnological perspective: a review. Pak J Biol Sci 23(11):1351–1356

Seid H, Birhane E, Kebede F, Hadgu KM, Negussie A, Norgrove L (2016) Response of selected indigenous dryland agroforestry tree species to salinity and implications for soil fertility management. Agrofor Syst 90:1133–1142

Seth MK (2003) Trees and their economic importance. Bot Rev 69(4):321–376

Shackleton S, Chinyimba A, Hebinck P, Shackleton C, Kaoma H (2015) Multiple benefits and values of trees in urban landscapes in two towns in northern South Africa. Landsc Urban Plan 136:76–86

Smith SE, Read DJ (2008) Mycorrhizal Symbiosis. Academic, San Diego

Smith J, Jones R, Brown T (2023) Role of parasitoid wasps in regulating wood boring beetle populations on *Acacia tortilis* in the Serengeti. J Ecol Conserv 45(2):112–125

Sow S, Ranjan S, Kumar N, Gitari NH, Dayal P, Kumar S (2024) Sustainable fodder production in South Asia through silvopastoral systems. Curr Sci 126:1217–1224

Sulieman HM (2018) Exploring the spatio-temporal processes of communal rangeland grabbing in Sudan. Pastoralism 8:1–11

Taha D, El Hajjaji S, Mourabit Y, Bouyahya A, Lee LH, El Menyiy N, Bourais I (2022) Traditional knowledge, phytochemistry, and biological properties of *Vachellia tortilis*. Plan Theory 11(23):3348

Tasisa BY, Nemomissa S (2019) Patch enclosure and localized effects of selected Acacia species on herbaceous richness and soil properties of rangelands in Somali regional state in Ethiopia. J Rangeland Sci 9(4):319–332

Teketay D (1996) Germination ecology of twelve indigenous and eight exotic multipurpose leguminous species from Ethiopia. For Ecol Manag 80(1–3):209–223

Tengnas B (1994) Agroforestry extension manual for Kenya. International Centre for Research in Agroforestry, Nairobi

Thomas FM, Blank R, Hartmann G (2002) Abiotic and biotic factors and their interaction as causes of oak decline in central Europe. For Pathol 32:277–307

Thwe-Thwe-Win HT, Watanabe A, Goto S (2015) Current genetic structure of teak (*Tectona grandis*) in Myanmar based on newly developed chloroplast single nucleotide polymorphism and nuclear simple sequence repeat markers. Trop Conserv Sci 8(1):235–256

Tiruneh AZ (2017) Effects of scattered *Acacia tortilis* (Forssk) hayne on soil properties in different land uses in Central Rift Valley of Ethiopia. J Sustain For 36(2):164–176

Tomar OS, Minhas PS, Sharma VK, Singh YP, Gupta RK (2003) Performance of 31 tree species and soil conditions in a plantation established with saline irrigation. For Ecol Manag 177(1–3):333–346

Turner-Skoff JB, Cavender N (2019) The benefits of trees for livable and sustainable communities. Plants People Planet 1(4):323–335

Tuskan GA, Difazio S, Jansson S, Bohlmann J, Grigoriev I, Hellsten U, Rokhsar D (2006) The genome of black cottonwood, *Populus trichocarpa* (Torr. & Gray). Science 313(5793):1596–1604

Uni D, Lerner D, Smit I, Mzimba D, Sheffer E, Winters G, Klein T (2023) Differential climatic conditions drive growth of *Acacia tortilis* tree in its range edges in Africa and Asia. Am J Bot 110(2):e16132

Yadav P, Kant R, Kothiyal P (2013) A review on *Acacia tortilis*. Int J Pharm Phytopharmacol Res 3(2):93–96

Transformative Role of Clonal Forestry in Agroforestry Success

11

Susmita Shil, Ankita Bhardwaj, Umakanta Dash, Manojkumar S., Akshay F. Madiwalar, and Asha K. Raj

Abstract

Tree improvement has revolutionized agroforestry in terms of producing hybrid clones with a wide range of functional traits such as cold and drought tolerance, disease resistance, and higher biomass production (timber and fodder) within a limited period. Understanding the yield potential of woody components has helped farmers select tree species of improved planting material for their agricultural fields. Species like poplar and eucalyptus have proved a boon to farmers as these species have strengthened the economic status of farmers through agroforestry. Assessment of promising clones and their performance in particular regions have made agroforestry successful. The expansion of clonal plantations has reduced pressure on natural forests. The crown architecture of clonal trees has recorded higher foliage efficiency than that of seedling-raised trees, which is an important determinant of under-story light availability for agricultural crops. Clonal plants also have higher physiological efficiencies, providing better photosystem protection from cellular oxidative stress in the hot, dry climate due to higher antioxidant activities. Hybrid clonal evaluation has been investigated in waterlogged soils, riparian buffers, and degraded soils of managed agricultural

S. Shil (✉) · A. Bhardwaj · M. S.
Department of Silviculture & Agroforestry, College of Forestry, Kerala Agricultural University, Thrissur, Kerala, India

U. Dash
Indian Agricultural Research Institute, Dirpai, Assam, India

A. F. Madiwalar
College of Forestry, Sirsi, University of Agricultural Sciences, Dharwad, Karnataka, India

A. K. Raj
AICRP on Agroforestry, Department of Silviculture and Agroforestry, College of Forestry, Kerala Agricultural University, Thrissur, Kerala, India

R. K. Kalia, R. Pathak (eds.), *Tree Biology and Biotechnology*, https://doi.org/10.1007/978-981-96-0002-1_11

landscapes. Clonal agroforestry is the promising and most sustainable way to produce food and wood in the current scenario of climatic adversities. The upcoming market for farmers is on the way to fetch income in terms of carbon credits for offsetting carbon emissions. The current challenge is to produce site-specific clones that ultimately determine the quality and quantity of production. Hence, in this chapter, we summarize the paradigms of clonal forestry in the different areas of interest in agroforestry.

Keywords

Tree improvement · Industrial Agroforestry · Clone · Short rotation forestry · Tree Breeding

11.1 Introduction

The journey of tree improvement through clonal forestry is marked by significant milestones, technological advancements, and impactful outcomes that have collectively revolutionized forestry and agroforestry practices. Except for poplars, willows and *Cryptomeria japonica*, the application of cloning techniques to forestry is a relatively new approach that gained some popularity throughout the world in the 1960s, despite its prevalence in horticulture for centuries (Leakey 1991). The early twentieth century marked the advent of systematic tree breeding, driven by advances in genetics and the principles of Mendelian inheritance. Foresters began to apply scientific methods to the selection and breeding of trees, focusing on characters e.g., growth rate, wood quality, and disease resistance. The period saw the formation of seed orchards and progeny testing programs, which laid the groundwork for more structured and effective tree improvement efforts (Libby and Rauter 1984). The mid-twentieth century witnessed a significant leap in tree improvement with the introduction of clonal forestry. Pioneering work in vegetative proliferation techniques, i.e., grafting, cuttings, and tissue culture, enabled the mass production of genetically identical clones from superior tree specimens. This innovation allowed foresters to capture and perpetuate the desirable traits of selected trees with unparalleled precision and consistency. One of the earliest successes in clonal forestry was the development of hybrid poplar clones, which combined the fast growth of *Populus deltoides* with the disease resistance of *P. nigra*. These hybrids demonstrated remarkable growth and adaptability, becoming a cornerstone of commercial forestry in North America and Europe. In commercial forestry, clonal forestry has become a keystone of tree improvement. Companies and research institutions have developed extensive clonal libraries of species such as *Eucalyptus*, *Populus*, and *Pinus*, tailored to specific site conditions and industrial requirements. These clones are rigorously tested for performance across different environments, ensuring their adaptability and robustness (Stettler et al. 1996).

Clonal forestry offers numerous benefits, such as enhanced use of genetic variability, as the variation between subjected clones encompass different additive variabilities along with the full expression of non-additive genetic variation. It may be because the clones maintain their full genetic significance. Additionally, a significant benefit is the ability to leverage genotype × environment interactions by deploying the most site-specific adapted clones (White et al. 2007). Similarly, concerns include potential future outbreaks of insect pests and diseases on specific clones, erosion of genetic variability in case the clones are closely linked or in limited numbers, and clone maturation affecting multiplication or productivity. Therefore, key strategies to mitigate risks and maximize the benefits of clonal forestry include developing and deploying a wide range of genetically varied clones, ensuring sustainable breeding support, implementing effective management, and maintaining constant research and development efforts. The clonal plantations of various conifers e.g., pines and spruce, and hardwoods like *Gmelina arborea*, casuarinas, poplars, *Acacia mangium*, and *eucalyptus* have been raised by several countries resulting in significant progresses in timber production and quality. Extensive commercial plantations using genetically improved *eucalyptus* clones have been established in Brazil, China, Portugal, South Africa, and Zaire, and they show higher productivity. The productivity of clonal plantations viz., *Eucalyptus grandis* × *E. uropyhylla* hybrids was about 55 m^3 per ha per year, while some clones have achieved the highest productivity up to 100 m^3 per ha per year (Bertolucci et al. 1995). Clonal plantations of *E. urophylla* and other eucalypts have been developed in China to meet the demands of pulp, paper, and timber. Similarly, India has had notable success with Poplar and clonal Eucalyptus plantations. Although several researchers have created protocols for the vegetative propagation of many species, the benefits could not be fully realized in the development, testing, and extensive distribution of robust clones in plantations. It must be appraised urgently to enhance the productivity and superiority of wood from the tree species suitable for cost-effective cloning (Lal 1993).

Besides greenery, agroforestry plantations supply significant timber and pulpwood in India. Almost 90% of annual wood production comes from sustainably managed plantations outside forests in Haryana and Punjab. Indian farmers recognize the advantages of trees and therefore have been maintaining naturally grown trees on their farms and fields. These trees provide a hedge against crop failure risks and offer lump-sum incomes when needed. However, due to small land holdings and general poverty, the benefit is not significant. Absentee landowners also tend to lease their lands annually rather than grow trees due to market demand uncertainties and regulatory hurdles for tree felling and timber transport. Despite this, farmers have shown interest in agroforestry plantations, like poplars and clonal eucalyptus, for significant benefits and income. The farmers remain ready to purchase good-quality plantlets. Key factors encouraging agroforestry include high productivity, better returns, sustainable market requirements, availability of improved planting stock, etc. Lower bank interest rates, benefits from treating agroforestry plantations as carbon sinks, and their integration with industries may boost the progression of clonal plantations having high yielding potential (Lal 2004).

11.2 Clone Production Methods

The production of clones is the basic step in commercial forestry or agroforestry to maximize productivity. Vegetative propagation is the only way of producing the clones. An effective, economically viable technique is essential for a successful programme. Clones can be produced by employing macropropagation and micropropagation approaches.

11.2.1 Macropropagation

It is a commonly followed technique for vegetative propagation of most plant species. It is a comparatively inexpensive and feasible way of clone production, even with minimal facilities. Still, macropropagation techniques are widely followed in small and medium-scale and community nurseries despite their seasonal limitations. Macropropagation includes cutting, grafting, budding, and layering techniques. As cutting is the widely adopted technique for plant propagation, the discussion in the current section is restricted to the cutting technique alone. Since the beginning, cuttings have been the most commonly used technique for clone production till today for many tree species like *Eucalyptus, Poplar, Casurina, Melia*, etc. (Tomar et al. 2023). Once a superior tree is identified, its parts, like stem, branch, or root can be used for multiplication based on the regenerative capacity of the species. The use of stem cuttings with apical dominance is well in practice for many fast-growing commercial timber tree species (Tomar et al. 2023). Branch cuttings are mostly preferred for fruit-yielding plant production (Ak et al. 2021). Lignified roots are also being used in some species; for instance, Harivel et al. (2006) used 15–20 cm long roots of *Faidherbia albida* while Meunier et al. (2008) used 10–20 mm diameter roots for the successful regeneration of *Spathodea campanulata*. Similarly, more than 16 cm long roots can be used for *Malus domestica* (Robinson and Schwabe 1977). The propagule's rooting success significantly relies on the nature of the mother plant species, age, size and position of shoot or root, maturity of the stem, and growth regulator treatment (Pal 1992).

Appropriate sources for obtaining cutting, growth media, season of cutting, and auxin treatment vary widely depending on the species. In the case of *Eucalyptus* sp., *Melia dubia, Acacia,* and *Dalbergia sissoo* younger stem, less lignified softwood cuttings are used for propagation (Seenivasan et al. 2017). As mentioned by Ezekiel (2010), around 90% of tropical trees are suitable for softwood cutting, and it is a highly adopted technique in community village nurseries. In some species like *Ficus carica, Poplar, Gliricidia sepium,* etc., lignified hardwood cuttings are commonly used. These hardwood cuttings require relatively more labour and auxin treatment than softwood cuttings (Land and Cunningham 1994). Cuttings are always recommended to obtain from the young plants of 2–3 years old at the optimal physiological condition of the mother plant with good nutritional status (Harfouche et al. 2007). Generally, spring is the best season for softwood cutting while it is winter for hardwood cutting and late spring or mid-summer (late growing season) for semi-hardwood cuttings (Land and Cunningham 1994). Auxins (IAA, NAA) are mostly used for rooting in a maximum number of plants. The ontogenic age of the cutting decides the concentration of auxin to be applied. The application rate and cost of Auxin can be reduced by mixing it with phenolic compounds during application. Researchers have recently explored ways to enhance root growth using plant growth-promoting rhizobacteria (PGPRs). However, this approach has yet to be extensively tested and widely adopted (Koyama et al. 2019).

11.2.1.1 Mini-cutting Technique: An AppealingMass-Production Technology

Conventional cutting techniques yield lower rooting success over time due to the ontogenic aging of mother plants. To counteract this issue, commercial plant production factories adopted 1990 Brazilian mini-cutting technology for mass production (Hihashi et al. 2015). It yields a large number of propagules with good rooting efficiency and a root system in a cost-efficient way. Mini cuttings are the young shoot cuttings obtained from the axillary sprouts of mother plants, propagated by stem cuttings of superior trees under the most favourable condition (Seenivasan et al. 2017). The mother plants are very intensively planted in a controlled mist chamber. They are irrigated and fertilised frequently to provide optimum nutrients for maintaining better plant nutrient status. Alternatively, mother hedge plants can also be grown in suitable hydroponic media (Bindumadhava et al. 2011). After 60 days, mother plants were ready for shoot harvest. The mini-cuttings harvested were treated with fungicide and planted in the root trainer containing coir pith or other suitable media in a mist chamber with optimum irrigation and temperature regime. This technology had elevated production capacity up to five times the conventional hedge methods (Palanisamy et al. 2020). It produced genetically highly uniform propagules with a meagre topophysis effect (Bindumadhava et al. 2011). It was a very feasible technology for some species like *Eucalyptus, Melia, Casurina* than the micro-propagation technique. Temperate species like *Sequoia sempervirens, Araucaria angustifolia, Prunus persica* were also reported to propagate using the mini-cutting technique (de Jesus et al. 2020; Pereira et al. 2018). The good rooting quality of mini-cuttings is attributed to a lower degree of juvenility and higher nutritional status of the plant maintained by regular application. It is also known for producing tap roots in propagules, which may not be possible with conventional cutting methods (Seenivasan et al. 2017). Moreover, cuttings obtained from clonal mini hedges require minimal rooting hormone or no hormone for propagation. More than 90% survival was reported in D. sissoo, M. dubia, E. camadulensis, T. populnea. In contrast, more than 75% survivability was recorded in *Casuarina junghuhniana,* Acacia hybrid and *Gmelina arborea* with this technology (Seenivasan et al. 2017).

11.2.2 Micropropagation Approaches

The totipotency of plants is highly exploited in the plant production industry through micropropagation. It has the advantage of producing a large number of disease-free plants in a smaller area. Thus, it is a highly practiced technique in the commercial plant production process. Fundamentally it involves four stages (Rout et al. 2006) i.e., (i) initiation: sterilized explants are placed in the suitable medium growing media under suitable light, humidity and temperature; (ii) shoot multiplication: explant multiplies and produces a shoot as a result of organogenesis under the influence of auxins and cytokinin; (iii) root induction: shoots produce roots in-vivo with increasing auxin concentration or roots can be produced ex-vivo; (iv) acclimatization: plantlets produced in-vitro are gradually hardened off for field transplantation.

The root induction stage is expensive; it accounts for 35–75% of the cost of the production (Wozny and Miler 2016). Some of the trees like cherry, plum, pear, and apple are challenging to produce good rooting percentages using in-vitro rooting technique (Fig. 11.1). Furthermore, in-vitro rooted plants acclimatized in the controlled conditions suffer physiologically and reduced survival percentages in the field conditions (Modgil et al. 2009). In many tree and medicinal crops, the ex-vitro rooting technique has proved to have maximum rooting and survival rates. Many investigations are being carried out for numerous crops to standardize this technology around the world as it fastens the propagation process by discarding the in-vitro root development stage. Ex-vitro rooting refers to the rooting of the shoot outside the test tube (Fig. 11.1) and combines the root induction and acclimatization processes (Modgil et al. 2009). Shoots produced in the in-vivo conditions are treated with an appropriate quantity of auxins and planted in a suitable medium under greenhouse conditions, inducing root production within a few weeks. It reduced production costs, enhancing the plants' rooting and survival percentage compared to in-vitro approaches (Arencibia et al. 2017; Shekhawat and Manokari 2015).

Fig. 11.1 In-vitro and ex-vitro rooting conditions

11.2.3 Bioreactors for the Future Automated Plant Production

In recent decades, bioreactors have gained more popularity in the mass production of plant materials. Although the use of bioreactors for plant material production was recognized in the 1980s, this technique has gained momentum in recent years due to its enhanced potential for automation. Initially, bioreactors were developed for microbial culture, followed by plant cell culture for secondary metabolite production, and later, they were employed for plant propagation (Vidal and Sanchez 2019). Conventional tissue culture technology is costly and labour intensive, requiring more manpower in culture test tube monitoring and frequent maintenance. Bioreactors are aseptic, closed systems that supply nutrients to the growing cell in liquid or gas. It enables intensive culturing with efficient monitoring and control of the system's environment, such as nutrient medium circulation, temperature, aeration, agitation, pH, etc. (Leathers et al. 1995). Bioreactor culture involves placing explants inside the bioreactors and supplying nutrients in the form of liquid or gas in contract with the explant temporarily or permanently. Liquid or gas form of nutrient media enables easy nutrient change automatically (Murthy et al. 2023). Bioreactor systems are of two kinds, continuous immersion system (CIS) and temporary immersion system (TIS). Among them, TIS is more suitable for plant shoot propagation (Fig. 11.2), wherein explants and nutrient liquid media remain in contact, preventing continuous immersion and oxygen deficiency in the culture media (Etienne and Berthouly 2002). It also enables air circulation in the system and prevents hyperhydricity of culture plants. This technique has resulted in better production of shoots and roots in many commercially important tree species like teak, eucalyptus, olive, apple, pistachio, poplar, etc., with 100% whole plant production in many species and acclimation percentage of more than 70% (Vidal and Sanchez 2019). TIS also paved the way for healthy propagule production of black lapacho

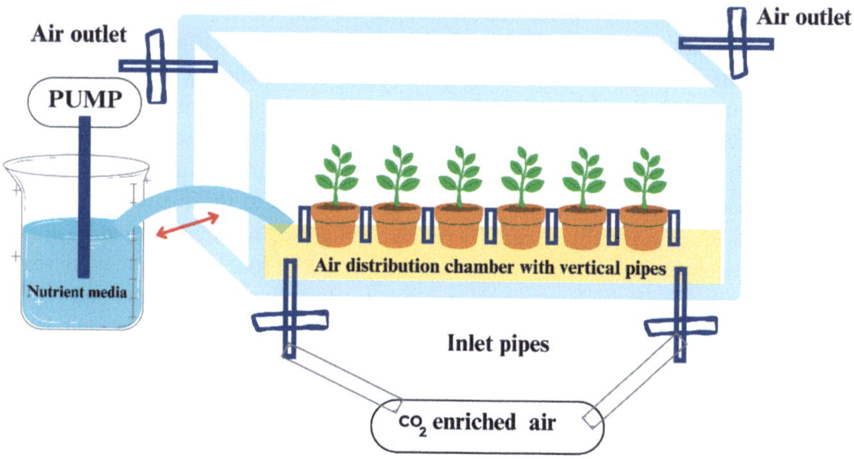

Fig. 11.2 Temporary immersion type bioreactor with additional forced ventilation

and myrobalan, which otherwise infected largely in the semi-solid agar medium in conventional tissue culture technique (Nasri et al. 2019; Duarte et al. 2016). However, bioreactors are used cost-effectively for mass multiplication of propagules for commercialization with automation. Still, the initial cost of its establishment is high and the technology is not standardized for all the species.

11.3 Physiological Functioning of Tree Species or Clones Under Stress Conditions

Both abiotic and biotic stresses have a vital role in plant productivity. Factors such as drought, waterlogging, extreme temperatures, and changes in soil conditions are grouped under abiotic stress, while various pests and diseases cause biotic stress (Suzuki et al. 2014). In a shifting global climate, variations in temperature and precipitation are expected to rise (IPCC 2008), resulting in reduced yield and productivity losses. Under stress conditions, each plant/crop/clone responds with changes in its physiological function or trait. The changes in physiological traits will help plants to survive and tolerate the stress conditions (Suzuki et al. 2014). The variation in clones for different traits can be exploited to develop biotic and abiotic stress-resistant clones from different tree species for higher productivity. Numerous studies have explored how clones respond to various biotic and abiotic stresses.

11.3.1 Impact of Combined Stressors on Clone Development, Yield, and Physiological Features

11.3.1.1 Adverse Interactions of Stresses

Identification and reaction of a particular stress on the plants are crucial, especially when these specific stresses have a negative impact on plant growth and reproduction. Drought and high temperatures are among the major factors restricting growth and yield of plants globally (Prasad et al. 2011). The combined effect of the stresses induces numerous functional changes that hamper the quality and production. Therefore, to satisfy the global food demands of a growing population, it is essential to find cultivars that have improved tolerance against these stresses. The combined impact of drought and heat stress are usually additive, indicating their nature of independence in the systems that plants use to respond to these stresses. Certain growth traits are specifically altered by one type of stress; however, they may not be applicable to another trait. It is evident that the amount of root biomass may increase due to drought; similarly, heat stress may positively influence leaf area and reproductive tissues. Although the combination of these stresses (drought and heat) affects the reproductive tissues contrarily. Similarly, the process of photosynthesis is particularly sensitive to these stresses. The stomatal closure limits the photosynthesis process by reducing the accessibility of carbon dioxide during drought conditions (Chaves et al. 2003). The negative effect of these stresses depends on the plant species or clones; for example, elevated temperatures increased photosynthetic rate, while drought reduced photosynthesis, stomatal conductance, PSII efficiency, and leaf water potential in European oaks (Arend et al. 2013). The negative impact of drought worsened under combined stress conditions, indicating potential benefits from elevated temperatures, but it had negative impacts when combined with drought.

Antioxidant, proline and bioactive compounds play important roles in overcoming stress conditions. Hence, variations in these compounds in different clones respond differently. Heavy metals have been found to cause greater harm to plant growth under the influence of other abiotic stresses, and variations in growth responses can be observed in different clones. Madiwalar et al. (2023) reported varied growth performance of Eucalyptus clones under heavy metal-affected soil conditions. This variation in different Eucalyptus clones can be exploited for phytoremediation of heavy metal-affected soil. A research study examined chilling stress on two rubber tree clones, GT-1 and Wenchang-217, using physiological methods in a controlled light chamber exhibited significant changes in physiological parameters between the clones. Wenchang-217 had higher malondialdehyde (MDA) levels and lower superoxide dismutase, peroxidase, and catalase as compared to GT-1. Stronger cold tolerance of GT-1 suggests that an activated oxidative quenching enzyme system is crucial for cold tolerance in rubber trees (Ping et al. 2012).

11.3.1.2 Favourable Interactions of Stresses

Certain amalgamations of stresses can be more favourable to the plants than individual stresses. For instance, drought stress reduces stomatal conductance, which can enhance plant tolerance to ozone (O_3) stress (Low et al. 2006). Various compounds like the presence of glycine, betaine and trehalose in the plant play a vital role in protecting plants under stress conditions and could be crucial to the plant response specifically to the particular combination of stresses. The increase in these substances under the stress combination was associated with maintaining a lower sodium to potassium ions ratio. This resulted in improved cell water status and photosynthetic performance compared to when plants experience salt stress alone (Allakhverdiev et al. 2003). Similarly, other studies also reported that the combined effect of heat and salt stress inhibited the accumulation of H_2O_2 and the oxidation of proteins (Rivero et al. 2013). It has also been demonstrated that boron inhibits the harmful impact of salt stress while the elevated CO_2 concentrations may be advantageous (del Carmen Martinez-Ballesta et al. 2008). The higher concentration of carbon dioxide reduces stomatal conductance, preventing the diffusion of ozone in the leaves, and increases water-use efficiency by lowering the transpiration rate (Brouder and Volenec 2008). The differential response of different clones to both biotic and biotic stresses is due to the changes in the physiological function or trait of each clone, which can be exploited to screen the superior clones from different tree species for higher yield and productivity.

11.4 Field Performance of Tree Clones

Monoclonal plantations have replaced organized clonal forestry as a prevalent form of land use since the early 1900s. Notwithstanding its lengthy history, large-scale industrial plantations only began using vegetative and clonal propagation for poplar and tropical and subtropical eucalyptus in the 1970s and 1980s (Libby and Ahuja 1993). Clonal plantations are beneficial for the superior genetic traits to quickly and immediately increase their genetic diversity (Saigal and Kashyap 2002). Vegetatively reproduced clonal plants are homogeneous and realistic, possessing entire genetic characteristics of the parental plant. They are created from a single mother tree with the most desired genetic traits (Abbas et al. 2024; Bhardwaj et al. 2024).

Chaukiyal et al. (2015) conducted an experiment to assess the morphological traits of pathologically tested nine clones of *D. sissoo* and observed a significant inter-clonal variation viz., a moderate increase in plant height, root length, collar diameter, number of branches and leaves per plant in all the clones. Similarly, Nautiyal et al. (2003) assessed thirty-three clones of *D. sissoo* for field performance and reported that plant height, collar diameter, flowering and seed production had highly significant variation among clones. Kumar and Banerjee (1986) also observed significant differences among growth traits in *D. sissoo* species. Mathur (2003) observed increased growth performance with increasing leaf number per plant. Plantations of clonal Eucalyptus have been the most economical land use alternative for reforestation and agroforestry. Its productivity has been reported to be about 20–25 m^3 per ha per year while in some cases a growth rate of 50 m^3 per ha per year has been observed under rainfed conditions also (Lal 2003, 2011). Several experiments have testified to the significant growth performance of *Eucalyptus* clones in the world, which have yielded mixed results in terms of survival, growth performance and productivity. An on-farm experiment conducted by Kumar et al. (2022) to investigate the performance of barley and cowpea under Eucalyptus clones revealed that both crops' phenotypic and functional features performed better when grown as a single crop in the agroforestry system with eucalyptus than when grown in combination. On the other hand, C-413 outperformed the other eucalyptus clones in terms of tree growth metrics, whereas C-83 clone's interspaces showed superior growth parameters and dry and green fodder output. Likewise, Tomar et al. (2020) carried out a study to evaluate nineteen clones of poplar and reported significant inter-clonal variation in terms of total height, collar diameter, number of branches and internodal length after 1 year. Dash et al. (2024) did an experiment to evaluate the performance of turmeric under G-48 clone of *P. deltoides* and found that yield was higher in sole turmeric cropping, but it was compensated by enhanced quality parameters (higher curcumin) under agroforestry system. Poplar trees (G-48) also attained satisfactory growth viz., mean height (21. 11 m), diameter (29.41 cm), crown spread (4.35 m), volume (0.59 m^3 $tree^{-1}$), green timber yield (4.82 q $tree^{-1}$).

Rakocevic et al. (2024) examined the effects of two environments, viz., monoculture and agroforestry, on photosynthesis, architecture, biomass production, and carbon and nitrogen savings in yerba-mate clones. Under the agroforestry system, higher photosynthesis was observed in female plant leaves than in the leaves of male plants. The carbon and nitrogen ratio in leaves was greater in monoculture than agroforestry, suggesting that monoculture represents a more environmentally degraded ecosystem. Crown attributes of 4-year-old trees of *Pinus radiata* developed from seedlings and clonal propagation in an agroforestry experiment with and without an understorey of lucerne showed no difference in tree heights (Bandara et al. 2004). However, the study reported larger DBH and total biomass in the clonal trees grown under no-understorey treatment than in seedling trees. Biomass distribution within the crown varied significantly between treatments. Clonal trees exhibited greater branch basal area and more medium-sized branches compared to seedling trees. Additionally, clonal trees had longer internodes and a higher crown shape ratio than their counterparts. The variations in the crown may lead to differences in the distribution of foliage area in the crown, potentially affecting the fraction of solar radiation intercepted. The performance of clones of various tree species under agroforestry and plantation systems is summarised in Table 11.1.

11.5 Rooting Patterns of Clonal Trees

Survivability and rooting rates of plantlets are crucial factors of consideration in clonal production. The root growth influences the above-ground growth and development of the plants. In Indonesia, Rosyidah and Nurjanto (2020) found that teak clones (WG-1, D-15, F-71 and WG-2) inoculated with mycorrhizal strains (Mixture of Acaulospora and Glomus) after 1 month of planting showed better root growth. Therefore, utilization of appropriate mycorrhizal inoculum and timing of inoculation are considered determinant factors for root formation and growth of teak clonal seedlings in agroforestry systems. The extensive root system of clones is highly valued for phytoremediation purposes. Based on rooting ability, clonal evaluation of some species has been carried out to standardize wind firm tree clones for windbreak agroforestry systems. The superior clones of casuarina (CJ-8, CJ-6, CJ-18, CJ-17 and CJ-9) have shown greater wind speed reduction under windbreaks ranging from 58.45 to 67.60% (Buvaneswaran et al. 2018). Poplar is the highly preferred tree species for agroforestry in both temperate and tropical regions. Besides above-ground biomass allocation and carbon sequestration studies, some disproportionate studies have evaluated below-ground biomass on a clonal basis. A study by Jha (2017) evaluated *P. euramericana* clone I-214 under the forestry system (FRS) and agroforestry system (AGF). The AGF recorded a maximum below-ground biomass (130 kg per tree) than FRS (120 kg per tree). Another study conducted in Quebec Canada, evaluated four clones of poplar in monoclonal and polyclonal plantations and found that root fraction in monoclonal was most distant compared to polyclonal plots (Elferjani et al. 2014). The root and shoot ratios were also higher in monoclonal compared to polyclonal plots. The researchers attributed that mixing clones increased above-ground biomass due to lesser competition among the clones, resulting in over-yielding under polyclonal plantations. Significant variation was observed for above and below-ground biomass partitioning and fine root morphology in monoclonal and polyclonal mixtures. The six juvenile spruce clones (*Picea*

Table 11.1 Performance of clones of various tree species under agroforestry and plantation system

Species	Clones	Site of development	Performance	Reference
Eucalyptus hybrid clone of *E. grandis* and *E. urophylla*	–	Brazil	The areas with the highest volume were those with eucalyptus hybrid clones cultivated in sunflowers and beans; after 16 months, this volume was 41.8% greater than that of eucalyptus hybrid clones grown conventionally	Westman (2010)
Poplar	Max-1, Koreana and Hybrid-275	Germany	Comparing Max-1 to the clones Hybrid-275 and Koreana, Max-1 demonstrated the highest re-sprouting, lowest mortality, and highest overall MAI	Langhof and Schmiedgen (2023)
Theobroma cacao	Inifap-, Inifap-, Inifap-9, and Neocriollo	Mexico	The average tree height in Agroforestry conditions was higher than in open conditions	Cocoletzi Vásquez et al. (2022)
Eucalyptus tereticornis	ITC Bhadrachalam	Central India	The maximum above-ground dry biomass was discovered for individual trees in the agri-silviculture system, while the highest biomass per hectare was observed in the compact block plantation	Nirmal and Handa (2021)
Poplar	S7C8, G-48, W-39 and Kranti	Uttarakhand, India	In comparison to clones W-39 and G-48, poplar clones S7C8 and Kranti recorded greater values of diameter at breast height and tree height. Clone S7C8 recorded the highest values of the various biomass components of trees compared to clones G-48 and W-39; clone Kranti yielded substantially more biomass in the stem, roots, and overall biomass. The agroforestry system trapped the most carbon stock with the clone Kranti	Banga et al. (2017)
Theobroma grandiflorum	32, 42, 44, 46, 47, 48, 51, 56, 57, 61, 62, 63, 64, 174, 215, 1074	Brazil	The clones 42, 44, 46, 47, 57, 61, and 64 exhibited satisfactory results concerning both the average fruit count and the morpho-agronomic attributes of the fruits	Alves and Chaves (2023)

Populus deltoides	G-48, AM-48, AM-49, AM-41, BR-510, AM-50, FS-18, FS-190, FS-155, L-90, L-87, L-89, S7C1, S7C15, S7C20, S7C4, L-200-84, Bahar, Udai and Kranti	Uttar Pradesh, India	While the largest diameter and basal area were found in L-87, the maximum height and volume were found in Udai. In Udai, poplar biomass, carbon sequestration, and carbon stock reached their peak.	Tomar et al. (2024)
Eucalyptus camaldulensis	P-13, 2136, P-50, P-23, P66, 526, 2070, 2023, P-32, P-14, 413, K-25, 2021, P45, 2013, IFGTB–4	Uttar Pradesh, India	The highest DBH was recorded in clone P-13 and maximum height was found in 3018	Srivastav et al. (2020)
Eucalyptus tereticornis	288, 07			
Eucalyptus hybrid	3018			
A clonal hybrid of *eucalyptus urograndis*		Brazil	The silvopastoral system showed greater financial viability than the plantation	Campos de Jesus et al. (2022)
Melia dubia	Sharad var., Bahumukhi var., Kartik var., Amar var., Varsha var., Shashi var. and Kshitiz var.	Madhya Pradesh, India	The lone plantation growth performance of *M. dubia* demonstrated the greatest height gain in Amar var. Similarly, Amar var. showed the greatest height gain under the agroforestry system.	Berry et al. (2022)

sitchensis) recorded higher fine root surface area when grown in polyculture or mixed plots (Donnelly et al. 2016).

11.6 Promotion of Short Rotation Clones Through Contract Tree Farming

The advantages owed by the short rotation clones are enormous and could easily be accepted by the farmers. Short rotation tree materials are the main raw material for lignocellulose-based pulp, paper, bio-energy, and charcoal industries. Promoting short-rotation tree clones for commercial farming requires a diversified strategy to make farmers and other stakeholders understand its advantages and implementation. Contract tree farming acts as a kind of transfer technology for small landholders without exposing farmers to market risks (Baumann 2000). Even if farmers in the contract might not get more economic benefits, it definitely assures the farmer's technological knowledge (Boulay 2010). Contract tree farming can be defined as a system in which a processing or marketing firm directly purchases the trees or its products from the farmer as per the contract terms and conditions arranged in advance. Contract farming in the tree has gained attention as a developmental strategy in the 2000s itself (Baumann 2000). Several frameworks for contract farming exist in the system; among them, best-suited models for tree clone promotion are discussed here.

11.6.1 Centralised Model

It involves a large, centralized marketing firm or processing unit that supports production and purchases from a large number of sellers, including farmers, processors and packagers (Eaton and Shepherd 2001). It is the most followed contract farming model, especially with the abundance of small-land holders. It is well suited for the tree clones which are popularised in the region. The buyer's involvement in the production aspect varies largely, from providing just planting material to providing services in all aspects of crop management. This model does not provide any insurance against calamities or loan services. It may hinder the involvement of low-income farmers with small-land holding as they need to wait until crop maturity to earn. The quality of the produce is strictly checked in this model; buyers can ensure the quality of produce by providing technical support and guaranteeing the purchase of produce.

Boulay (2010) has documented the centralised contract farming system in Thailand between pulp and paper industries and farmers, wherein pulp and paper industries provide *E. camaldulensis* planting stock at the rate of 1 baht (the Thai currency) per unit for normal seedlings, 2.5 baht for rooted cuttings and 7 baht for tissue cultured plants to the contracted farmers. At the time of harvest, most industries were guaranteed a minimum purchasing price based on the diameter of the thin end of the log. It was 500 for <1-inch-thick log, 700 for >1.5-inch log, and 1200 for

>2-inch log. Though, the company didn't have any command over the type of planting material to be used by the farmer and seed origin seedlings were available in the companies at cheaper rates, more than 70% of contract farmers chose eucalyptus clones for planting as it was available at subsidized rate from industries and profitable.

11.6.2 Nucleus Estate Model

It slightly differs from the centralized model. However, it works in a similar manner to the centralized model; additionally, the processing or marketing firm (buyer) also owns the estate and grows the produce (Eaton and Shepherd 2001). It is one of the best-suited models to promote new tree clones as the farm owned by the firm can act as a demonstration plot to showcase technology (performance of superior new clones) to the farmers and encourage them to absorb the technology. Besides, in the large estate area, the firm also ensures the supply of some minimum quantity of raw material. Like in the centralised model, the main drawback is the unavailability of insurance and credit to small-scale farmers who adopt tree farming.

One such kind of nucleus estate model was adopted in the Tree Biotechnology Programme (TBP) in Uganda in 2002, which aimed to reduce deforestation and promote sustainable forest management in the region through the introduction and popularisation of fast-growing tree clones. TBP identified exotic Eucalyptus clones based on the local condition's performance. Further, plantations of these improved clones were established strategically in some busy areas so that local people could notice the difference between clonal and non-clonal eucalyptus growth. Later, with the help of the Gatsby club, the program encouraged local people to take up private enterprise in clonal nurseries by disseminating clones, technologies and some business support. Consequently, several certified eucalyptus clonal nurseries have been established in the country. The seedling production in the nursery was further boosted by the Sawlog Production Grant Scheme (SPGS) for promoting commercial forestry and by degazetting government land for leasing in the country (Calabrese 2021).

11.6.3 Multipartite Model

Farmers are typically involved in tandem with statutory organizations and private firms in a multipartite arrangement. Separate organizations looking into the aspects of production, management, processing, marketing, and finance provision are involved in multipartite contract farming (Eaton and Shepherd 2001). In this model, financial institutes offer loans and credits to the farmers; technical guidance is provided by the service industries or research institutes and at the end industries procure the product as per the pre-determined condition. These models are highly suitable for promoting tree clone cultivation even in low-income small landholders' farms.

In southern India, Tamil Nadu tri-and quad-partite models of contract tree farming have revolutionised quality pulp wood production by promoting selected clones of *Casurina*. The models were expanded to 11,284 ha at 4389 farmer fields in the state (Parthiban et al. 2011). The tri-partite model involves three partners: farmers, the paper industry, and financial institutes. Industry supplies selected *casuarina* clones to the farmers at a subsidized rate with an agreement of buying produce back at a minimum rate of Rs. 2000 per tonne. In the case of quad-partite model, the research institute is an additional entity that is involved in identifying and providing clones to the industries for mass multiplication and technical advice to the farmers. Industry harvests produce from the farmer's field according to the contract. Due to the cultivation of superior *casuarina* clone, farmers earned an average total revenue of Rs. 3 lakhs at the end of 3 years. The superior clone has almost doubled the farmer's profit and also benefited the industry by improving pulp recovery by up to 49% with a good kappa number (a specific character acquired by pulp testing).

Another successful tri-partite model developed by the WIMCO-NABARD Scheme promoted and popularised poplar clones in the poplar growing areas of India in 1984. Under this scheme WIMCO Seedling Limited promoted and supplied poplar clones, viz., G-3 and G-48, to farmers and provided technical guidance under buy-back agreement. Following the WIMCO's report sent to financial institutes on farmers' agreements, the local banks rendered credits for the costs associated with plantation maintenance to farmers, allowing them to participate in the credit and also a crop insurance coverage program. On average, farmers earned a net profit of 1.15 lakh per acre. After popularising the clones and cultivating profit, farmers started growing poplars without any credit facilities and also started growing intercrops in poplar fields (Deshpande 2005).

11.7 Conclusion and Future Perspective

The transformative role of clonal forestry in agroforestry success is undeniable. Through centuries of tree improvement, from historical practices to modern technological advancements, the development of new clones and varieties has significantly enhanced the productivity, resilience, and sustainability of forestry and agroforestry systems. The implementation of advanced breeding techniques, the exploration of both native and non-native species, and the growing preference among farmers for clonal planting material have all contributed to this progress. Vegetative and reproductive clone production methods have enabled the mass propagation of superior genotypes, ensuring consistency and quality in planting material. These clones have demonstrated remarkable physiological functioning under various abiotic and biotic stresses, making them adaptable to diverse environmental conditions. Comparative studies of clone performances in height, diameter, and volume have shown their superiority over traditional varieties in agroforestry and conventional forestry settings. The rooting systems of clonal and seedling trees have revealed significant differences, with clonal trees often exhibiting more robust and extensive root systems. This characteristic supports soil health and enhances the overall stability of agroforestry systems. The promotion of short-rotation clones through models like contract tree farming has further incentivized the adoption of clonal forestry, providing economic benefits to farmers and contributing to sustainable land management.

Looking ahead, the future of clonal forestry in agroforestry holds immense promise. Continued research and innovation will be crucial in addressing emerging challenges and maximizing the benefits of clonal forestry. As climate change continues to impact global ecosystems, breeding and selecting clones that can thrive under changing climatic conditions will be imperative. Expanding farmer education and engagement initiatives will be critical for the widespread adoption of clonal forestry. By embracing future opportunities and addressing emerging challenges, clonal forestry can continue to drive sustainable agricultural practices, enhance environmental health, and improve the livelihoods of farmers around the world.

References

Abbas G, Chauhan V, Dash U, Keprate A, Madiwalar AF, NikhilBalu R, Banoo R (2024) Effect of organic manures on growth and yield of *Raphanus sativus* L. under Melia based agroforestry system: a case study from North Western Himalaya. Biol Forum Int J 16(1):01–06

Ak BE, Hatipoglu IH, Dikmetas B (2021) Propagation of fruit trees. In: Pakyurek M (ed) Recent headways in pomology. Iksad, Turkey, pp 55–92

Allakhverdiev SI, Hayashi H, Nishiyama Y, Ivanov AG, Aliev JA, Klimov VV, Murata N, Carpentier R (2003) Glycinebetaine protects the D1/D2/Cytb559 complex of photosystem II against photo-induced and heat-induced inactivation. J Plant Physiol 160:41–49

Alves RM, Chaves SF (2023) Selection of *Theobroma grandiflorum* clones adapted to agroforestry systems using an additive index. Acta Sci Agron 45:e57519

Arencibia AD, Gomez A, Poblete M, Vergara C (2017) High-performance micropropagation of dendroenergetic poplar hybrids in photomixotrophic Temporary Immersion Bioreactors (TIBs). Ind Crops Prod 96:102–109

Arend M, Brem A, Kuster TM, Gunthardt-Goerg MS (2013) Seasonal photosynthetic responses of European oaks to drought and elevated daytime temperature. Plant Biol 15:169–176

Bandara GD, Mead DJ, Whitehead D (2004) Understorey vegetation and the crown architecture of Pinus radiata seedling and clonal trees in an agroforestry system. N Z J For Sci 34(2):139–157

Banga A, Yadava A, Sah VK (2017) Growth, biomass and carbon sequestration potential of different poplar (*Populus deltoides* Bartr.) clones in agroforestry system with wheat (*Triticum aestivum*) varieties in Tarai belt of Uttarakhand. J Environ Sci Technol 5(3):32–36

Baumann P (2000) Equity and efficiency in contract farming schemes: the experience of agricultural tree crops (Vol. 111). Overseas Development Institute. London, p 48

Berry N, Saravanan S, Shukla A (2022) Growth Performance of *Melia dubia* Cav. varieties under agroforestry system in Jabalpur district, Madhya Pradesh. Adv Biores 13(2):21–25

Bertolucci FL, Demuner BJ, Garcia SL, Ikemori YK (1995) Increasing fiber yield and quality at Aracruz. In: Eucalpyt plantations: improving fibre yield and quality. Proceedings of the CRC-IUFRO conference, pp 19–24

Bhardwaj A, Shil S, Ashwath MN, Dash U, Kumar V, Vats P (2024) Interaction between below-ground and above-ground resources in tree-crop systems. In: Jatav HS, Rajput VD, Minkina T, Hullebusch EDV, Dutta A (eds) Agroforestry to combat global challenges: current prospects and future challenges, Sustainable development and diversity. Springer Nature, pp 249–274

Bindumadhava H, Tamak J, Mahavishnan K, Upadhyay AP, Varghese M, Sharma N (2011) Clonal propagation in *Eucalyptus camaldulensis* using minicutting technique. Curr Sci 25:1578–1585

Boulay A (2010) Contract tree farming and smallholders: drivers of adoption in Thailand. PhD thesis, Australian National University

Brouder SM, Volenec JJ (2008) Impact of climate change on crop nutrient and water use efficiencies. Physiol Plant 133:705–724

Buvaneswaran C, Kumar KV, Velumani R, Senthilkumar S (2018) Experimental design for evaluation of clones of *casuarina* for windbreak agroforestry system. J Tree Sci 37(1):11–18

Calabrese L (2021) Promoting commercial forestry in Uganda. The experience of the tree biotechnology programme. GATSBY Africa

Campos de Jesus LD, Luana Pasa D, Antônio de Farias J, Chagas de Souza S (2022) Performance of the clone *Eucalyptus urograndis* in a conventional and silvopastoral system. Floresta 52(2):222–230

Chaukiyal SP, Bhatia P, Kumar M, Dobhal P, Kumar S (2015) Performance of *Dalbergia sissoo* clones in relation to morphological traits under nursery conditions. J Exp Zool India 18(1):43–48

Chaves MM, Maroco J, Pereira J (2003) Understanding plant responses to drought from genes to the whole plant. Funct Plant Biol 30:239–264

Cocoletzi Vasquez E, Hipolito-Romero E, Ricano-Rodriguez J, Ramos-Prado JM (2022) Ecophysiological plasticity of *Theobroma cacao* L. clones in response to the structure and microclimate of agroforestry systems in Mexico. Bot Sci 100(4):960–976

Dash U, Gupta B, Bhardwaj DR, Sharma P, Kumar D et al (2024) Tree spacings and nutrient sources effect on turmeric yield, quality, bio-economics and soil fertility in a poplar-based agroforestry system in Indian Himalayas. Agrofor Syst 98(2):911–931

de Jesus JS, Dorleans VR, Ribeiro DP, Sodre GA, Barbosa RM (2020) Mini-cuttings of forest and fruit species. Cientifica 48(1):67–75

del Carmen Martinez-Ballesta M, Bastias E, Carvajal M (2008) Combined effect of boron and salinity on water transport: the role of aquaporins. Plant Signal Behav 3:844–845

Deshpande CS (2005) Contracting farming as means of value-added agriculture. Department of Economic Analysis and Research (Occasional paper-42), National Bank for Agriculture and Rural Development (NABARD), Mumbai, pp 19-42

Donnelly L, Jagodziński AM, Grant OM, O'Reilly C (2016) Above-and below-ground biomass partitioning and fine root morphology in juvenile Sitka spruce clones in monoclonal and polyclonal mixtures. For Ecol Manag 373:17–25

Duarte E, Sansberro P, Luna C (2016) Micropropagation of *Handroanthus heptaphyllus* (Vell.) Mattos from seedling explants. Afr J Biotechnol 15:1292–1298

Eaton C, Shepherd A (2001) Contract farming: partnerships for growth, FAO agricultural services bulletin 145. Food & Agriculture Organisation, p 161

Elferjani R, DesRochers A, Tremblay F (2014) Effects of mixing clones on hybrid poplar productivity, photosynthesis and root development in northeastern Canadian plantations. For Ecol Manag 327:157–166

Etienne H, Berthouly M (2002) Temporary immersion systems in plant micropropagation. Plant Cell Tissue Organ Cult 69:215–231

Ezekiel A (2010) Viable options and factors in consideration for low-cost vegetative propagation of tropical trees. Int J Plant Biol 6(2):187–193

Harfouche A, Baoune N, Merazga H (2007) Main and interaction effects of factors on softwood cutting of white poplar (*Populus alba* L.). Silvae Genet 56:287–294

Harivel A, Bellefontaine R, Boly O (2006) Aptitude a la multiplication vegetative de huit especes forestieres d'interet au Burkina Faso. Bois Forets Tropiques 288(2):39–50

Hihashi EN, de Silveira RLVA, Gonçalves AN (2015) Monitoramento nutricional e fertilizaçao em macro, mini e microjardim clonal de Eucalyptus. In: Nutricao e fertilizacao florestal. IPEF, Piracicaba, p 427

IPCC (2008) In: Kundzewicz ZW, Palutikof J, Wu S (eds) Climate change and water, Technical paper of the intergovernmental panel on climate change. Cambridge University Press, Cambridge/New York

Jha KK (2017) Root structure and below-ground biomass of hybrid poplar in forestry and agroforestry systems in Mediterranean France. Not Sci Biol 9(3):422–432

Koyama R, Aparecido Ribeiro Júnior W, Mariani Zeffa D, Tadeu Faria R, Mitsuharu Saito H, Simoes Azeredo Goncalves L, Ruffo Roberto S (2019) Association of indole butyric acid with *Azospirillum brasilense* in the rooting of herbaceous blueberry cuttings. Horticulture 5(4):68

Kumar PS, Banerjee AC (1986) Provenance trials of *Acacia nilotica*. J Tree Sci 5(1):53–56

Kumar A, Kumar M, Poonia PK, Dhillon RS, Daneva V, Dagar H (2022) Performance of field crops and growth of Eucalyptus clones under Eucalyptus based agri-silvicultural system in semiarid regions of western Haryana. Biol Forum Int J 14(3):1030–1035

Lal P (1993) Economics of clonal forestry plantations. In: Proceedings of Workshop on 'Production of genetically improved planting material for afforestation programmes', Field Document No.7, FAO, Project RAS/91/004. FAO, Los Banos, pp 108–115

Lal P (2003) Clonal eucalyptus plantations in India. In: Turnbull JW (ed) Eucalypts in Asia. Proceedings of an international conference held at Zhanjiang, China from April 7–11, 2003, ACIAR Proceedings No.111, Canberra

Lal P (2004) Integrated development of farm forestry plantations and wood based industries. Indian For 130(1):71–78

Lal P (2011) Clonal agroforestry plantations in India. Indian J Ecol 38:6–10

Land SB Jr, Cunningham M (1994) Rooted cutting macropropagation of hardwoods. In: Proceedings of the southern regional information exchange group biennial symposium on Forest genetics: applications of vegetative propagation in forestry. US Department of Agriculture, Forest Service, Southern Forest Experiment Station, Huntsville

Langhof M, Schmiedgen A (2023) 13 years of biomass production from three poplar clones in a temperate short-rotation alley cropping agroforestry system. Biomass Bioenergy 175:106853

Leakey RR (1991) Towards a strategy for clonal forestry: some guidelines based on experience with tropical trees. In: Tree breeding & improvement, pp 27–42

Leathers RR, Smith MAL, Aitken-Christie J (1995) Automation of the bioreactor process for mass propagation and secondary metabolism. In: Aitken-Christie J et al (eds) Automation and environmental control in plant tissue culture. Springer, pp 187–214

Libby W, Ahuja MR (1993) Clonal forestry II -genetics and biotechnology. Springer, Pp, p 277

Libby WJ, Rauter RM (1984) Advantages of clonal forestry. For Chronicle 60(3):145–149

Low M, Herbinger K, Nunn AJ, Häberle KH, Leuchner M, Heerdt C, Werner H, Wipfler P, Pretzsch H, Tausz M et al (2006) Extraordinary drought of 2003 overrules ozone impact on adult beech trees (*Fagus sylvatica*). Trees 20:539–548

Madiwalar AF, Dhillon GP, Singh A, Singh P, Singh B (2023) Eucalyptus clones respond differentially for heavy-metals phytoextraction and carbon sequestration in tree biomass and soil with distillery effluents irrigation in North-Western India. Proc Indian Natl Sci Acad 89(1):77–100

Mathur R (2003) Physiological and biochemical responses of some promising clones of *Dalbergia sissoo* Roxb. to varying water stress conditions. PhD thesis, F. R. I. (Deemed University), Dehradun, India, 250p

Meunier Q, Bellefontaine R, Monteuuis O (2008) La multiplication vegetative d'arbres et arbustes medicinaux au benefice des communautes rurales d'Ouganda. Bois et Forets des Tropiques 295(2):71–82

Modgil M, Sharma T, Thakur M (2009) Commercially feasible protocol for rooting and acclimatization of micropropagated apple rootstocks. Acta Hortic 839:209–214

Murthy HN, Joseph KS et al (2023) Bioreactor systems for micropropagation of plants: present scenario and future prospects. Front Plant Sci 14:1159588

Nasri A, Baklouti E et al (2019) Large-scale propagation of Myrobolan (*Prunus cerasifera*) in RITA® bioreactors and ISSR-based assessment of genetic conformity. Sci Hortic 245:144–153

Nautiyal S, Pal M, Sagta HC, Rawat RS, Negi DS (2003) Clonal variation in growth, flowering and seed production of *Dalbergia sissoo* in a clonal seed orchard. J Trop For Sci 15(1):97–108

Nirmal A, Handa AK (2021) Biomass and volume models for clonal *Eucalyptus tereticornis* coppice under agroforestry systems in central India. Trees 1(1):1

Pal M (1992) Clonal propagation for yield improvement in forest plantations. J Indian Pulp Pap Tech Assoc 4(1):61–64

Palanisamy C, Ramanathan S et al (2020) Large scale multiplication of *Casuarina junghuhniana* Miq. clonal plants through mini-cutting technique. J Agric Sci Technol 10(2):98–105
Parthiban KT, Seenivasan R et al (2011) Designing and augmenting pulpwood supply chain through contract tree farming. Indian J Ecol 38:41–47
Pereira MDO, Angelo AC et al (2018) Roots of mini-cutting of clones of *Sequoia sempervirens* in different seasons of the year. Cerne 24:452–460
Ping L, Jun-Jun H, Yan-Li Y, Xiao-Hong D, Ru-Xiong C, Li-Feng W (2012) Differential responses of two rubber tree clones to chilling stress. Afr J Biotechnol 11(70):13466–13471
Prasad PVV, Pisipati SR, Momcilovic I, Ristic Z (2011) Independent and combined effects of high temperature and drought stress during grain filling on plant yield and chloroplast EF-Tu expression in spring wheat. J Agron Crop Sci 197:430–441
Rakocevic M, Batista ER, Matsunaga FT, Wendling I, Marcheafave GG et al (2024) Canopy architecture and diurnal CO_2 uptake in male and female clones of yerba-mate cultivated in monoculture and agroforestry. Ann Appl Biol 184(2):210–225
Rivero RM, Mestre TC, Mittler R, Rubio F, Garcia-Sanchez F, Martinez V (2013) The combined effect of salinity and heat reveals a specific physiological, biochemical and molecular response in tomato plants. Plant Cell Environ 37(5):1059–1073
Robinson JC, Schwabe WW (1977) Studies on regeneration of Apple cultivars from root cuttings propagation aspects. J Hortic Sci 52(2):205–220
Rosyidah U, Nurjanto HH (2020) The use of mycorrhizae in production of clonal seedlings of superior teak, a prospective tree species for increasing agroforestry system productivity. IOP Conf Ser Earth Environ Sci 449(1):012043
Rout GR, Mohapatra A, Jain SM (2006) Tissue culture of ornamental pot plant: a critical review on present scenario and future prospects. Biotechnol Adv 24:531–560
Saigal S, Kashyap D (2002) Review of company-farmer partnerships for the supply of raw material to wood-based industry. Ecotech Services (India)
Seenivasan R, Chezhian P et al (2017) Mini-cutting technique for mass multiplication of pulp wood species. In: Parthiban KT, Seenivasan R (eds) Plantation and agroforestry pulpwood value chain approach. Scientific Publisher, New Delhi, p 517
Shekhawat MS, Manokari M (2015) Efficient *in-vitro* propagation by *ex vitro* rooting methods of *Artemisia absinthium* L., an ethnobotanically important plant. Chin. J Biol 2015:73405
Srivastav A, Tomar A, Agarwal YK (2020) Performance of Eucalyptus clones in Trans-Ganga region of Uttar Pradesh, India. Indian J Agrofor 22(1):43–47
Stettler RF, Bradshaw HD, Heilman PE, Hinckley TM (1996) Biology of Populus and its implications for management and conservation. NRC Research Press, Ottawa
Suzuki N, Rivero RM, Shulaev V, Blumwald E, Mittler R (2014) Abiotic and biotic stress combinations. New Phytol 203(1):32–43
Tomar A, Yadav A, Srivastav A, Agarwal YK, Mishra C, Vishwakarma P (2020) Growth performance of different clones of Poplar (*Populus deltoides* Marsh.) in Nursery. Bull Environ Pharmacol Life Sci 9(4):09–12
Tomar A, Singh BK et al (2023) Clonal forestry: the future of tree growing. In: Singh S, Tomar A (eds) Clonal forestry: principles and practices. Narendra Publishing House, Delhi
Tomar A, Singh BK, Khan FA, Srivastava A, Singh S (2024) Growth, biomass and carbon sequestration of *Populus deltoides* clones. Environ Ecol 42(2):541–546
Vidal N, Sanchez C (2019) Use of bioreactor systems in the propagation of forest trees. Eng Life Sci 19(12):896–915
Westman E (2010) Growth response of eucalyptus hybrid clone when planted in agroforestry systems: an approach to mitigate social land conflicts and sustain rural livelihood. Second cycle, A1E. Umea: SLU, Department of Forest Ecology and Management.
White TL, Adams WT, Neale DB (2007) Forest genetics. CAB International, Oxfordshire, pp 466–477
Wozny A, Miler N (2016) Leds application in *ex vitro* rooting and acclimatization of chrysanthemum (Chrysanthemum × Grandiflorum/Ramat./Kitam.). Electron J Pol Agric Univ 19(4):1–8

Agroforestry and Trees Outside Forests: Its Implications

12

Suresh Ramanan S., A. Arunachalam, Ankit Verdiya, and A. K. Handa

Abstract

The cultivation of trees and agricultural crops in combination is an ancient practice termed agroforestry. There are several examples of these traditional land-use practices. The retention or growing of trees with other components for multiple welfare purposes from the same piece of land is the main principle of agroforestry. The worldwide concerns in tree planting and agroforestry operations during the past decades prompted the beginning and spread of different terms like community forestry, social forestry, urban forestry, tree outside forest (TOF), etc. According to the Food and Agriculture Organisation, TOF may include agricultural land, agroforestry systems coupling meadows and pasture, built-on land, barren land, in lines along rivers, canals, and roads. In other words, agroforestry enhances the tree cover outside the forests and works as a surrogate for natural forests, sequestrating carbon and helping pressure off natural forests. In India, other than forest and wooded lands, most of the trees are either naturally or intentionally grown on roadsides, along canals, agricultural bunds and community land. In some situations, trees are also grown for boundary demarcation and most importantly all these areas of tree growing fall under the scientific classification of agroforestry. In this review, various types of land cover such as forest, other wooded land, and other land with tree cover, along with a comparison between agroforestry and TOF and the concept of TOF, including community forestry, social forestry, and urban forestry has been discussed thoroughly.

S. Ramanan S. (✉) · A. Arunachalam · A. Verdiya · A. K. Handa
ICAR-Central Agroforestry Research Institute, Jhansi, Uttar Pradesh, India
e-mail: suresh.s@icar.gov.in

R. K. Kalia, R. Pathak (eds.), *Tree Biology and Biotechnology*,
https://doi.org/10.1007/978-981-96-0002-1_12

Keywords

Agroforestry · Ecosystem services · Forest · Human well-being · Sustainable development · Trees outside forests · TOF

12.1 Introduction

Agroforestry, the combination of trees and shrubs in cultivated landscapes, is being referred to as trees outside forests (TOF) (Rizvi et al. 2020; Dabral et al. 2023). Agroforestry practice combines crop cultivation, rearing of livestock, and growing of trees on a similar piece of land, creating a sustainable system that offers numerous environmental, economic, and social benefits. By promoting biodiversity, enhancing soil potency, and reducing the requirement of chemicals, agroforestry contributes significantly to environmental sustainability (Ramanan and Arunachalam 2021). Fundamentally, the concepts of agroforestry and TOF have been used interchangeably in most literature.

On introspecting, forests are the beginning point representing the most fertile land-based ecosystems in the world and are indispensable for life on earth. Forests cover about 4.06 billion ha of area globally, which accounts for 31% of the land area of the world. Interestingly, more than 50% of global forests, i.e., 815, 497, 347, 310 and 220 million ha are limited to five countries, viz., the Russian Federation, Brazil, Canada, the USA and China, respectively (FAO 2020). Globally 16 billion people depend on forests directly or indirectly for goods, socio-economic, cultural and ecological services (FSI 2021). Agricultural expansion due to increasing food demand for the growing human population led to deforestation, fragmentation of forests and consequently to the loss of forest biodiversity and ecosystem services (Pental 2021) and resulted in a faster increase in the global croplands (Zomer et al. 2009, 2016). Coupled with this, increasing urbanization also impacts forest ecosystems. More than half of the global human populace lives in towns and cities (Ramachandra et al. 2017). It has been predicted that about 420 million ha of forest area worldwide has vanished due to deforestation since 1990, which is 3.2% of the global land. There is an urgent need to accelerate the recovery of degraded forests and their associated biodiversity and ecosystem services have insisted on the simplest solution, i.e., tree planting (Arunachalam et al. 2021; Patel et al. 2023). Trees offer enormous potential to restore and rehabilitate degraded lands and ecosystems and improve the ecosystem's provision, including goods and services. Trees significantly play an important role in improving the environment and uplifting humankind's economic conditions, specifically in the rural economy (Ren et al. 2005).

Trees are among the most important components of a forest but they are not found only in forests. The discrete trees and small groups of trees in non-forested areas are essential resources worldwide, collectively known as trees outside forests (TOF) (Dogra and Chauhan 2016). Conservation, expansion, and sustainable management of TOF can be a nature-based solution to current environmental concerns and issues. TOF are found both in urban and rural landscapes. These tree resources can be found in agricultural land, woodlots in the rural landscape, riparian zones, and in and around human settlements (FAO 2016). TOF offers several ecosystem services across four categories, i.e., provisioning (fruits, fuelwood, timber and medicines), regulating (climate parameter, air cleansing and soil erosion regulation), cultural (aesthetic value, recreation, spiritual and religious value and cultural heritage), and other secondary services (habitat provision, water cycling and oxygen production). The importance of trees has been well understood by several farming communities for centuries. Farming communities have been conserving and increasing the TOF ambit through traditional tree-based agricultural farming. In several developing countries, trees and forests have been an integral component of subsistence farming systems. Farm forests, hedgerows, and isolated trees are pivotal for the sustainability of agricultural landscapes. The present chapter reviews various land cover related to forests such as forests, other wooded land, and other land with tree cover, along with the comparison between agroforestry and TOF and the concept of TOF, including community forestry, social forestry, and urban forestry.

12.2 Comparison Between Agroforestry and Trees Outside Forests

The term 'trees outside forests' has been even more relevant in the present day. It is defined as 'all the trees growing outside recorded forest area irrespective of the patch size' (FSI 2021). The national agency for forest area monitoring also estimates TOF in the urban and rural areas separately. The trees in urban areas are sometimes collectively referred to as 'urban forestry'. Under the changing climatic and cultural scenarios of urban forestry, urban green spaces and lung spaces are attracting attention among dwellers due to social, cultural and ecological relevance. Studies have systematically documented the dearth of urban forests and green space in the cities. Therefore, there is no means to increase the TOF area in urban areas. Therefore, to achieve the 33% target, TOF in rural areas is very vital and it is the last resort for the nation to achieve the superfluous carbon sink of 2.5–3 billion tonnes of carbon dioxide equivalent through additional forest and tree cover by 2030, committed by the country under Paris Agreement of UNFCCC (Clark et al. 2021). Against this backdrop, there are a few concepts in TOF that need to be discussed.

12.2.1 Farm Forestry

The practice of cultivating and managing trees in the compact block on agricultural lands. Typical man-made plantations are examples of this practice.

12.2.2 Social Forestry

A forestry that aims at producing a flow of protection and recreational benefits for the community is termed social forestry. The term was coined by forest scientist Westoby at the ninth Common Wealth Forestry Congress held in 1968 in New Delhi (Rakatama and Pandit 2020). A precise definition of social forestry is difficult; however, it is usually implicit in the growing of trees coupled with bamboo, grasses, and legumes for rural development. There are also opinions that the social forestry concept encompasses farm forestry, community forestry, extension forestry and agroforestry. The difference between all these three terminologies is listed in Table 12.1.

12.2.3 Agroforestry

Fundamentally, all three programmes aimed at planting trees outside the forest area and the choice of species was according to the situation. The results were convincing that planting trees was beneficial both in economic and ecological terms (Cernea 2019). Unfortunately, both social forestry and farm forestry are government-driven programmes. It is here that agroforestry differs, yet it is also the concept of planting

Table 12.1 Difference between farm forestry, social forestry and agroforestry

Aspect	Farm forestry	Social forestry	Agroforestry
Definition	Individual owners growing trees on their lands	Government-led program for community forestry	Intentional integration of trees into agricultural systems
Primary goal	Economic benefits for individuals	Social, environmental, and economic benefits for communities	To enhance agricultural productivity and sustainability
Ownership	Private ownership	Often community-owned or managed	Private or communal ownership
Scale	Typically small-scale	Can be small to large-scale	Can vary from small to large-scale
Purpose	Timber, fuelwood, fodder, and other products	Environmental conservation, fuelwood, timber, and social benefits	Enhancing soil fertility, water management, biodiversity conservation, and climate resilience
Management	Managed by individuals	Managed by communities or government agencies	Managed by farmers or landowners
Emphasis	Individual economic benefits	Community and environmental benefits	Integration of trees with agricultural practices for multiple benefits for enhancing productivity
Examples	Planting trees for timber or fruit production on own lands	Afforestation and reforestation projects in communal areas	Alley cropping, silvopastoral, windbreaks, and other integrated systems

trees outside forest areas. Agroforestry, in layman's words, will be combining agriculture and forestry. However, agroforestry needs a detailed explanation. On the corollary, agroforestry is a relatively new terminology for the traditional way of cultivating trees in the agricultural landscape in various forms.

The significant foundational work of agroforestry was carried out by the IDRC, Canada, under the leadership of Dr. John Bene. The International Development Research Centre (IDRC) came out with some major recommendations to improve the productivity of agricultural lands, and one of them was to focus on a combined production system that includes forestry, crops and animal husbandry. This paved the way for the establishment of agroforestry as a science. After establishing the World Agroforestry Centre or International Centre for Research in Agroforestry (ICRAF), its works on agroforestry showcased the role of trees in woodland (Nair 1993). Today, agroforestry research has entered the fifth phase, focusing more on dynamics and socio-ecological systems at the landscape, livelihood, and policy levels. Yet, there are misconceptions about the definitions of agroforestry among different stakeholders. The recent publication from the World Agroforestry Centre (WAC) titled 'Sustainable development through trees on farms: agroforestry in its fifth decade' systematically brings in analogy to elucidate the development of agroforestry and enlighten the readers on the definition of agroforestry (van Noordwijk 2019). It is stated that the initial years of agroforestry research were mainly devoted to proving the existence of agroforestry. Historically, farmers have incorporated trees in farmlands globally, and a recent study at a global scale also proved that more than 40% of agricultural land has 10% of tree cover. Therefore, the scientific works on agroforestry have always attempted to provide a framework for the existing practice in the farmer's field. In the 1970s, agroforestry was explained as follows: 'Agroforestry is a collective name for land-use systems and technologies where woody perennials (trees, shrubs, palms, bamboo, etc.) are deliberately used on the same land management unit as crops and/or animals, in some form of spatial arrangement or temporal sequence'. In agroforestry systems, there are both ecological and economic interactions between the different components. Afterwards, the explanation was modified to highlight the contribution of agroforestry: 'Agroforestry is a dynamic, ecologically based, natural resource management system that, through the integration of trees on farms and in agricultural landscapes, diversifies and sustains production for increased social, economic and environmental benefits for land users at all levels'. Besides this, van Noordwijk (2019) suggested that agroforestry is a combination of agriculture and forestry; it is a land use that combines aspects of both, including the agricultural use of trees.

In this way, agroforestry is a forestry-centric area of study; however, it may involve different disciplines. Despite all the similarities and dissimilarities, Table 12.2 examines the intricacies between agroforestry and TOF.

Understanding the popularity and acceptance of terms like 'agroforestry' and 'trees outside forests' (TOF) provides valuable insights into their recognition and use across different regions. Google, being the most widely used search engine globally, offers a rich dataset to analyse the usage of terms and familiarity. By examining search trends, one can infer how these concepts are received and

Table 12.2 Theoretical comparison between agroforestry and trees outside forests

Aspect	Agroforestry	Trees outside forests (TOF)
Definition	Deliberate incorporation of trees or shrubs with crops or livestock on the same land	Trees grown outside of traditional forest areas, including urban areas, along roadsides, or in agricultural fields
Primary purpose	To optimize the benefits of land use by combining agriculture and forestry practices	To provide environmental, social, and economic benefits outside of designated forest areas
Typical locations	Rural agricultural landscapes, farms, and pastoral lands	Urban areas, rural non-forest landscapes, roadsides, parks, and other non-forest areas
Integration with agriculture	Trees are intentionally incorporated with crops or livestock to create a symbiotic relationship	Trees may or may not be intentionally integrated with agricultural practices
Biodiversity impact	Enhances biodiversity by creating diverse ecosystems combining trees, crops, and animals	Contributes to biodiversity in non-forest areas, but the impact varies based on the context and management
Environmental benefits	Improves soil health, water management, and carbon sequestration, while reducing erosion	Provides shade, improves air quality, enhances urban aesthetics, and contributes to microclimate regulation
Economic benefits	Provides multiple income sources (e.g., timber, fruits, crops) and enhances farm resilience	May offer timber, fruits, and other products, but economic benefits are often secondary to environmental and social roles
Scale and intensity	Typically managed on a larger scale, with intentional design and management practices	Can be on a smaller, more scattered scale, with less intensive management compared to agroforestry
Management practices	Requires specific management practices to balance the needs of crops, trees, and livestock	Management varies widely depending on the location and purpose of the trees, often less intensive
Policy and governance	Often supported by specific agricultural and forestry policies	May fall under various policy frameworks, including urban planning, environmental protection, and agricultural policies

prioritized in various countries. For terms such as ‘agroforestry’ and ‘trees outside forests’, Google Trends can show how frequently these terms are searched and how their popularity fluctuates (Proulx et al. 2014; Jun et al. 2018). If one term shows a consistently higher search volume compared to the other, it indicates greater general interest or recognition. For instance, ‘agroforestry’ consistently outranks ‘trees outside forests’, suggesting that the term is more commonly used or understood in public discourse (Fig. 12.1). The Google Trends also allows users to break down search data by region. This capability is crucial for understanding how the acceptance of these terms varies by country. However, in this analysis, we have taken up the entire world and present a panoramic view (https://trends.google.com/trends/explore/TIMESERIES/1711973400?hl=en-S&tz=-330&date=all&q=%22Melia+d

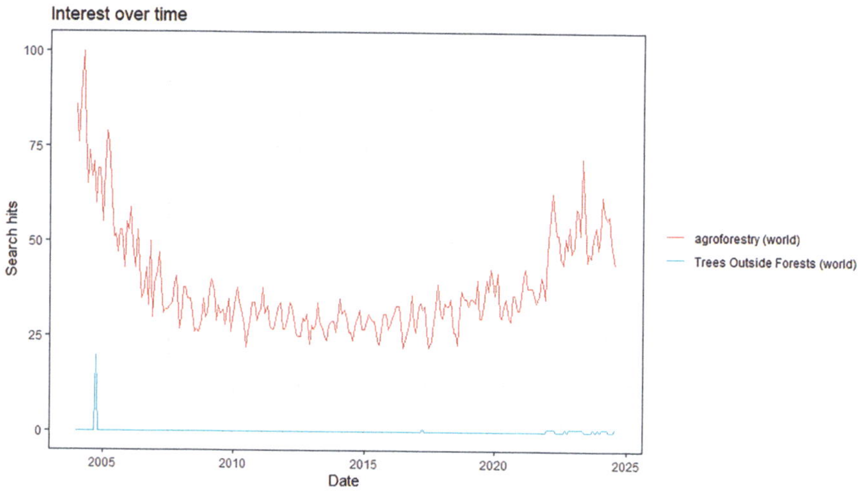

Fig. 12.1 Empirical comparison of 'agroforestry' and 'trees outside forests' using Google trends

ubia%22,%22Melia+azedarach%22,%22Melia+volkensii%22,%22Melia+composita%22&sni=3).

While Google Trends offers valuable insights into the popularity of search terms, it has limitations when comparing terminologies. The data is based on relative search volumes, not absolute numbers, which can skew comparisons if search volumes are low overall. Additionally, Google Trends data is influenced by various factors such as regional differences in internet usage, language variations, and varying levels of public awareness, which can affect the accuracy of comparisons. The tool does not account for the context in which terms are used or the nuances of their meanings, potentially leading to misinterpretations. Search trends might reflect current events or specific campaigns rather than long-term trends. Consequently, while Google Trends can indicate general interest levels, it should be supplemented with other data sources to comprehensively analyse terminology usage and acceptance. More specifically, it does not have a dataset before 2005. In this context, we also attempted to understand the academia mindset on the popularity and acceptance of scientific terms like 'agroforestry' and 'trees outside forests' using Google Books database. Google Books (https://books.google.com/) provides a valuable resource for analysing the usage of words and phrases over an extensive period, making it a suitable tool for examining the evolution of terminology related to sustainability from 1900 to 2019. By leveraging the vast corpus of digitized texts available on Google Books, researchers can track the frequency and context of terms such as 'sustainability', 'agroforestry', and 'environmental science' over time. These are typically referred to as n-grams. Studies have demonstrated the utility of Google Books for such analyses (Kousha and Thelwall 2015; Pechenick et al. 2015; Ramanan et al. 2020). These studies show how the database can reveal shifts in terminology and conceptual focus within the literature. For example, an increase in the frequency of sustainability-related terms may reflect growing public and academic interest in environmental issues and sustainable practices. Conversely, a decrease or shift in terminology might indicate changes in research focus or public awareness.

The advantage of using Google Books lies in its extensive historical coverage and the breadth of material it encompasses, including books, journals, and other published works. This allows researchers to capture a broad spectrum of usage and identify trends across different periods. However, it is essential to consider that Google Books data represents only a portion of the published literature and may not fully capture all nuances of term usage due to variations in publication practices and digitization efforts over time. Overall, Google Books serves as a robust tool for tracing the development and prevalence of sustainability-related terminology, offering insights into how concepts have evolved and been perceived across different years.

A detailed view of the occurrence of various terms related to sustainability within the books available in Google Corpora has been given in Fig. 12.2. The data is articulated as a percentage of occurrence relative to the number of books published each year. This approach allows for a nuanced understanding of how frequently terms such as 'agroforestry' and 'trees outside forests' have appeared in published literature over time.

From the analysis, it is evident that the term 'agroforestry' is used more frequently than 'trees outside forests'. This higher frequency indicates a stronger acceptance and recognition of the term within the literature. The term 'agroforestry' has a broader scope and application, encompassing various practices that integrate trees and shrubs into agricultural landscapes. This broad applicability and the associated benefits likely contribute to its more frequent use. The term prominence suggests that it has been more widely adopted in both academic and practical contexts, reflecting its significant role in discussions around sustainable agricultural practices.

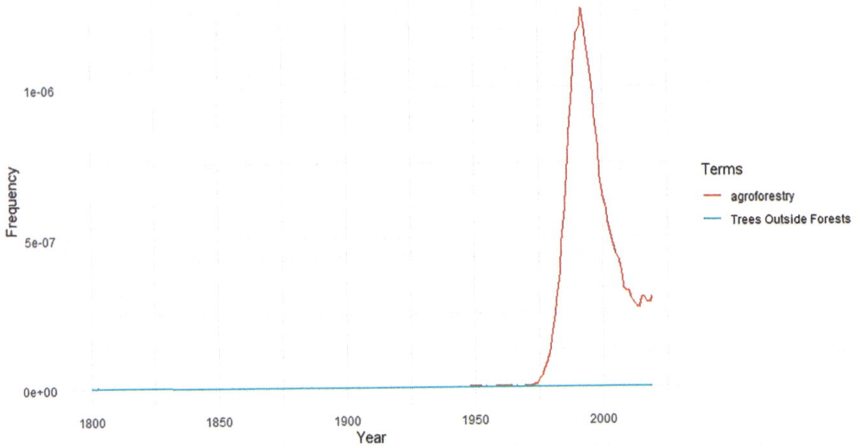

Fig. 12.2 The screenshots depict the frequency of words (y-axis) agroforestry and trees outside forests (case insensitive) in the Google Books from 1950 to 2022 (x-axis)

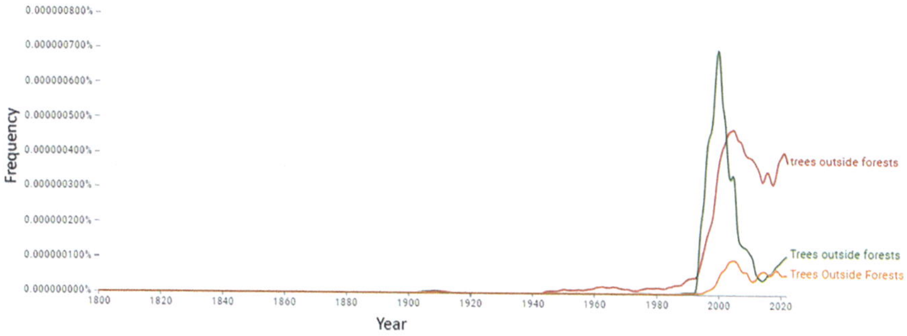

Fig. 12.3 The screenshots depict the frequency of words (y-axis) trees outside forests (including variation) in the Google Books from 1950 to 2022 (x-axis)

Figure 12.3 provides additional insight by highlighting the variations in the usage of the term 'trees outside forests'. The term appears in several forms, including 'trees outside forests', 'Trees outside Forests', and 'Trees Outside Forests', among others. This variation indicates a degree of inconsistency in how the term is referenced in academic literature. Despite these variations, the term has seen notable usage, particularly around 2000. This period marks a time when there was a growing emphasis on integrating trees into non-forest environments, such as urban and peri-urban parts, to address environmental and ecological concerns.

Analysis of the term 'trees outside forests' reveals that, while there is a significant presence of this term in the literature, it has not reached the same frequency level as 'agroforestry'. This discrepancy might be attributed to the more specialized nature of 'trees outside forests' compared to the broader concept of 'agroforestry'. The focus on trees outside traditional forest boundaries may appeal to specific sectors like urban planning and conservation, which could explain the observed usage patterns (Regmi and Garforth 2010; Thomsen et al. 2016).

In contrast, 'agroforestry' experienced its peak in usage during the 1990s. This period saw a heightened interest in sustainable agricultural practices and integrated land use, which likely contributed to the increased prominence of the term. The peak in the 1990s reflects a time when the concept of agroforestry gained considerable traction in both academic research and practical applications, driven by growing awareness of environmental sustainability and the benefits of integrating trees into farming systems. Additionally, search outcomes from Google Trends further corroborate these findings. The data reveals that 'agroforestry' is one of the most frequently used terms in comparison to 'trees outside forests'. Fig. 12.1 from the Google Trends analysis shows a clear trend where 'agroforestry' consistently ranks higher in search frequency. This higher search volume indicates greater public and academic interest in the term, reinforcing the notion that 'agroforestry' is more widely recognized and utilized.

The prominence of 'agroforestry' in search trends and its frequent appearance in academic literature underscore its significance in the field of sustainable land use. The term's broad application and relevance to a wide range of practices make it a central concept in discussions about integrating trees into agricultural systems. The sustained interest in 'agroforestry' reflects its role in addressing contemporary environmental challenges and promoting sustainable agricultural practices (Verchot et al. 2007; Prasad et al. 2014; Abbas et al. 2017). On the other hand, the term 'trees outside forests', while important, does not match the same level of recognition as 'agroforestry'. The variations in the terminology and the specific focus on trees in non-forest environments may limit its broader appeal compared to the more inclusive concept of agroforestry. However, the presence of the term in academic literature and its increasing usage over time suggest that it remains a relevant and valuable concept, particularly in discussions related to urban greening and sustainable land management (Singh et al. 2013; Paul et al. 2017).

Overall, the analysis of both Google Books data and Google Trends highlights the differing levels of acceptance and usage of these terms. 'Agroforestry' emerges as the more widely recognized and utilized term, reflecting its broader applicability and significance in the field of sustainability. In contrast, 'trees outside forests' has seen considerable usage, particularly in the early 2000s, but does not achieve the same level of prominence as 'agroforestry'. The variations in terminology and the specific focus of 'trees outside forests' may influence its relative frequency and recognition in academic and public discourse.

12.3 Agroforestry and Trees Outside Forests: Definition Insights

The FAO produces two key reports related to global forest resources, viz., Global Forest Resource Assessment (GFRA) (FAO 2020), published every 5 years, and State of the Worlds' Forests (SOFO), released biennially (FAO 2022). These reports are based on data provided by individual countries, which conduct their own periodic assessments (Fischer et al. 2011). The GFRA provides a comprehensive dataset on various types of land cover related to forests. It categorizes land into three main types: forest, other wooded land, and other land with tree cover. These are defined as follows:

12.3.1 Forest

This category refers to land spanning more than 0.5 hectares and meets specific tree height and canopy cover criteria. Specifically, it includes areas where trees are taller than 5 m and have a canopy cover of higher than 10%, or where trees are able to reach the thresholds in their natural environment. Notably, forest land does not include areas mostly utilized for crop cultivation or urban purposes. Additionally, it encompasses windbreaks, shelterbelts, and tree corridors that cover more than 0.5 hectares of areas and have a width exceeding 20 m.

12.3.2 Other Wooded Land

This designation applies to land that does not qualify as forest but still has tree cover. It includes areas spanning areas higher than 0.5 hectares where trees are taller (more than 5 m) with a canopy cover ranging from 5% to 10%, or where the total cover of trees, shrubs, and bushes exceeds 10%. Similar to forest land, this category excludes areas mainly used for agriculture or urban development.

12.3.3 Other Land with Tree Cover

This includes land categorised as 'Remaining Land Area' that spans higher than 0.5 hectares and has a canopy cover of over 10%. The trees on this land are capable of attaining a height of 5 m at maturity. This category also covers areas under bamboo, orchards, and palms.

12.4 Conclusion

The definitions of Forest, Other Wooded Land, and Other Land with Tree Cover provided by FAO in its GFRA and SOFO reports are crucial for categorizing land cover types. These categories each have distinct criteria and implications for reporting, particularly when it comes to areas under agroforestry. The Forest category is defined as land spanning higher than 0.5 hectares with trees taller (more than 5 m) and a canopy cover exceeding 10%, or where trees can attain these thresholds in-situ. The category explicitly excludes land predominantly under crop cultivation or urban use. However, it does include certain agroforestry practices, such as wind-breaks, shelterbelts, and tree corridors, provided they cover more than 0.5 hectares and are wider than 20 m. This inclusion acknowledges that specific agroforestry practices can contribute to forest cover, but only if they meet particular spatial and structural criteria. The Other Wooded Land category encompasses land that does not qualify as forest but still has significant tree cover, including covering areas higher than 0.5 hectares having trees taller than 5 meters with a canopy cover between 5% and 10%, or where the total cover of trees, shrubs, and bushes exceeds 10%. Like the Forest category, it also excludes land used for crop cultivation or urban development. While this category may include some agroforestry systems, particularly those with lower tree density, it largely fails to capture agroforestry areas that are more integrated with agricultural activities, leading to potential underreporting.

The other land with tree cover are categorised as 'Remaining Land Area' covering higher than 0.5 hectares area with a canopy cover exceeding 10%. The trees under this area can attain a height of 5 m at maturity. It also reports areas under bamboo, orchards, and palms. While it could include some agroforestry systems, especially those involving tree crops like orchards, it does not fully account for the diversity and complexity of agroforestry practices. The focus on tree cover rather than the integration of trees with crops limits the category's applicability for comprehensive agroforestry reporting. The FAO's definitions effectively categorize different land cover types, but they have limitations in fully capturing the complexity and diversity of agroforestry systems. Agroforestry, which integrates trees into agricultural landscapes, often exists at the intersection of forestry and agriculture, making it challenging to fit within the existing framework.

12.5 Way Forward

To address this challenge, it is essential to reconsider and refine current reporting mechanisms to ensure they more accurately capture the full extent and significance of agroforestry. Developing more nuanced definitions or establishing a distinct category for agroforestry within global land use assessments could bridge this gap. Such an approach would not only improve the precision of global evaluations but also elevate the visibility of agroforestry's vital contributions to forestry and agriculture. By acknowledging and integrating these contributions, we can better reflect the growing recognition of agroforestry as a crucial practice for enhancing biodiversity, climate resilience, and sustainable agriculture. Furthermore, it's important to standardize the usage and definition of terms like TOF across global literature to facilitate more consistent and comprehensive reporting on this important land use practice.

Acknowledgement The authors would like to thank the support from the Indian Council of Agricultural Research and the Trees Outside Forests in India Program.

References

Abbas F, Hammad HM, Fahad S et al (2017) Agroforestry: a sustainable environmental practice for carbon sequestration under the climate change scenarios-a review. Environ Sci Pollut Res 24:11177–11191

Arunachalam A, Ramanan SS, Handa AK (2021) Administering agroforestry at the district level. Curr Sci 121:473–474

Cernea M (2019) Alternative social forestry development strategies. In: Ives J, Pitt DC (eds) Deforestation. Routledge, London, pp 159–190

Clark B, DeFries R, Krishnaswamy J (2021) India's commitments to increase tree and Forest cover: consequences for water supply and agriculture production within the Central Indian Highlands. Water 13(7):959

Dabral A, Shankhwar R, Martins-Ferreira MAC et al (2023) Phenotypic, geological, and climatic spatio-temporal analyses of an exotic *Grevillea robusta* in the Northwestern Himalayas. Sustain For 15(16):12292

Dogra AS, Chauhan SK (2016) Trees outside forests in India: socio-economic, environmental and policy issues. In: Forestry technologies – a complete value chain approach. Scientific Publishers, pp 84–102

FAO (2016) In: Salbitano F, Borelli S, Conigliaro M, Chen Y (eds) Guidelines on urban and peri-urban forestry. FAO

FAO (2020) Global forest resources assessment 2020. FAO

FAO (2022) The state of the World's forests 2022. FAO
Fischer C, Kleinn C, Fehrmann L et al (2011) A national level forest resource assessment for Burkina Faso – a field based forest inventory in a semiarid environment combining small sample size with large observation plots. For Ecol Manag 262:1532–1540
FSI (2021) India state Forest report 2021. FSI
Jun S-P, Yoo HS, Choi S (2018) Ten years of research change using Google trends: from the perspective of big data utilizations and applications. Technol Forecast Soc Change 130:69–87
Kousha K, Thelwall M (2015) An automatic method for extracting citations from Google Books. J Assoc Inf Sci Technol 66:309–320
Nair PKR (1993) An introduction to agroforestry. Springer Science & Business Media
Patel N, Sethi T, Ravisankar T et al (2023) Technical report on greening and restoration of wastelands with agroforestry (G.R.O.W). NITI Ayog
Paul C, Weber M, Knoke T (2017) Agroforestry versus farm mosaic systems – comparing land-use efficiency, economic returns and risks under climate change effects. Sci Total Environ 587:22–35
Pechenick EA, Danforth CM, Dodds PS (2015) Characterizing the Google Books corpus: strong limits to inferences of socio-cultural and linguistic evolution. PLoS One 10:e0137041
Pental D (2021) Challenges for India in agriculture and the pivotal role of R&D in meeting these. Dialogue – Sci Sci Soc 4:1–47
Prasad R, Newaj R, Chavan SB, Dhyani SK (2014) Agroforestry land use: a way forward for reducing carbon footprint in agriculture. Indian J Agrofor 16:15–23
Proulx R, Massicotte P, Pepino M (2014) Googling trends in conservation biology. Conserv Biol 28:44–51
Rakatama A, Pandit R (2020) Reviewing social forestry schemes in Indonesia: opportunities and challenges. Forest Policy Econ 111:102052
Ramachandra TV, Aithal BH, Kulkarni G, Vinay S (2017) Green spaces in Bengaluru: quantification through geospatial techniques. Indian For 143:307–320
Ramanan SS, Arunachalam A (2021) Introspection of agroecology for food systems from agroforestry perspective. Indian J Agrofor 23:141–145
Ramanan SS, George AK, Chavan SB et al (2020) Progress and future research trends on Santalum album: a bibliometric and science mapping approach. Ind Crop Prod 158:112972
Regmi BN, Garforth C (2010) Trees outside forests and rural livelihoods: a study of Chitwan District, Nepal. Agrofor Syst 79:393–407
Ren HC, Lu YL, Liu C et al (2005) Eco-environment contribution of agroforestry to agriculture development in the plain area of China – Huai'an Prefecture, Jiangsu Province as the case study area. J Environ Sci 17:327–331
Rizvi RH, Handa AK, Sridhar KB et al (2020) Mapping agroforestry and trees outside Forest. Jointly published by the ICAR, Central Agroforestry Research Institute (CAFRI), Jhansi and World Agroforestry (ICRAF), South Asia Regional Programme, New Delhi
Singh G, Singh B, Rathore TS (2013) Effects of agroforestry land use on microclimate modification and productivity in dry areas. Indian J Agrofor 15:68–78
Thomsen P, Bühler O, KrisTOFfersen P (2016) Diversity of street tree populations in larger Danish municipalities. Urban For Urban Green 15:200–210
van Noordwijk M (2019) Sustainable development through trees on farms: agroforestry in its fifth decade. ICRAF, Nairobi
Verchot LV, Van Noordwijk M, Kandji S et al (2007) Climate change: linking adaptation and mitigation through agroforestry. Mitig Adapt Strateg Glob Chang 12:901–918
Zomer RJ, Trabucco A, Coe R, Place F (2009) Trees on farm: analysis of global extent and geographical patterns of agroforestry. ICRAF Working Paper Agroforestry Centre
Zomer RJ, Neufeldt H, Xu J et al (2016) Global tree cover and biomass carbon on agricultural land: the contribution of agroforestry to global and national carbon budgets. Sci Rep 6:29987

Tree Health Management in Rajasthan: Insects-Pests, Diseases and Protection Strategies for Some Important Tree Species

13

Shiwani Bhatnagar and Neelam Verma

Abstract

Tree health management is crucial for sustaining biodiversity and ecological balance. Understanding the impact of insects, pests and diseases on key tree species in the region is an essential requirement. The major threats such as termites, leaf miners and fungal infections that significantly affect tree health and ecosystem stability are mostly prevalent in the region. Through a comprehensive literature review, this paper highlights the susceptibility of major key tree species found in Rajasthan, including Babul (*Vachellia nilotica*), Khara Jal (*Salvadora persica)*, Tendu (*Diospyros melanoxylon*) Guggal (*Commiphora wightii*), Gundha (*Cordia dichotoma)* and Palash (*Butea monosperma*) to various biotic stressors. It explores effective protection strategies encompassing integrated pest management (IPM), bio-control, and cultural practices tailored to the local environment and socio-economic conditions. The present chapter also underscores the importance of community engagement and policy interventions in promoting sustainable tree health management practices against detrimental factors, ensuring their resilience and longevity in the face of evolving environmental challenges.

Keywords

Tree health · IPM · *Acacia nilotica* · *Salvadora persica* · *Diospyros melanoxylon* · *Cordia dichotoma* · *Butea monosperma*

S. Bhatnagar (✉) · N. Verma
ICFRE-Arid Forest Research Institute, Jodhpur, Rajasthan, India
e-mail: shiwani@icfre.org

R. K. Kalia, R. Pathak (eds.), *Tree Biology and Biotechnology*,
https://doi.org/10.1007/978-981-96-0002-1_13

13.1 Introduction

Tree health management in Rajasthan, India, is a critical area of concern given the region's diverse ecological landscapes and the pivotal role of trees in sustaining both environmental equilibrium and human livelihoods by providing a diverse array of economically significant products such as timber, pulp, resins, gums, fibre and food. The arid and semi-arid climate of Rajasthan presents unique challenges to tree species, making them vulnerable to a range of biotic and abiotic stressors such as insects, pests, diseases, and adverse climatic conditions. The state boasts a plethora of tree species that contribute significantly to its biodiversity and ecosystem services. Iconic species like Babul (*Acacia nilotica*), Khara Jal (*Salvadora persica)*, Tendu (*Diospyros melanoxylon*) Guggal (*Commiphora wightii*), Gundha (*Cordia dichotoma)* and Palash (*Butea monosperma*) are integral to socio-economic fabric of the state, providing resources such as timber, fodder, and medicinal products, while also playing crucial roles in soil conservation and water management.

Despite their ecological and economic importance, these trees face numerous threats to their health and survival. Insects like termites and pests such as leaf miners, leaf defoliators, wood borers, etc., pose substantial risks, while diseases caused by fungi and bacteria further exacerbate the challenges. These threats not only compromise the health of individual trees but also threaten broader ecological stability and the resilience of local ecosystems. Effective tree health management strategies are therefore, imperative to mitigate these risks and ensure the sustainable management of tree resources of the state. This necessitates a holistic approach that integrates scientific research, community engagement, and policy interventions tailored to the region's specific environmental conditions and socio-economic contexts.

Thus there is a need to explore the various dimensions of tree health management in Rajasthan, emphasizing the need for comprehensive strategies that address current challenges and future sustainability concerns. By enhancing our understanding of these issues and promoting effective management practices, we can safeguard the invaluable tree species of Rajasthan for generations to come.

13.2 Insect-Pests and Diseases of Trees

Diseases and insect pests present significant challenges to forests and the forest industry by affecting the growth of trees and the production of both wood and non-wood products. While specific statistical data on their impact is limited, it is well-documented that various insects and diseases severely damage both naturally regenerating forests and plantations. Trees are threatened by a diverse array of pathogens, including bacteria (Shahbaz et al. 2022), nematodes (Khan 2020), viruses (Pathak and Silawat 2015), and fungi (Pathak et al. 2015). Insects and other invertebrates also pose risks, with defoliators and borers causing direct harm to trees, while some insects act as disease vectors, exacerbating the challenges faced by forest ecosystems.

13.2.1 Babul (*Vachellia nilotica*)

Vachellia nilotica, often referred to as 'Babul' or 'Kikar', is a tree that falls under the subfamily Mimosoideae within the Fabaceae (Leguminosae) family. *Vachellia* generally develops into a moderately sized tree with a wide, spreading crown. It thrives in tropical and subtropical regions and is native to the Indian Subcontinent. Additionally, it can be found across Tropical Africa, Burma, Sri Lanka, Saudi Arabia, Egypt, West Sudan, and East Sudan. In India, natural babul forests are primarily found in Rajasthan, Gujarat, Haryana, Maharashtra, Karnataka and Andhra Pradesh (Raj et al. 2016). This versatile plant, which was introduced to Queensland from India in the late 1890s for its ornamental appeal (Bolton 1989), has a wide range of uses (Kaur et al. 2005).

13.2.1.1 Insect/Mites and Their Management

The major insects/mites affecting the species (Table 13.1) and their management are listed below:

Leaf feeder: Dichlorvas (0.076%) can be sprayed if a severe infestation of leaf feeder is observed (Senthilkumar and Murugesan (2015) or application of phosphamidon 0.05% @ 0.6 ml/litre as a spray (ICFRE 2020) may be used.

Borer: Sapling can be cut at ground level, and a small amount of crude oil such as Kerosene or 0.1% solution of Dichlorvos can be introduced into the tunnel. Once applied, the hole should be sealed with damp soil (Joshi and Jamaluddin 2007) to manage borers.

Table 13.1 Major insects/mites affecting *Vachellia nilotica* species

Insects/mites	References
Celosterna scabrator	Brunck (1994), Roychoudhury and Mishra (2020)
Oxyrchachis tarandus	Dhileepan (2015)
Heliothis armigera	(ICFRE 2020)
Achaea janata	Dhileepan (2015)
Selepa celtis	Dhileepan (2015)
Euproctis lunata, E. fraternal	Beeson (1941), Mathur and Singh (1959)
Indarbela quadrinotata	Dhileepan (2015)
Eumeta crameri	Dhileepan (2015)
Mylloceros spp.	Dhileepan (2015)
Sinoxylon dekkancnse	Satyavir and Verma (1996)
Acmaeodera aurifer	Satyavir and Verma (1996)
Homoeocerus signatus	Satyavir and Verma (1996)
Aphis craccivora	Satyavir and Verma (1996)
Agrilus sp.	Satyavir and Verma (1996)
Bemisia tabaci	Garg et al. (2017)
Caryedon serratus	Singal and Toky (1990)

13.2.1.2 Diseases and Their Management

The major causal agents (Table 13.2) of diseases and their management are summarized as follows:

Removal of infected trees and fruiting bodies of fungus, along with improving soil aeration, provides partial protection. For effective root rot management, utilizing a combination of *Trichoderma*, *Pseudomonas fluorescens*, and PGPR mix can be considered. These beneficial microorganisms work synergistically to enhance soil health and plant resilience, offering a comprehensive approach to manage root rot. Soil application of neem cake is also beneficial; it can be done once every 6 months, along with fertilizers. Tridemorph at the rate of 2 ml or Hexaconazole at 1 ml per 100 ml of water is applied for root feeding (https://agritech.tnau.ac.in, retrieved on 4.8.24).

13.2.2 Guggal (*Commiphora wightti)*

Commiphora wightii (Arnott) Bhandari, commonly known as guggulu or Indian bedellium, is a medicinal plant adapted to arid environments. It is classified under the genus Commiphora within the Burseraceae family, which comprises approximately 185 species (Reddy et al. 2012). This particular species is renowned for its guggul gum-resin, a valuable substance in Ayurvedic medicine. Guggul products are well-regarded in the market, and local communities also use the gum to create *Dhupa* for religious ceremonies (Rakholiya 2010).

C. wightii is distributed across several regions in Rajasthan, such as Jaipur, Jalore, Jaisalmer, Jodhpur, Kota, Udaipur, Sawai Madhopur, Ajmer, Alwar, Bundi, Bikaner, Sikar, Churu, Nagaur, Pali, Sirohi and Jhunjhunu (Tiwari et al. 2001). However, the species is experiencing a significant population decline in its natural habitat, resulting in its classification as a threatened species (Reddy et al. 2012). It is also listed on the International Union for Conservation of Nature Red List of Threatened Plants (Natesh and Mohan Ram 1999). Furthermore, plants cultivated in nurseries or plantations often suffer from termite infestation and root rot, leading to yellowing leaves and eventual plant death (Reddy et al. 2012).

Table 13.2 Major pathogens affecting *Vachellia nilotica* species

Pathogens	References
Ganoderma lucidum	Luna et al. (2006)
Curvularia lunata	Singh and Jamaluddin (1995)
Macrophomina phaseolina	Srivastava and Kalyani (1990)
Oidium sp.	Bhat and Hedge (1990)
Haplosporella bakeriana	Verma et al. (1997)
Fomes badius	ICFRE (2020)

13.2.2.1 Insect/Mites and Management

The major insects/mites affecting the species (Table 13.3) and their management are listed below:

Termites: They can be effectively managed by using calcium disulphide or kerosene to destroy the termitorium (Rohini and Pandiarana 2013) and by drenching the soil with chloropyriphos @ 4 ml/litre of water (https://vikaspedia.in, retrieved on 10.8.24).

Mealy bugs: For plants with minor infestations, manual removal of bugs can be effective, especially in miniature conservatories or home gardens. Alternatively, a water jet can be used to dislodge pests. For more persistent issues, dichlorvos 76 EC @ 2 ml with 2 gm of fish oil rosin soap per litre water can be mixed and applied to the plants. Additionally, sticky bands can be placed around the plant's stems to trap mealy bug crawlers (Tanwar et al. 2007).

Spodoptera litura: Profenophos 50% EC @ 1000 ml per ha, or Deltamethrin 2.8 EC @ 750 ml per ha is applied (https://agritech.tnau.ac.in, retrieved on 4.8.24) to manage *S. litura*.

Bemisia sp.: Diafenthuron 50 WP @ 0.1% or Acetamiprid 20 SP @ 0.5 g per litre can be sprayed (Jain et al. 2022, https://agritech.tnau.ac.in/forestry/forest_disease_gmelina.html, retrieved on 23.7.24) for its management .

Euproctis lanata: The application of NSKE @ 5% (Nizamani et al. 2016) or spray of Chlorpyriphos @2 ml/l (Mani et al. 2008) can be utilized for the management of *E. lanata*.

13.2.2.2 Disease and Their Management

The major pathogens (Table 13.4) diseases and their management is summarized as follows:

Cercospora leaf spot: One of the following treatments per acre may be applied to against the pathogens: (https://agritech.tnau.ac.in/crop_protection/):

Table 13.3 Major insects/mites affecting the *Commiphora wighti* species

Insect/mites	References
Odontoterm obesus	Orwa et al. (2009), Ahmed et al. (2016)
Microtermis mycophagus	DMAPR Annual report, Anand (2012–13)
Ferrisia vergata	DMAPR Annual report, Anand (2012–13)
Spodoptera litura	DMAPR Annual report, Anand (2012–13)
Bemisia sp.	Dalal (1989), Kurian and Sankar (2007)
Euproctis lanata	Dalal (1989), Kurian and Sankar (2007)
Miresa albipuncta	Beeson (1941)

Table 13.4 Major pathogens affecting the *Commiphora wightti* species

Pathogens	References
Ascochyta commiphorae	Ramakrishnan and Sundaram (1954)
Phoma glomerata (Corda)	Sharma and Gour (1987)
Alternaria alternata	Majumdar et al. (1988)
Sclerotium rolfsii	Orwa et al. (2009)
Rhizoctonia bataticola (Taubenh.) Butler and *Rhizoctonia solani*	AFRI annual report (2007–2008)
Cercospora sp.	Dalal (1989), Kurian and Sankar (2007)
Xanthomonas axonopodis pv. commiphorae (Xac)	Kurian and Sankar (2007), Samanta and Mandal (2013)

- Mancozeb @ 400 g
- Copper oxychloride @ 500 g
- Carbendazim 50% WP @ 200 g
- Propiconazole @ 200 ml

Xanthomonas bacterial blight: The application of copper-based bactericides (Gent and Schwartz 2005) may be used to manage the disease.

Sclerotium root rot: Soil drenching with a 0.2% solution of Bavistin or Dithane M-45 at monthly intervals (https://agritech.tnau.ac.in, retrieved on 4.8.24) may be used to manage the disease.

Phoma disease: Soil solarisation (Deb et al, 2020), Dazomet (Molina-Vargas 2013) and benzimidazoles (Deb et al. 2020) application may be done to manage Phoma diseases effectively.

Rhizoctonia root rot: The treatment combining Bavistin (2 g per kg seed) with 20% neem extract can be used to manage the disease (Kumar et al. 2011).

Alternaria leaf spot: Natural plant defences may be enhanced by applying soil microbes like Trichoderma and Mycorrhiza, combined with compost or farm-yard manure to combat the disease. Additionally, difenoconazole (0.025%), azoxystrobin (0.023%), or bavisin (0.1%) may be sprayed (Kumar 2018).

13.2.3 Tendu (*Diospyros melanoxylon*)

Tendu leaves (*Diospyros melanoxylon*) hold significant commercial value in India as they are utilized as wrappers for *bidis* (mini-cigar packed with tobacco flakes). Their ideal thickness, texture, colour, and flexibility make them the preferred choice over leaves from other tree species. This tree species is severely affected by certain insect-pests, causing substantial losses in state revenues and foreign exchange for the country. This tree is found in the states of Karnataka, Andhra Pradesh, Maharashtra, Rajasthan, Tamil Nadu and Madhya Pradesh. It thrives as a deciduous forest tree, exhibiting broad adaptability to challenging soil and climatic conditions and displaying resilience against both biological and environmental stresses.

13.2.3.1 Insect/Mites and Management

The major insects/mites affecting the species (Table 13.5) and their management are listed below:

Root borer: Insecticidal solution viz., Dichlorvos (0.0125%), may be injected to manage the insect-pest.

Gall-forming insect: The process involves plucking and disposing of the old leaves. Dimecron, Dimethote and spraying a 0.03% solution of cypermethrin or fenvalerate (Sumicidine) twice in March–April, with a 15-day interval between applications, can be effective. Moreover, pruning should be performed during the first week of March (ICFRE 2020).

Defoliators: Natural enemies like *Nomuraea rileyi* against *Hypocala rostrata* (Dadwal et el. 1986) can be used. The spray of systemic insecticide can also be done (ICFRE 2020).

13.2.3.2 Diseases and Their Management

The major pathogens, diseases (Table 13.6) and their management are summarized as follows:

The disease can be managed through timely applications of Bordeaux mixture or any copper-based fungicide. Natural plant defences of the plant may be enhanced by applying soil microbes like Trichoderma and Mycorrhiza. Leaf spots can be managed by spraying azoxystrobin (0.023%), or bavisin (0.1%) (Kumar, 2018).

13.2.4 Khara Jal (*Salvadora persica*)

Salvadora persica (L), often referred to as miswak or khara jal, is a versatile plant known for its natural oral hygiene benefits. It is a tree species adapted to salt-affected lands in arid regions such as Rajasthan, Gujarat, Haryana, Punjab and Western Uttar Pradesh (Arya et al. 2014). As a facultative halophyte, it selectively stores excess salts in its fully grown leaves, as well as in the bark. The shedding of these parts

Table 13.5 Major insects/mites affecting the *Diospyros melanoxylon* species

Insect/mites	References
Hypocala rostrata	Kumar (1991)
Miresa albipuncta	Kumar (1991)
Euthelia lamdabilis	Kumar (1991)
E. nais	Kumar (1991)
Lamida carbonifera	Kumar (1991)
Phycita sp.	Kumar (1991)
Trioza obseleta; Trioza spp.	Kumar (1991)
Cossus cadambae	Kumar (1991)
Stromatium barbatum	Kumar (1991)
Xyleborus noxius	Kumar (1991)
Plocaederus ferrugineus	Kumar (1991)

Table 13.6 The major pathogen/diseases affecting the *Diospyros melanoxylon* species

Pathogens/diseases	References
Pseudocercospora tessellata	Braun et al. (2020)
Cladosporium sp.	Parida (2020)
Alternaria species	Parida (2020)
Pestalotia diospyri	Mehta et al. (2020)
Stereum lobatum	Kumar et al. (2017)
Penicillium sp.	Kitagawa and Glucina (1984)
Crown gall	Kumar et al. (2017)
Dematophora root rot	Kumar et al. (2017)

Table 13.7 Major insects/mites affecting the *Salvadora persica* species

Insects/mites	References
Colotis amata	Bhatnagar et al. (2014), Haldhar et al. (2015)
Catopsilia crocale	Kumar and Ahmed (1996)
Colotis ephiae	Baumer (1983)
Resseliella salvadorae	Moawia et al. (2012)
Eriohyes sp.	Kant and Arya (1971)

helps remove excess salts from the plant (Amonkar and Karmakar 1978). *S. persica* is also cultivated in saline coastal areas (Makwana et al. 1988). The plant has a variety of uses as its twigs and young stems are used as toothbrushes, shoots are fed to camels, plant ash is processed into salt, and the bark has antimicrobial properties that help to prevent plaque and support oral health. Additionally, it is used to relieve back pain, chest ailments, and stomach issues and the seed oil is applied topically to alleviate rheumatism (Arya et al. 2014).

13.2.4.1 Insect/Mites and Management

The major insects/mites affecting the species (Table 13.7) and their management are listed below:

Eriohyes sp.: Pruning the sections of host plants that are heavily affected by galls is effective in managing mites and miticides viz., Abamectin 1.9% EC@1 ml per litre, Diafenthiuron 25% WW+ Pyriproxyfen 5% WW@ 2 ml per litre can be utilized for effective management (Bhatnagar et al. 2023).

Catopsilia crocale: Christopher and Mathavan (1993) reported *Apanteles glomeralus* (L.) as its biocontrol agents.

Resseliella salvadorae: *Platygaster salvadorae* parasitizes the gall maker and can be utilized as its potential bio-control agent (Mani 1973).

13.2.4.2 Diseases and Their Management

The major pathogen (Table 13.8) diseases and their management are summarized as follows:

Spraying of mancozeb at a concentration of 2.5 g per litre or Bavistin at 1 g per litre water is reported to be highly effective in managing the Cercospora leaf spot, the major disease of the plant species (ICFRE 2020). The phytoplasma disease

Table 13.8 Major pathogens affecting the *Salvadora persica* species

Pathogens	References
Phytoplasma	Kumar et al. (2012)
Cercospora udaipurensis	https://infonet-biovision.org/trees/toothbrush-tree, retrieved on 11.11.24; https://apps.worldagroforestry.org/treedb/AFTPDFS/Salvadora_persica, retrieved on 12.11.24
Placosoma salvadorae	https://infonet-biovision.org/trees/toothbrush-tree, retrieved on 11.11.24; https://apps.worldagroforestry.org/treedb/AFTPDFS/Salvadora_persica, retrieved on 12.11.24
Sephogloeum salvadorae	https://infonet-biovision.org/trees/toothbrush-tree, retrieved on 11.11.24; https://apps.worldagroforestry.org/treedb/AFTPDFS/Salvadora_persica, retrieved on 12.11.24

causing pathogen gets transmitted by leafhopper (https://www.invasive.org, retrieved on 11.11.2024). Phytoplasma disease can be managed through effective management of insect vectors, removal of infected plants, and the use of plant growth-promoting microbes (Trivedi et al., 2016).

13.2.5 Gundha (*Cordia dichotoma*)

Cordia dichotoma is a well-regarded traditional medicinal plant prevalent throughout India. Its fruit, rich in polysaccharides, produces a jelly-like, sticky substance upon ripening. This plant is highly valued for its antibacterial, antiviral, and antitussive properties, playing a crucial role in various treatments. The fruit in India is used to treat conditions such as respiratory distress, cold cough, catarrh and fevers (Basu et al. 1984; Vohora 1986). *C. dichotoma* flourishes across diverse forest ecosystems, from dry deciduous and moist deciduous forests to tidal forests. In Maharashtra, it is particularly prolific in the moist monsoon forests (Hussain and Kakoti 2013). *Eriophyes cordiae* has been reported to be the major mites (Mani 1959; Kant and Sharma 1975) found on the species. Pruning the sections of host plants that are heavily affected by galls is highly effective in managing the mites besides this miticides, viz., Abamectin 1.9% EC @1 ml per litre, Diafenthiuron 25% WW+ Pyriproxyfen 5% WW@ 2 ml per litre can also be utilized for effective management of the mites (Bhatnagar et al. 2023).

13.2.5.1 Disease and Their Management

Leaf spot disease caused by *Alternaria* sp. (Singh et al. 2017)) and rust caused by *Aecidium* sp. (Kumar and Khurana 2016) are major diseases of the tree species. Palmarosa oil (0.025%) causes total inhibition of both the fungal mycelium growth and spore germination of the *Alternaria* (Kamalalakshmi 1996). The application of neem oil (Karthikeyan et al. 2005) and Eucalyptus oil (5%) is also effective in the management of *Alternaria* sp. (Supraja and Tiwari 2022). Hexaconazole 5% EC and thiauzamide 24% SC are reported to be effective against this pathogen (Kumar and Singh 2017). Studies reported that the rust in *C. dichotoma* can be effectively managed by Tebuconazole 50% EC (Bajoriya et al. 2023).

13.2.6 Palash (*Butea monosperma*)

Butea monosperma (Lam.) Kuntze (Family: Fabaceae), widely recognized as the Flame of the Forest or Palash is native to South Asia, including India, Bangladesh, and Myanmar. This deciduous tree is renowned for its vibrant orange-red flowers that bloom in clusters during the spring. It features trifoliate leaves and produces flat pods containing seeds. *B. monosperma* holds cultural importance in Hindu mythology and is closely associated with the arrival of spring and the Holi festival in India. The flowers are a source of orange dye used in herbal gulal (Mazumdar et al. 2011). Beyond its ornamental appeal, the tree is valued in Ayurvedic medicine for its diverse medicinal values such as aphrodisiac, depurative tonic, astringent and diuretic (Nadkarni 2002), and exhibits anti-fertility, aphrodisiac, and analgesic effects (Anon 1988). Locally, the roots are applied to treat conditions such as filariasis, night blindness, helminthiasis, piles, ulcers, and tumors (Mengi and Deshpande 1999), while the flowers are used to address diarrhoea (Bhalla and Walter 1999).

13.2.6.1 Insect/Mites and Management

The major insects/mites (Table 13.9) and their management are summarized as follows:

White grubs: Thorough summer ploughing is done to ensure that pupae are exposed to intense sunlight and are more vulnerable to bird predation. Seed treatment with imidacloprid 17.8SL @ 2.0 ml per kg seed and application of Quinalphos 25EC @ 3.2 litres per ha in the root zone 3 weeks after the adult emergence are highly effective (https://krishi.icar.gov.in, retrieved on 5.8.24).

Spodoptera litura: Pheromone-baited traps and light traps can be employed. Spodoptera NPV (Nuclear Polyhedrosis Virus) is useful against this pest (Senthilkumar and Murugesan 2015). Emamectin benzoate 5 SG @ 5 g/10 litres. water is also effective in managing the insect-pest (https://ctri.icar.gov.in/for_controlPests.php, retrieved on 12.11.24)

Odontotermes obesus: https://agritech.tnau.ac.in (retrieved on 4.8.2024) reported following measures:

- Application of undecomposed farmyard manure (FYM) or compost should be avoided.
- Termite mounds within and around the nursery or field should be eliminated, targeting the queen and associated colonies. Their eradication is also helpful.
- Spray of neem oil @5% is also effective.
- Application of calcium at the trunk base also helps in deterring termite attacks.
- The base and up to 2 meters height on the trunk with 5% neem oil is swabbed for effective management.

Myllocerus: https://agritech.tnau.ac.in (retrieved on 4.8.2024) reported following measures:

- Adult weevils should be collected and destroyed.
- In regions prone to endemic issues, 15 kg per ha of carbofuran 3G is applied for 15 days after planting.

Taragama siva: http://eagri.org (retrieved on 5.8.2024) reported following measures:

Table 13.9 Major insects/mites affecting the *Butea monosperma* species

Insects/mites	References
Anoplaenemis phasiana	Vaibhav et al. (2016), Anuragi et al. (2023)
Aonidiella orientalis	Vaibhav et al. (2016), Anuragi et al. (2023)
Cyclopelta obscura	Vaibhav et al. (2016)
Dasychira mendosa	Vaibhav et al. (2016)
Holotrichia serrata	Vaibhav et al. (2016)
Euproctis (Nygmia)varion	Vaibhav et al. (2016)
Megalurothrips distalis	Vaibhav et al. (2016)
Acoclita crytiolitha	Vaibhav et al. (2016); Anuragi et al. (2023)
Endaeus buteae	Vaibhav et al. (2016)
Belippa laleana	Vaibhav et al. (2016), Anuragi et al. (2023)
Beara dichromella	Vaibhav et al. (2016), Anuragi et al. (2023)
Cyclopelta obscura	Anuragi et al. (2023)
Cyclopelta siccifolia	Anuragi et al. (2023)
Myllocerus cardoni	Vaibhav et al. (2016)
Myllocerus discolor	Vaibhav et al. (2016)
Odontotermes obesus	Vaibhav et al. (2016), Anuragi et al. (2023)
Deltas eucharis	Anuragi et al. (2023)
Paratachardina pseudolobata	Vaibhav et al. (2016)
Spodoptera litura	Vaibhav et al. (2016)
Taragama siva	Vaibhav et al. (2016)
Parasa lepida	Vaibhav et al. (2016), Anuragi et al. (2023)
Aulacaspis sp.	Singh et al. (2013, Anuragi et al. (2023)
Giaura sp.	Anuragi et al. (2023)
Holotrichia serrata	Anuragi et al. (2023)
Lampideo boeticus	Anuragi et al. (2023)
Pseudaulacaspis cockerelli	Singh et al. (2013)
Amblyrrhinus poricollis	Singh et al. (2013), Anuragi et al. (2023)
Peltotrachelus sp.	Singh et al. (2013), Anuragi et al. (2023)
Physomerus grossipes	Singh et al. (2013), Anuragi et al. (2023)

- Collecting and destroying larvae from the plants may be one of the best strategies.
- Employment of light traps to lure and elimination of adult insects is more effective.
- Application of insecticides from the organophosphate group by spraying is also effective.

Lampides boeticus: https://agritech.tnau.ac.in (retrieved on 4.8.2024) reported following measures:

- Use Neem oil at a concentration of 2%.
- Apply Emamectin benzoate 5% SG @ 220 grams per hectare.
- Apply NSKE @5% twice, subsequently Triazophos @0.05%.

Pseudaulacaspis cockerelli: Buss and Dale (2016) suggested the following measures:

Table 13.10 Major pathogens affecting the *Butea monosperma* species

Pathogens	References
Erythricium salmonicolor	Vaibhav et al. (2016)
Pseudodiplodia buteae	Vaibhav et al. (2016)
Xanthomonas buteae	Vaibhav et al. (2016)
Phytoplasma	Vaibhav et al. (2016)
Phomopsis buteae	Vaibhav et al. (2016)

- Horticultural oils and dormant oils are recommended.
- Soil drenching or spray on the trunk with systemic pesticides like dinotefuran can effectively manage scale insects throughout the season.
- Application of acetamiprid on foliage or bark can also be done to manage the infestation.

13.2.6.2 Diseases and Their Management

The major pathogens (Table 13.10) diseases and their management are summarized as follows:

Xanthomonas sp., Pseudomonas fluorescens and Bacillus substilis: Neem oil @5% can be used (Ambadkar et al. 2015) or spray of copper hydroxide or copper hydroxide-mancozeb or Acibenzolar-S-methyl can be done (Gent and Schwartz 2005) to manage the disease.

Phomopsis sp.: Regular application of protective fungicides in the field reduces the levels of disease-causing agents and prevents infections through any wounds that may occur (http://www.extento.hawaii.edu, retrieved on 5.8.2024). Dithane Z-78 (0.2%) or Bordeaux mixture (1%) can be used to check the disease (Vinodkumar 2012).

Erythricium: Copper oxychloride, copper carbonate and Bordeaux mixture can be used (Thankamma et al. 1986; Kueh 1986; Ram et al. 1982) to check the disease. Spray of calixin @0.1% a.i or Contaf @0.001% (Mohanan 2011) can also be done to manage the disease.

13.3 Conclusion

Effective tree health management requires a comprehensive approach to address the diverse challenges posed by insects-pest and diseases. The unique climatic and environmental conditions of Rajasthan create a complex landscape for tree health, necessitating targeted strategies for different species. Implementing integrated pest management (IPM) practices is crucial. This includes regular monitoring, manual removal of insect-pests, the strategic application of insecticides like organophosphates and physical barriers such as sticky bands. Understanding the life cycles and behaviours of pests enables more precise and effective interventions. Disease management should focus on early detection and prevention. Employing proper sanitation, pruning infected branches, and using disease-resistant tree varieties can help mitigate the spread of pathogens. Additionally, maintaining soil health and avoiding overwatering can reduce susceptibility to diseases. For the protection of important tree species in Rajasthan, a combination of preventive measures and responsive actions is essential. Regular inspections, timely treatments, and cultural practices tailored to specific tree species enhance resilience against both biotic and abiotic stressors. By adopting a holistic and adaptive approach to tree health management, stakeholders can improve the vitality of tree species, ensuring the sustainability of forest ecosystems, and supporting the broader environmental and economic well-being of the region. Continued research and collaboration will be vital in refining these strategies and addressing emerging challenges.

13.4 Way Forward

To advance tree health management in Rajasthan, a strategic and multi-faceted approach is essential. The following actions are recommended to enhance the management of insect- pests, diseases and protection strategies for important tree species:

- Establish comprehensive monitoring systems to track pest populations, disease outbreaks, and tree health indicators. Implementing data collection protocols will facilitate early detection and enable more accurate forecasting and response strategies.
- Develop and promote IPDM programs that combine biological, physical, chemical, and cultural management methods. Encourage using pest-resistant tree varieties, natural predators, biofertilizers and organic treatments to minimize chemical use and environmental impact.
- Conduct research to identify and characterize natural predators, parasitoids, and bioinoculants that can effectively manage pests and diseases affecting trees in Rajasthan. Implement pilot release programs to evaluate their effectiveness and adaptability in local conditions before broader deployment.
- Research should focus on developing resistant tree varieties, effective management measures, and innovative treatment methods tailored to the region's unique conditions.
- Educate people about the significance of tree health and the responsibility of a person to avert and manage tree pests and diseases. Engage local communities in conservation efforts and encourage them to report sightings of pests and symptoms of diseases.
- Provide training to forestry professionals, local communities, and landowners on best practices for tree health management. Workshops, field demonstrations, and educational materials can enhance knowledge and skills in pest and disease management.
- Continuous assessment of tree health management strategies and their adaptation based on emerging challenges and new information. Flexibility in management practices will help to address unforeseen issues and improve overall tree health outcomes.

References

AFRI Annual report (2007–2008) Arid Forest Research Institute, Jodhpur (Rajasthan)

Ahmed I, Nizamani, Rustamani MA, Khaskheli MI, Maree JM et al (2016) Management of Hairy Caterpillar, *Euproctis Fraterna* Moore of Jujube, *Ziziphus Mauritiana* Lam. Sci Int (Lahore) 28(1):416–422

Ambadkar CV, Dhawan AS, Shinde VN (2015) Integrated management of bacterial blight disease (oily spot) of pomegranate caused by *Xanthomonas axonopodis* pv. *punicae*. Int J Plant Sci 10(1):19–23

Amonkar DV, Karmakar SM (1978) Ion regulation in *Salvadora persica*. J Biol Sci 21:13–18

Anon (1988) Butea monosperma (Lam.) Taub. The wealth of India – raw materials, vol II: B, Rev edn. Council of Scientific and Industrial Research, New Delhi

Anuragi H, Ramanan SS, Arunachalam A, Rajarajan K (2023) Palash (*Butea monosperma*) [monograph]. ICAR-Central Agroforestry Research Institute, Jhansi, pp 1–128

Arya R, Lohara RR, Meena RL (2014) Survival and biomass production of *Salvadora Persica* on various types of salt affected soils under arid conditions in Rajasthan and Gujarat. Ann Arid Zone 53(1):43–48

Bajoriya DK, Mishra KK, Yadav VK (2023) Evaluation of newer fungicides for the management of wheat stem rust caused by (*Puccinia graminis* f. sp. *tritici*). J Cereal Res 15(2):277–283

Basu NG, Ghosal PK, Thakur S (1984) Structural studies on a polysaccharide fraction from the fruits of *Cordia dichofoma* Forst. Carbohydr Res 131:149–155

Baumer M (1983) Notes on trees and shrubs in arid and semi-arid regions. FAO, EMASAR Phase II, Rome

Beeson CFC (1941) The ecology and control of the forest insects of India and neighbouring countries. Manager Publication, Government of India, p 767. (Reprinted 1961)

Bhalla V, Walter H (1999) Research Bulletin of the Punjab University. Science 48:87–94

Bhat MN, Hedge RK (1990) Foliar disease of some forest nurseries and plantations in Karnataka-new records. Indian J For 13:349–350

Bhatnagar S, Goran P, Singh S (2014) Life cycle of small Salmon arab *Colotis amata* (Lepiodoptera: Pieridae). Bioscan 9(3):1063–1066

Bhatnagar S, Singh S, Khan AU, Sharma N, Kumar B (2023) A comprehensive approach to manage *Eriophyes prosopidis* mite induced flower galls in *Prosopis cineraria* (L.) Druce. Int J Environ Climate Change 13(11):1496–1507

Bolton MP (1989) Biology and ecology of Prickly *Acacia*. Unpublished report. Richmond Prickly *Acacia* Field Day, Brisbane Australia: Queensland Department of Lands, pp 21–25

Braun U, Nakashima C, Bakhshi M, Zare R, Shin HD, Alves RF, Sposito MB (2020) Taxonomy and phylogeny of cercosporoid ascomycetes on Diospyros spp. with special emphasis on *Pseudocercospora* spp. Fungal Syst Evol 6:95–127. https://doi.org/10.3114/fuse.2020.06.06

Brunck F (1994) Pests and diseases of *Acacia nilotica* (Les ravageurs et maladies de *l'Acacia nilotica*.). Bois Forêts Tropiques 239:59–65

Buss EA, Dale A (2016) Managing scale insects on ornamental plants. ENY-323 (MG005), series of the Department of Entomology and Nematology, UF/IFAS Extension, Gainesville, pp. 1-6

Christopher MSM, Mathavan S (1993) Role of a hymenopteran parasite and a bacterium in the regulation of a lepidopteran pest in tropical ecosystem. Int J Trop Insect Sci 14:477–482

Dadwal VS, Jamaluddin, Soni KK (1986) Note on *Nomuraea rileyi* (farlow) Samson, an Entomogenous Fungus on Tendu Defoliators, *Hypocala rostrata*. Indian Forester 112(1):88–89

Dalal KC (1989) Germplasm collection and therapeutic, botanical, agriculture, chemical and clinical aspects of the endangered indigenous plant guggal (*Commiphora wightii* (Arn.) Bhand). A status cum review report, 8th workshop at Faizabad, December 1989

Deb D, Khan A, Dey N (2020) *Phoma* diseases: epidemiology and control. Plant Pathol 00:1–15. https://doi.org/10.1111/ppa.13221

Dhileepan K (2015) New biocontrol opportunities for prickly acacia: exploration in India. Final report. Published by Meat & Livestock Australia Limited PO BOX 1961 North Sydney NSW 2059

DMAPR Annual Report (2012–13) Maiti S (ed) Annual report, Directorate of Medicinal and Aromatic Plants Research, Boriavi, Anand-387–310, Gujarat, India

Garg VK, Patel Y, Jaga PK, Sharma SK, Bansal GR (2017) Status of white fly (*Bemisia tabaci* Gennadius) in Vindya Plateau. Int J Agric Environ Biotechnol 10(5):581–587

Gent DH, Schwartz HF (2005) Management of *Xanthomonas* leaf blight of onion with a plant activator, biological control agents, and copper bactericides. Plant Dis 89(6):631–639

Haldhar SM, Bhargava R, Singh RS, Krishna H, Sharma SK (2015) First report of *Colotis amata* (Lepidoptera: Pieridae) on *Salvadora persica* (Capparales: Salvadoraceae) in Rajasthan, India: incidence and morphometric analysis. Fla Entomol 98(2):442–445

https://infonet-biovision.org/trees/toothbrush-tree, retrieved on 11.11.24

https://www.invasive.org, retrieved on 11.11.2024

https://ctri.icar.gov.in/for_controlPests.php, retrieved on 12.11.24

https://apps.worldagroforestry.org/treedb/AFTPDFS/Salvadora_persica, retrieved on 12.11.24

Hussain N, Kakoti BB (2013) Review on ethnobotany and psychopharmacology of *Cordia dichotoma*. J Drug Deliv Ther 3(1):110–113

ICFRE (2020) A user manual on Forest insect pests and diseases. Indian Council of Forestry Research and Education, Dehradun

Jain DK, Swami H, Kumar H, Singh B, Chouhan BS, Chhangani G (2022) Management of whitefly *Bemisia tabaci* (Genn.) and yellow mosaic virus disease infesting Blackgram, *Vigna mungo* (Linn). Legume Res Int J 1:1–6

Joshi KC, Jamaluddin (2007) Handbook of diseases, insect pests and their control measures in forest nurseries/plantations. Madhya Pradesh, Tropical Forest Research Institute, Jabalpur, p 58

Kamalalakshmi C (1996) Studies on leaf blight disease of Jathimalli *Jasminum grandiflorum* L. incited by *A. alternata* (Fr.) Keissler.agg. MSc thesis, Tamil Nadu Agricultural University, Coimbatore, India

Kant U, Arya HC (1971) Anatomy of the leaf gall on *Salvadora persica*-D induced by *Eriophyes*. Marcellia 37(1–2):47–57

Kant U, Sharma SL (1975) Studies on *Cordia Dichotoima* leaf galls induced by *Eriophyes cordiae*. In: All India symposium on form, structure and function in plants, p 31

Karthikeyan M, Jayakumar V, Radhika R, Bhaskaran R, Velazhahan DA (2005) Induction of resistance in host against the infection of leaf blight pathogen (*Alternaria palandui*) in onion (*Allium cepa* var *aggregatum*). Indian J Biochem Biophys 42:371–377

Kaur K, Michael H, Arora S, Harkonen P, Kumar S (2005) In vitro bioactivity-guided fractionation and characterization of polyphenolic inhibitory fractions from *Acacia nilotica* (L.) Willd. ex Del. J Ethnopharmacol 99:353–630

Khan MR (2020) Nematode infestation, a potential threat to Indian forests. Indian Phytopathol 73:397–414

Kitagawa H, Glucina PG (1984) Persimmon cultivation in New Zealand. Science Information Publishing Centre, Wellington, pp 1–74

Kueh TK (1986) Cocoa diseases. Annual report of the research branch, Department of Agriculture, Sarawak, for the year 1984. Ministry Agriculture Community Development, Kuching, pp 83–85

Kumar S (1991) Insect pests of tendu and their management. Indian J MFP 1:92–98

Kumar V (2018) *Alternaria* diseases of litchi. NRCL TB-13, ICAR-National Research Centre on Litchi, Muzaffarpur, p 28

Kumar S, Ahmed SI (1996) *Salvadora persica* – a new host record for *Catopsilia crocale* Cram (Lepidoptera: Pieridae). Indian For 122(1):83–84

Kumar N, Khurana SMP (2016) *Aecidium* sp. causing serious leaf rust on *Cordia dichotoma* G. Forst in Haryana-First Record. Int J Curr Microbiol App Sci 5(8):55–58

Kumar P, Singh S (2017) *In Vitro* evaluation of fungicides and plant extracts against *Alternaria solani* (Ellis) causing early blight in tomato (*Lycopersicon esculentum* Mill.). Int J Curr Microbiol App Sci 6(9):820–827

Kumar S, Sharma S, Pathak DV, Beniwal J (2011) Integrated management of *Jatropha* root rot caused by *Rhizoctonia bataticola*. J Trop For Sci 23(1):35–41

Kumar S, Singh V, Lakhanpaul S (2012) Detection and characterization of a phytoplasma associated with witches'-broom disease of *Salvadora persica* in India. J Gen Plant Pathol 78(4):294–297

Kumar S, Lal N, Nath V (2017) Tendu (*Diospyros melanoxylon* (roxb)). In: Ghosh SN, Singh A, Thakur A (eds) Underutilized fruit crops: importance and cultivation. Part II. Jaya Publishing House, New Delhi, pp 1327–1338

Kurian A, Sankar MA (2007) In: Peter (ed) Medicinal Plant: Vol. 2, Horticulture series II. New India Publishing Agency, New Delhi, p 113

Luna RK, Kaushal P, Banyal R (2006) Study on mortality of Kikar (*Acacia nilotica*) in Punjab. Indian Forester 132:281–296

Majumdar VL, Agnihotri JP, Sharma ML (1988) Mycoflora of gum guggulu *Commiphora wightii* affected by die back. Bull Med Ethn Res 9:136–138

Makwana MT, Patolia JS, Iyengar ERR (1988) *Salvadora* plant species suitable for saline coastal wasteland. Trans Indian Soc Desert Technol 2:121–131

Mani MS (1959) Cecidotheca Indica. Agra University l. Res (Science) 891

Mani MS (1973) Plant galls of India. Macmillan, pp 1–354

Mani M, Kulkarni NS, Banerjee K, Adsule PG (2008) Management of caterpillar on grapevine. Extension folder no. 4. National Research Centre for grapes

Mathur RN, Singh B (1959) A list of insect pests of forest plants in India and the adjacent countries arranged alphabetically according to the plant genera and species, for the use of forest officers Part 5-List of insect pests of plant genera D to F. Indian Forest Bulletin Delhi (NS) Ent 171(4):1–165

Mazumdar PM, Das MK, Das S (2011) *Butea monosperma* (LAM). Kumtze – a comprehensive review. Int J Pharma Sci Nanotechnol 4(2):1390–1393

Mehta N, Jain A, Rajkumar M (2020) Impact of pruning of *Diospyros melanoxylon* Roxb. (Tendu) bushes on yield and quality of leaves in Maharashtra. J Pharm Phytochem 9(1):1360–1365

Mengi SA, Deshpande SG (1999) Anti-inflammatory activity of *Butea frondosa* leaves. Fitoterapia 70(5):521–522

Moawia EM, Ensaf SI, Sharma RM (2012) First record of *Resseliella salvadorae* (Rao) (Diptera: Cecidomyiidae) and its parasitoid from stem and leaf galls of *Salvadora persica* L. Sudan J Threatened Taxa 4(13):3215–3217

Mohanan C (2011) Pink disease in teak plantations and its management. Final report of project No. KFRI 579/09, KFRI research report No. 386, KFRI, Peechi pp19

Molina-Vargas LF (2013) Mechanism of action of isothiocyanates. A review. Agronomía Colombiana 31:68–75

Nadkarni KM (2002) Indian Materia Medica, vol 1. Popular Prakashan, Bombay, pp 223–225

Natesh S, Mohan Ram HY (1999) An update of green medicine. J Indian Bot Soc 78:13–23

Nizamani IA, Rustamani MA, Khaskheli MI, Maree JM, Rajput LB, Nizaman SA (2016) Management of hairy caterpillar, *Euproctis fraterna* Moore of Jujube, *Ziziphus mauritiana* Lam. Sci Int (Lahore) 28(1):416–422

Orwa C, Mutua A, Kindt R, Jamnadass R, Anthony S (2009) Agroforestree database: a tree reference and selection guide version 4.0. http://www.worldagroforestry.org/sites/treedbs/treedatabases.asp. Accessed 13 July 2024

Parida S (2020) Seasonal incidence of major diseases of 'Kendu' Plants (*Diospyros melanoxylon* Roxb.) and their management. Int J Bot Stud 5(3):165–169

Pathak H, Silawat SC (2015) Viral diseases of four tree species in forest nurseries of Indore, India. J Bombay Nat Hist Soc 112(3):191–193

Pathak H, Maru S, Satya HN, Silawat SC (2015) Fungal diseases of trees in forest nurseries of Indore, India. J Plant Pathol Microbiol 6:297

Raj A, Haokip V, Chandrawanshi S (2016) *Acacia nilotica*: a multipurpose tree and source of Indian gum Arabic. South Indian J Biol Sci 1(2):66–69

Rakholiya KB (2010) Pseudomonas – the causal agent of gummosis of guggal. Int J Plant Prot 3(2):400–401

Ram A, Bastos STG, de Figueiredo JM (1982) Chemical control of *Corticium salmonicolor* (Berk.) & Br., causal agent of pink disease of cocoa. Revista Theobroma 12(4):241–247

Ramakrishnan TS, Sundaram NV (1954) Notes on some fungi from South India-IV. Indian Phytopathol 7:140–151

Reddy CS, Meena SL, Krishna P, Charan PD, Sharma KC (2012) Conservation threat assessment of *Commiphora wightii* (Arn.) Bhandari. An economically important species. Taiwania 57(3):288–293

Rohini N, Pandiarana N (2013) Package of practices for cultivation of Guggal (*Commiphora wightii*). https://www.krishisewa.com/crop-production/149-guggul.html

Roychoudhury N, Mishra RK (2020) Babul borer, *Celosterna scabrator* and its control measures. Van Sangyan 7(12):13–16

Samanta J, Mandal K (2013) In planta detection of *Xanthomonas axonopodis* pv. *commiphorae* using fyuA and rpoD genes. Indian J Exp Biol 15(6):470–476

Satyavir, Verma SK (1996) Insect pest of agroforestry leguminous trees in India. Ann Arid Zone 35(4):349–359

Senthilkumar N, Murugesan RS (2015) Insect pests of important trees species in South India and their management information. Institute of Forest Genetics and Tree Breeding, Tamilnadu, p 133

Shahbaz E, Ali M, Shafiq M, Atiq M, Hussain M, Balal RM, Sarkhosh A, Alferez F, Sadiq S, Shahid MA (2022) Citrus canker pathogen, its mechanism of infection, eradication, and impacts. Plants (Basel) 12(1):123

Sharma ML, Gour HN (1987) A new leaf-spot of *Commiphora wightii* a medicinal plant, caused by *Phoma* sp. Curr Sci 56(11):538–539

Singal SK, Toky OP (1990) Carryover of bruchid, *Caryedon serratus* (Olivier) (Coleoptera) from field to stores through seeds of *Acacia nilotica* (L.) Willd. in India. Trop Pest Manag 36:66–67

Singh B, Jamaluddin (1995) *Acacia nilotica*: new host of *Curvularia*. Indian For 121(2):161

Singh J, Monobrullah M, Jaiswal AK (2013) New record of insect pests attacking Butea monosperma – a commercial host for culturing lac insect, *Kerria lacca* (Kerr). Indian J Entomol 75(3):225–231

Singh A, Bhatnagar S, Singh S, Rathore LS (2017) Foliar disease infection on some forest trees. J Agric Ecol 3:38–41

Srivastava KK, Kalyani KB (1990) Charcoal root rot in *Acacia nilotica* nursery- a new disease from India. Myforest 26:163–164

Supraja P, Tiwari S (2022) Efficacy of some selected essential oils against Alternaria leaf spot of brinjal (*Solanum melongena L.*). Pharma Innov J 11(9):2580–2586

Tanwar RK, Jeyakumar P, Monga D (2007) Mealy bugs and their management, Technical bulletin. National Centre for Integrated Pest Management, New Delhi

Thankamma L, Nair JM, Deo AD, Kumar D (1986) Prophylactic brush on application of Emulsicop for pink disease management in Eucalyptus. Indian For 112(11):1016–1018

Tiwari DN, Kumar K, Tripathi A (2001) *Guggal Utthan Center* for sustainable development and poverty alleviation, Allahabad, India, p 117

Trivedi P, Trivedi C, Grinyer J, Anderson IC, Singh BK. (2016) Harnessing host-vector microbiome for sustainable plant disease. Front Plant Sci 7:1423

Vaibhav DL, Thamilarasi K, Ghosh J, Mohanasundaram A, Sharma KK (2016) Monograph on Palas, *Butea monosperma* (Lam.) Taburt, IINRG, Ranchi, India

Verma RK, Jamaluddin, Soni KK (1997) Die back of leguminous trees in young plantations caused by *Haplosporella* in Central India. Indian For 123:652–655

Vinodkumar (2012) Brinjal diseases and their control. www.agropedia.iitk.ac.in/content/brinjal-diseases-and-their-control. Retrieved on 23.7.24

Vohora SB (1986) Unani joshandah drugs for common cold, catarrh, cough and associated fevers. J Ethnopharmacol 16:201–211

Insect Torments of Timberland: Monitoring to Microbial Control and Beyond

14

A. A. Safeena Majeed, Kishore Pujar, Priyanka, and V. Rathna

Abstract

Insects are an essential part of the forest ecosystem, serving various critical functions. Annually, around 68 million ha area of forests around the globe suffer insect damage, which is more than double the area destroyed by wildfires. Some groups of foliage, root and bark feeders are considered major forest pests, apart from termites and wood-boring insects, causing irreversible damage to moist timber. Additionally, global climate change could increase the risk of new pest invasion pathways and outbreaks by vertical and horizontal tree migration, host expansion, mismatch of time and food web etc. Forest pest's management is a complex process requiring a solid framework such as integrated pest management (IPM), with minimal adverse impacts on other resources while maintaining the forest's health and vitality. Management options for these non-native forest pests include classical biocontrol, microbial insecticides, inoculative biocontrol, silvicultural management to promote tree health, resistance tree breeding, push–pull strategy, RNA interference, gene silencing, etc. The present chapter reviews the insect nuisances of timberland and their monitoring with the aim of forest pest management using ecologically sound, economically viable, socially and environmentally acceptable strategies.

A. A. Safeena Majeed (✉)
Indian Institute of Science, Bangalore, Karnataka, India

K. Pujar · Priyanka
Department of Agricultural Entomology, University of Agricultural Sciences, GKVK, Bangalore, Karnataka, India

V. Rathna
ICFRE – Institute of Forest Productivity, Ranchi, Jharkhand, India

R. K. Kalia, R. Pathak (eds.), *Tree Biology and Biotechnology*,
https://doi.org/10.1007/978-981-96-0002-1_14

Keywords

Microbe · Microbial control · Defoliators · Wood feeder · Forest entomology · Tree breeding · Pesticides

14.1 Introduction

Insects play important roles in the forest ecosystem, performing vital functions such as pollination, nutrient cycling, and eliminating sick and weak trees. They are also an important component of the food chain. Many forest insects are predatory and feed on forest insect pests, regulating the pest population under check. However, around 68 million ha of forest, twice the area affected by wildfires, is prone to insect attacks annually (FAO 2005). Some groups of foliage, root and bark feeders are considered major forest pests, apart from termites and wood-boring insects, causing irreversible damage to timber. Among the pests of forest crops, the xylophagous (wood feeding) termites (Rhinotermitidae) cause major havoc, especially in Eucalyptus plantations, causing 80% of seedling deaths. The white grub, *Holotrichia* sp., is an occasional major pest, causing wilting and lodging of trees. Another major defoliator, *Hyblaea puera*, and trunk borer, *Alcterogystia cadambae,* account for a significant damage to Eucalyptus and teak, followed by the sapling borer, *Sahyadrassus malabaricus*. In recent years, invasive whiteflies (*Aleurodicus disperses* and *A. rugioperculatus*) have damaged teak plantations. Webber *Atteva fabriciella* in Malabar Tree of Heaven (*Ailanthus triphysa*), the terminal shoot borer *Hypsipyla robusta* in mahogany and the bagworms *Pteroma plagiophleps* in Albizia and *Ptero maplagiophleps* are gaining the pest status. Additionally, global climate change could increase the risk of new pest invasion pathways and outbreaks by vertical (elevational) and horizontal (latitudinal) tree migration, host expansion, mismatch of time and food web, etc. For example, high global temperature favouring Dendroctonus ponderosae Hopkins' overwintering survival, resulting in tree mortality.

14.2 Forest Pests

These pests can be broadly categorized as follows based on their nature of damage:

14.2.1 Wood Feeders

Wood borers are a serious threat to timber in urban and forest areas. Among the pests of forest crops, the following are the major wood feeding:

14.2.1.1 Termites

They belong to the family Rhinotermitidae. The insect initially feeds on the fallen woods, builds a colony, and later migrates to damage standing trees. Their internal feeding causes hollow stems and eventually results in the lodging of the damaged tree. Termites are a major menace in Eucalyptus plantation, especially during the first year of transplanting, causing around 80% of seedling deaths in *E. tereticornis* and *E. grandis* (Nair and Varma 1981).

14.2.1.2 Wood Beetles

A majority of the tree borers belong to the order Coleoptera, commonly known as wood beetles such as death watch beetle, common furniture beetle, false powder post beetles, powder post beetles, old house borer, anoplophera longhorn beetles, wharf borer etc. The beetles lay eggs on the leaves and/or wood, and grub emerges inside the stem region to form tunnels through feeding. In certain instances, the markings of tunnels can be seen or the frass present at the entry hole is an indication of the pest introduction.

14.2.1.3 White Grub

Holotrichia sp. lays eggs in the soil or organic matter. Emerged grubs start scraping the roots and later entirely feeds on roots, resulting in the trees' wilting and lodging. It is considered an occasional major pest of Eucalyptus (*E. grandis*) and teak plantations.

14.2.2 Defoliators

The major defoliator of the forest ecosystem is the teak defoliator, *Hyblaea puera* and trunk borer, *Alcterogystia cadambae.* These two are a major pest of Eucalyptus and teak. The sapling borer, *Sahyadrassus malabaricus,* is of minor significance as pest; however, their serious infestation is reported in plantations with luxurious weed growth, sometimes even causing stem breakage of the saplings with 20% infestation and is rated as a minor pest.

The webber *Atteva fabriciella* is the major pest of the Malabar tree of heaven, *Ailanthus triphysa,* which webs on the leaves and lives inside the web. The larvae scrape from the inside and cause damage to the crop. The terminal shoot borer *Hypsipyla robusta,* a major pest of mahogany, *Suietenia macrophylla,* feeds on the emerging shoots and causes an early break to the plant growth. Bagworm, *Pteroma plagiophleps* is a major pest of *Albizia, Paraserianthes falcataria,* whose larvae majorly feed on the leaves.

14.3 Insect Pests of Timber

Apart from termites, timber is prone to attack by numerous wood borers, some causing extensive damage right from larval to adult stages. Though their damage is enormous and irreversible, it often goes unnoticed on timber and furniture in homes till the very last stage. Proper curing and treatment of these timbers can reduce the attacks. Comparatively, moist timber attracts more infestation than dry timber.

Some wood-boring insects support the decomposition of dead trees, thus keeping the forest ecosystem clean.

14.4 Forest Entomology and Pest Management

Forest entomology is a branch of insect science that deals with the study of forest insects which damage trees and timber and their effective management tactics. Pest management involves two basic strategies, viz., 'prevention' and 'suppression' to reduce the forest insect pests. Early detection of pest infestations, practices to promote predatory populations, and proper forest management are pivotal in minimizing forest insect pests.

The efforts intended to reduce future infestations by manipulating stand, forest and landscape conditions is known as 'prevention'; those to reduce current infestations by remedial measures are known as 'suppression'. Though only insecticides can control the pest, they reappear when the effect fades and can develop resistance to these chemicals. Hence, an integrated technique that utilises all the available pest management strategies as economically as possible with the least negative impacts on the environment is known as Integrated Pest Management (IPM).

14.5 Factors Affecting Pest Population Dynamics

Details on the life history of pests are needed to develop accurate surveys and management strategies that have the least interference with natural pest mortality factors. The major factors affecting pest population dynamics are given below:

14.5.1 Climate

Temperature influences the geographical distribution, dispersal, developmental rate, reproduction, voltinism, overwintering and survival. High temperatures as a result of climate change can reduce the generation time and improve the overwintering survival of pests (e.g., Mountain pine beetle, *Dendroctonus ponderosae*) resulting to tree mortality (Amman et al. 1990). Climate change can induce pest invasions and outbreaks by tree migration, time lags, host expansion, food web mismatch etc. All insects have a certain temperature, rainfall and other climatic threshold, beyond which they cannot survive. These limits are prerequisite in the development of models to predict the impact of pests (Powell and Bentz 2014), their expansion (Lesk et al. 2017) and invasion to new areas.

14.5.2 Natural Enemies

Natural enemies like predators, parasitoids and pathogens exert a top-down effect on herbivore populations and are the major natural mortality factors (Price et al. 2011). Understanding their dynamics and effects on pest populations is valuable in the classic of biological control (Bauer et al. 2015) and in formulating IPM programs.

14.5.3 Host-Pest Interaction and Host Quality

The availability of host trees exerts a considerable bottom-up effect, thus forming a base for resistance breeding and IPM. Although trees have developed vigorous growth or regrowth mechanisms to tolerate the infestations, seldom this makes the host more susceptible to herbivores, where other defence mechanisms come into the picture. For example, the fast-growing *Eucalyptus* tolerate defoliation due to their fast growth, while the slow-growing *Eucalyptus* species use physical and phytochemical properties of leaves. However, both biotic and abiotic stresses condition the trees, reduce the vigour and host defense mechanism, and make them more susceptible.

14.5.4 Dispersal/Movement

The inward (immigration) and outward (emigration) movement of insect pests resulting in overcrowding and depopulation are important in determining the distribution ranges of forest pests. Although wind-assisted long-range dispersal flights of beetles are seldom seen, short-range dispersal is often in forest ecosystems. Unintentional anthropogenic activities like the movement of infested wood materials (Jacobi et al. 2012), packaging and shipping containers also spread pests (e.g., gypsy moth and emerald ash borer) outside their native habitats. However, regulatory and phytosanitary actions have considerably reduced the risk (Haack et al. 2014).

Understanding the factors affecting pest dispersal leads to accurate prediction of the rate and direction of spread, aiding effective IPM strategies. All these require accurate identification and monitoring of the pest population to know when and how to initiate the management tactics.

14.6 Integrated Pest Management (IPM)

It is the maintenance of pest population below tolerable levels by utilizing all the available tactics in an ecologically sound, economically efficient and socially acceptable manner by minimizing risks to humans, non-target beneficial organisms, and the environment. IPM is a system that, in the context of the associated environment and population dynamics of the pest species, utilizes all suitable techniques and methods in a compatible manner as possible and maintains pest populations at levels below those causing economic injury. The concept of pest management programmes in forest ecosystems recently has rationalized from targeting a single pest to considering IPM as an integral part of sustainable forest management, including ecological processes like carbon sequestration, and soil and water quality conservation (Holvoet and Muys 2004). Hence, the key aim of IPM is to reduce the ill effects of pest with minimum negative effects on ecosystem services and function along with its major component is understanding the interaction between forest and pest, its biology, ecology and the factors affecting population dynamics.

14.6.1 Components of IPM

14.6.1.1 Monitoring

Monitoring is an inventory control that keeps a record of the presence, population density and impacts of insect pests and its natural enemies, thus helping in prioritizing the management practices based on the tree damage estimated from survey data. Monitoring generates qualitative and quantitative data on pests to decide when and how best management actions can be imposed. It is quite challenging in forests compared to agricultural systems due to difficulties in sampling visibility of damage and the cryptic nature of tricky insects (Ric et al. 2007). Monitoring can be done manually by directly counting the number of insects or species per unit plant sample or indirectly by quantifying the percent damage (defoliation, deformation or mortality). It can also be done with the help of traps (light, sticky, malaise, pheromone-baited and food lure traps), and even artificially stressed trees are used as traps.

Surveys to monitor and damage prediction are of two types: ground survey and aerial survey. The ground survey is done under small stands and in easily accessible areas where infestation symptoms are clear. It is a highly effective survey, and the symptoms are very strongly correlated with insect species. It provides additional information on the pest population and mortality but is arduous and time-consuming. The aerial survey was initially started to track tree damage and mortality, but gradually, it was utilized for impact assessment of insects on forest resources, monitoring pests and diseases, and the presence of new pests in the new areas (Ciesla 2000; Carnegie et al. 2008). The estimation of damage levels provides proper guidance for further survey or management (Johnson and Wittwer 2008). The aerial surveys are often followed by ground surveys to study damages more accurately.

Satellite technologies and laser-based landscape mapping (e.g. LiDAR) have been used to identify tree species, monitor forest insects, map tree distributions (Somers and Asner 2012) and stressful areas, map insect damage (Stone and Mohammed 2017), and assess defoliation, tree mortality, and foliar discolouration. Yet, there are a few downsides of remote sensing, such as the collection of incomplete information and time lag in processing the output into a format that is rapidly refining. Integration of multiple survey and monitoring tools also provides more reliable information for IPM programmes.

14.6.1.2 Models in IPM

Models elucidate the factors and processes affecting pest population, their dynamics and density-damage relations to make management predictions and formulate future research by understanding the gaps. Models improve the efficacy of sampling methods and predict pest dynamics under changing climates, as well as their distribution, dispersal pattern, and rate of spread, to elucidate the risk of invasion. Though field experiments generate good data on population dynamics, they are carried out on small spatial and temporal scales and often consider a single factor or a few factors, not the mechanisms of outbreak. Hence, Hunter and Dwyer (1997) developed a mathematical model combining field experiments to understand the dynamics of the western tent caterpillar *Malacosoma californicum*, larch budmoth *Zeiraphera diniana* and gypsy moth *Lymantria dispar* to study the host-pathogen interaction. Temperature-based phenology models forecast pest life history, while statistical regression models enumerate the relationship between pest density and damage to determine thresholds (Fry et al. 2008). Results of pest modelling offer integration of stand dynamics and forest inventory to develop management decisions.

14.6.1.3 Regulatory/Quarantine Controls

This is needed to prevent the introduction of exotic species and to restrict the spread of already established species to a new environment. It gives information on the regulated species and materials, the scope of quarantine and penalties for defiance. These are conversant by pest risk analysis, by computing the risks of introduction based on life history, host distributions, invasion pathways of the pest etc., as per the World Trade Organization agreement on sanitary and phytosanitary measures (SPS) (Yemshanov et al. 2009).

Prohibitions of host plants of high-risk pests are a simple and effective approach for preventing invasive species, allowing the importation only of low-risk plant genera. Many countries have banned the importation of live plants, allowing only planting material, while others have banned plants with soil. However, the SPS agreement circumvent countries using false prohibitions and phytosanitary treatment, while the pre-border biosecurity allows safe importation of high-risk commodities. The International Standards for Phytosanitary Measures No. 15 is an international trade standard established in 2002 for wood packing materials that requires heat treatment or fumigation with compliance (Haack et al. 2014).

14.6.1.4 Cultural Control

Cultural control may include several cultural practices, viz., sanitation, salvage, silvicultural control etc. The sanitation is a cultural operation involving the pruning of affected tree branches and the removal or treatment of infested trees to destroy pests within. Fallen trees are burned, debarked, chipped, solarized, or transported to mills to kill the broods by processing and milling lumber. The salvage involves harvesting and processing dead trees to recover at least a little economic worth and address the safety concerns of dead trees for forest visitors and workers. Though salvage is not a pest-suppressive approach with immediate results, it can limit future pest outbreaks and silvicultural control is the backbone of IPM in forestry, which starts with the selection of pest-free planting material based on site requirements and other cultural operations targeting the overall improvement of tree health by enhancing the innate host defence mechanisms. Apart from increasing natural enemies, mixed planting of different species reduces the pest population by impeding host searches by pests (Thompson et al. 2009). Maintaining optimum stand density by thinning to manage the stand structure and tree species composition ultimately influences the susceptibility against bark beetles (Fettig et al. 2007). In addition, proper log residue management and pre-emptive removal of hosts (Vannatta et al. 2012) are done to reduce the risks of future infestations. Lopping can be practiced against teak trunk borer, *Alcterogystia cadambae* and regular thinning and removal of overmature trees coupled with trapping is effective in preventing the outbreak of Sal borer, *Hoplocerambyx spinicornis*. Other methods include the withdrawal of susceptible tree cultivation and replacement with alternative tree species in high-risk endemic areas of pests, followed by mahogany and *Gmelina arborea* with *Ailanthus* spp. (Nair 2007).

14.6.1.5 Tree Breeding

The use of pest-resistant trees is a preventive and eco-friendly measure. Resistance is the genetic competence of trees to avoid, limit or withstand pest infestation. Tree resistance to insect pests is often polygenic and could also be due to favourable physical factors of the trees like hard stem characteristics, resin system, surface waxes or hairiness. Conventional breeding for resistance is constrained by the long reproductive cycle of trees. Traditional tree breeding has focused more on disease resistance and polygenic traits, since only a few resistance traits in trees are monogenic. However, plant biotechnology has made tremendous progress in the recent past, making it possible to transfer desirable genes across species (Strauss et al. 1991). For example, the development of transgenic trees with chitinase genes, proteinase inhibitor genes, resistant Bt genes from bacterium, baculovirus genes etc. (Nair 2007). Many transgenic trees with noble genes of various traits have been developed with high mortality to lepidopteran pests, eucalypt leaf beetles (Frankenhuyzen and Beardmore 2004), Bt genes against insect defoliators and eucalypt resistant genotypes against *Leptocybe invasa*. RNA interference technologies such as gene silencing also hold promise for managing invasive forest insects. The perennial nature of the trees is a major drawback. Hence, the stable expression of genes should be evaluated due to the potential risk of gene drift into the environment as biosafety concerns (Vettori et al. 2016).

14.6.1.6 Semiochemicals

These are the chemicals used by insects for communication. These chemicals could be explored to trap them, disrupt their mating, attract and kill insects, and inhibit their colonization on tree stands. Semiochemicals are advantageous in being environmentally friendly and are target-specific. Multiple semiochemicals can be combined and applied together for better control. They can be used for mass trapping of pests, their mating disruption and anti-aggregation. Mass trapping involves the phenomenon in which aggregation pheromones are used by insects to attract members of the population, irrespective of sex, for protection, reproduction and feeding. Mass trapping has been tried to reduce populations of both bark and ambrosia beetles. For example, Ipsenol has been used to trap the bark beetles of *Ips* sp. Frontalin has been used for the management of *Dendroctonus* beetle (Baake 1991). The anti-aggregation pheromones are emitted when trees are colonized by insects. Trees baited with formulations of these pheromones can repel the attacks. For example, verbenone has been used against bark beetles, *Ips* and *Dendroctonus* spp. (Baake 1991).

The mating disruption involves the usage of synthetic analogues of natural sex pheromones used by insects for mate finding, thus confusing them to reduce mating. It has huge success in agricultural and orchard crops in the management of lepidopteran pests (Cardé and Minks 1995) and least in other insect orders with limited results are available for pinhole borers, cerambycids and sawflies (Martini et al. 2002; Funes et al. 2011; Sweeney et al. 2017). For example, aerial application of synthetic pheromone *disparlure* suppressed the gypsy moth populations (Tobin and Blackburn 2007) and application of sex pheromone reduced the mating in spruce budworm (Rhainds et al. 2012).

The pheromone baited traps or stressed trees as traps for mass trapping is another usage of semiochemicals for population reduction to lure and kill a larger population of insects (e.g. bark beetles). However, mass trapping can also lead to attacks on adjacent healthy trees due to eliciting strong responses and can trap beneficial insects. Hence, timely replacement of old traps, trap modifications that suit local situations, and watchful placement of traps in areas where tree loss is acceptable (Ross and Daterman 1997, 1998) should be practiced.

Push–pull technology is another strategy that manipulates insects to colonize one area while ignoring the other, such as reducing the main crop's pest load by attracting them to other plants. It uses both attractants and repellents simultaneously. For this method to be effective, the target pest should respond strongly to attractants and repellents. Successful push–pull results are available for bark beetles (Seybold et al. 2018), but due to the spillover attacks, only the push (repellent) system has gained more popularity than the pull (attractant) system.

14.6.1.7 Biological Control

It is the reduction of pest population through the activity of biocontrol agents like parasitoids, predators and pathogens. There are four different strategies in biocontrol: classical, inoculation, inundation, and conservation. Classical biological control is the importation and release of coevolved exotic biological control agents for permanent establishment and pest management. Reportedly around 6158 species of predators and parasitoids were introduced worldwide for managing 588 forest pests, of which 172 pest species were successful (Kenis et al. 2017). For example, the usage of egg parasitoid *Anaphes nitens* against eucalypt weevils, *Gonipterus* spp. has been reported to be successful (Kenis et al. 2017).

Inoculation is the intentional release of biocontrol agents infrequently, as once a year, allowing natural multiplication and establishment so that future progenies will control the pests. Here, control is expected from the progenies, not from the release itself. In the process of inundation, biocontrol agents are used in large quantities for pest control immediately after release. This is also called biotic insecticides because, the control is achieved by the released organisms themselves, not on their progeny. Utilization of *Bacillus thuringiensis* subsp. *kurstaki* (Btk) and egg parasitoid, *Trichogramma minutum* against the spruce budworm, *Choristoneura fumiferana* is its best example. In contrast, the conservation process includes activities that protect or enhance populations of natural enemies by environmental manipulation. It involves several activities like providing alternate food sources (nectar and pollen producing plants) and hosts to the natural enemies, nesting and overwintering sites for birds, limited or targeted pesticide application, skip row treatment, maintaining non-treated buffer zone, use of selective pesticides etc. for environmental manipulation.

14.6.1.8 Biopesticides and Microbial Control

The application of disease-causing agents like fungi, viruses, bacteria, and nematodes directly to insect pests for their control is known as microbial control. The products derived from these microbes are called biopesticides.

The *Bacillus thuringiensis* (Bt) is a soil bacterium that kills insects by producing toxic crystal proteins called delta-endotoxin. It is an ideal and safe management tactic because of its host specificity by target binding and requirement of alkaline gut pH. *Bt* is more of a biopesticide than microbial control, where a huge number of commercial formulations are available. *Bt israelensis* (Bti) is used against mosquitoes, blackflies and fungus gnats. *Bt san diego/tenebrionis* has been utilized against tenebrionid beetles while *Bt. kurstaki* (Btk) has been used against tent caterpillars and spruce budworms. *Bt kurstaki* is the most widely used microbial insecticide for control of forest defoliators in Asia, Europe and North America (Cranshaw 2009). Similarly, baculoviruses are the major group of entomopathogenic viruses that replicate in the nucleus of insect cells after ingestion and causes quick mortality. They are promising, due to their host specificity, quick dispersal in the population and persistence in the environment. It has three major groups: NPV (Nucleopolyhedrosis viruses), where virons are encapsulated in a polyhedral protein matrix; GV (Granulosis viruses), with individually encapsulated virons; and CPV (Cytoplasmic viruses). Larvae of forest defoliators and sawflies are highly susceptible to baculovirus and can be applied either by ground or aerial application. The aerial application LdM-NPV resulted in higher larval mortality against the Cork oak gypsy moth, *Lymantria dispar,* at ultra-low volumes (Ruiu et al. 2021).

Fungi infect insects through the integument, penetrate inside, proliferate, spread, and drain the nutrients; subsequently, insects get killed. Most insect orders and all developmental stages are subjected to attack. Weather plays an important role in determining the effectiveness, because fungi need moisture and high humidity to germinate. *Beauvaria bassiana*, the white muscadine fungus, is the most common fungi used as microbial control. Application of *Entomophaga maimaiga* has been reported for the management of gypsy moth, *Lymantria dispar* in Japan (Ciesla 2011). Similarly, nematodes enter into the insect body through the natural openings like the mouth, anus, and spiracles and soft integumental areas in between two scelrites. The nematode, *Deladenus siricidicola*, is an ovarian parasite of the wood wasp, *Sirex noctilio,* rendering them unable to reproduce (Zondag 1969).

14.6.1.9 Pesticides

Although pesticides are the last rescue due to social and environmental concerns, they are an essential part of IPM. It involves spraying or dusting insecticides either manually with ground-based equipment like sprayers, blowers, tree and soil injection devices etc., or aerially with helicopters. Pesticides in forestry include organic and inorganic insecticides, insect growth regulators (IGRs), soaps, horticultural oils etc. Insecticides must be used with high precision for economic and ecological benefits, which otherwise create problems like resistance, resurgence and residue, in addition to pollution hazards. Intensively pesticide application targets seed orchards, nurseries, recreation sites and short-rotation woody crops, and the application of insecticides on individual trees is to seek protection from bark beetles, defoliators and woodborers. Due to the residue and other negative impacts of synthetic insecticides, the focus has been shifted towards finding safer, portable and long-lasting alternatives. Stem injection is a technique that requires a small quantity of systemic insecticides that are translocated in the entire plant system, offering protection from insect pests attacking multiple parts.

Other alternatives like integration of pesticides, trapping, *Bt* etc., with synthetic formulations of entomopathogenic microorganisms like the white muscadine fungus, *Beauveria bassiana,* against bark beetles and wood borers. Trapping the beetles using semiochemicals, treating them with microbial agents and then releasing these individuals back into field populations to contaminate the pest population (Lyons et al. 2012) and applying various suspensions of spores to the surfaces of fallen and standing trees (Davis et al. 2018) are also some of the alternatives.

14.7 Cost-Economics of Pest Management

Quantifying the benefits of pest management in forest ecosystems is tough, unlike in agricultural systems. Though the monetary value of forest goods like forest trees, timber and fibre can be estimated and compared to management costs to determine the net returns, other ecological services of forest cannot be calculated. However, comparison of growth in treated and untreated stands and use of models can predict the value of the stand, but such analyses are rare, and there are long gaps between management actions and harvests that make the analyses difficult. Hence, cost-benefit analysis mainly determines the tactic with the lowest cost range among available tactics.

14.8 Case Studies on Invasive Pests and Their Microbial Control

Despite these strict regulations and quarantine, many pests hold the record for their continental spread and establishment, threatening the native biodiversity, and are called invasive pests. Some of those discussed include their economic significance, invasion, distribution, nature of damage, extent of possible control, and even failures.

14.8.1 Cypress Aphid, *Cinara cupressi* Buckton (Hemiptera: Aphididae)

It is native to North America and is specifically associated with conifers belonging to the Cupressaceae and Pinaceae families. This aphid species complex is widespread globally (Watson et al. 1999). It poses a threat to *Widdringtonia* species, which is endangered in Africa and has been reported to damage the Chilean cedar, *Austrocedrus chilensis*, and Patagonian cypress, *Fitzroya cupressoides*. In both natural forests and plantations, the aphid feeds on small twigs and parts of the crown, often causing branch die-back. A secondary issue is the secretion of large quantities of honeydew, which encourages the growth of sooty mold (Eskiviski et al. 2005). This mold reduces photosynthesis, increases respiration, and produces chlorosis of foliage and stunted growth, especially in young trees (Penteado et al. 2000).

To mitigate the impact of *C. cupressi*, various governmental agencies have implemented pest management programs focused on biological control. An extensive survey conducted in southern Chile from 2007 to 2013 identified many strains of *Lecanicillium* and *Neozygites* species (Montalva et al. 2010). Research in multiple laboratories worldwide has shown high mortality rates of *C. cupressi* caused by *Lecanicillium attenuatum* (Loureiro et al. 2004; Montalva et al. 2017) and *Neozygites turbinata* and *Neozygites osornensis* are known to cause epizootics. Though they can be mass-produced as hyphal bodies and/or protoplasts, their field efficacy mostly depends upon the environment, especially humidity and temperature.

14.8.2 Eucalyptus Weevil, *Gonipterus platensis* Marelli (Coleoptera: Curculionidae)

Native to Australia, *Gonipterus platensis* has been inadvertently introduced to other regions, where it has become a significant pest of eucalyptus trees (Jeger et al. 2018). It is now recognized as part of a cryptic species complex known as *G. scutellatus* based on morphological and molecular data and is considered a quarantine pest in countries like France, Italy, Portugal, and Spain (Jeger et al. 2018).

The larvae feed on young leaves, leading to defoliation of the upper canopy, while adults consume the edges of mature leaves that hinders tree growth. Damage is evident through scalloped leaf edges, die-back of shoot tips, and the formation of epicormic shoots (Jeger et al. 2018). Initially, the damage appears as a scorched brownish hue on young foliage, eventually destroying twigs and buds. Severe defoliation stunts give the trees a stag-headed appearance. A survey during 2016–2018 in Chile identified natural infections of adult *G. platensis* by species of *Beauveria*, *Hirsutella*, and *Metarhizium*. In South Africa, three formulations of *Beauveria bassiana* and *Metarhizium acridum* spore suspensions were tested against *G. scutellatus* adults in the lab, with *B. bassiana* showing the highest efficiency, achieving 100% mortality and emerging as a promising strain for IPM programs.

14.8.3 Gypsy/Spongy Moth, *Lymantria dispar* L. (Lepidoptera: Erebidae)

Lymantria dispar is native to Eurasia and North Africa, where it is a major pest of deciduous trees, particularly in Europe (Tobin and Blackburn 2007), equally problematic in the Balkan Peninsula, with frequent outbreaks. Its range has been expanding ever since the accidental introduction to the United States from France due to the broad host range and the non-selective oviposition behaviour of females. It is now one of the most destructive forest pests globally, threatening both commercial and native forests (Paini et al. 2018). The ecological consequences of defoliation by *L. dispar* include changes in forest succession, watershed characteristics, and wildlife distribution patterns (Fajvan and Wood 1996). The urticating hairs and frass also pose significant health risks in urban areas. Among invasive pests, *L. dispar* has been the subject of extensive research on microbial control agents, including *Bt kurstaki*, nuclear polyhedrosis virus (LdMNPV), and *Entomophaga maimaiga*. The introduction of *E. maimaiga* from Asia to the United States led to epizootics, though infections by *I. fumosorosea* and occasional infections by *B. bassiana* have also been noted. Currently, *E. maimaiga* is the most significant host-specific fungal pathogen of *L. dispar* larvae in North America (Hajek et al. 2008).

14.8.4 Asian Longhorned Beetle, *Anoplophora glabripennis* Motschulsky (Coleoptera: Cerambycidae)

A highly destructive wood-boring insect, *Anoplophora glabripennis* targets and kills healthy trees and is native to Asia (Meng et al. 2015). In China, the insect has devastated over 90 million trees since the 1980s, particularly due to large-scale monoculture plantings of non-native Populus and Salix species (Pan 2005). In the last two decades, numerous accidental introductions have occurred in North America and Europe, largely due to international trade, with maple trees (*Acer* spp.) being a common host. The females bore into the cambium and lay eggs individually (Haack et al. 1997), which, upon hatching, feed beneath the bark along the phloem-cambium interface, eventually boring into the heartwood and creating galleries. Adults exit the galleries after chewing through the phloem and bark of leaf petioles and young twigs. Studies have shown that *Beauveria* and *Metarhizium* can reduce adult longevity and oviposition by impregnating non-woven fiber bands with *B. asiatica* and *B. brongniartii* (Dubois et al. 2004). Similarly, *M. anisopliae* isolates have also been found to reduce larval eclosion (Hajek et al. 2008).

14.8.5 Emerald Ash Borer, *Agrilus planipennis* Fairmaire (Coleoptera: Buprestidae)

The emerald ash borer, native to temperate North Eastern Asia, including China, Korea, and Russia, is a highly invasive beetle that has caused the death of millions of ash trees in North America, leading to significant ecological and economic damage (Herms and McCullough 2014). All North American and European ash species are vulnerable to *A. planipennis*, with black, green, and white ash being the most susceptible, while blue ash shows the most resistance in North America. Adult beetles feed on foliage; females lay eggs under the bark or in cracks. The larvae feed on the cambium and phloem, creating serpentine galleries that disrupt the flow of sap, water, and nutrients, leading to canopy decline and eventual tree death within 2–6 years (Knight et al. 2013).

Several entomopathogenic fungi, including *Beauveria bassiana*, have been identified as potential biocontrol agents for the emerald ash borer. Applications of *B. bassiana* strain GHA to tree trunks before adult emergence in summer have been found to reduce adult longevity, female fecundity, and delayed larval development (Liu and Bauer 2008). Conidial sprays of *B. bassiana* on ash bark before adult emergence have also been shown to be a significant mortality factor.

14.8.6 Oak Lace Bug, *Corythucha arcuata* (Say) (Hemiptera: Tingidae)

Corythucha arcuata, a species native to North America, is a major pest of oak trees in forests, urban areas, and rural settings worldwide (Mutun et al. 2009). After its initial detection in northern Italy (Bernardinelli and Zandigiacomo 2000), *C. arcuata* has spread throughout Europe, Turkey, Russia, and Iran (Dobreva et al. 2013). In Bulgaria, within 5 years of its introduction, it had invaded most of the country, affecting about 85% of the oak leaves (Simov et al. 2018). Both adults and nymphs feed on the underside of leaves, causing characteristic black spots, while the upper surfaces develop whitish blotches or stippling. Heavy infestations can lead to premature leaf drop due to reduced photosynthesis and respiration.

Despite the significant damage caused by this pest, operative phytosanitary procedures to control its spread are lacking and oils or contact insecticides are the only available options (Sparks et al. 2015). However, a study in Turkey tested ten entomopathogenic fungi against *C. arcuata* nymphs and adults under laboratory conditions (Sonmez et al. 2016). Fungi from the genera *Metarhizium*, *Beauveria*, *Isaria*, and *Myriodontium* (@1×10^7 per mL conidial concentration) were found to be effective and *B. bassiana* exhibited the highest pathogenicity, resulting in 80–90% mortality within 14 days (Sonmez et al. 2016).

14.9 Conclusion

Though trees have naturally evolved effective defense mechanisms against insect pests, pest epidemics in forest ecosystems are relatively rare. However, management interventions are essentially required when trees are under stressful conditions or when predators and other natural control mechanisms fail to keep the pest populations under check. The aim of forest pest management is to prevent their density from reaching damage thresholds, using ecologically sound, economically viable, socially and environmentally acceptable strategies. This requires the selection of efficient survey and monitoring tools, a proper understanding of life history and other pest population parameters and their interaction with forest resources to decide the time of intervention. IPM programs in forestry should be an integral part of ecosystem management and should be well-integrated with forest management activities. Though operational implementation of IPM in forests faces legal, organizational, and financial constraints, it is continuously developing through research and adaptive application. The most important aspect of pest management is that no matter how sophisticated the IPM system is, there is always room for area and pest-based improvement, allowing more accurate pest mitigation with fewer disagreeable ecological impacts.

References

Amman GD, McGregor MD, Dolph RI (1990) Mountain pine beetle, Forest Insect and Disease Leaflet 2:12. USDA Forest Service

Baake A (1991) Using pheromones in the management of bark beetle outbreaks. In: Baranchikov YN, Mattson WJ, Hain FP, Payne TL (eds) Forest insect guilds: patterns of interaction with host trees, General technical report NE-153. USDA Forest Service, Northeastern Research Station, Newton Square, pp 371–377

Bauer LS, Duan JJ, Gould JR, Van Driesche R (2015) Progress in the classical biological control of *Agrilus planipennis* Fairmaire (Coleoptera: Buprestidae) in North America. Can Entomol 147(3):300–317

Bernardinelli I, Zandigiacomo P (2000) First record of the oak lace bug, *Corythucha arcuata* (Say) (Heteroptera, Tingidae) in Europe. Inf Fitopatol 12:47–49

Cardé RT, Minks AK (1995) Control of moth pests by mating disruption – successes and constraints. Annu Rev Entomol 40:559–585

Carnegie AJ, Cant RG, Eldridge RH (2008) Forest health surveillance in New South Wales Australia. Aust For 71(3):164–176

Ciesla WM (2000) Remote sensing in forest health protection. USDA Forest Service Forest Health Technology Enterprise Team Report FHTET-00-03, Fort Collins

Ciesla WM (2011) Forest entomology: a global perspective. Blackwell Publishing Ltd Zondag/Wiley

Cranshaw WS (2009) Bacillus thuringiensis, Fact Sheet 5.556. Colorado State University Extension

Davis TS, Mann AJ, Malesky D, Jankowski E, Bradley C (2018) Laboratory and field evaluation of the entomopathogenic fungus *Beauveria bassiana* (Deuteromycotina: Hyphomycetes) for population management of spruce beetle, *Dendroctonus rufipennis* (Coleoptera: Scolytinae), in felled trees and factors limiting pathogen success. Environ Entomol 47:594–602

Dobreva M, Simov N, Georgiev G, Mirchev P, Georgieva M (2013) First record of *Corythucha arcuata* (Say) (Heterotera: Tingidae) on the Balkan Peninsula. Acta Zool Bulg 65:409–412

Dubois T, Li Z, Jaifu H, Hajek AE (2004) Efficacy of fiber bands impregnated with *Beauveria brongniartii* cultures against the Asian longhorned beetle, *Anoplophora glabripennis* (Coleoptera: Cerambycidae). Biol Control 31:320–328

Eskiviski E, Agostini J, Toloza R, Coll O (2005) Daños Producidos por el Pulgón del Pino Cinara Atlantica W. (Hemiptera: Aphididae) en Plantas Jóvenes de Pinus Taeda L; XI Jornadas Técnicas Forestales y Ambientales; INTA Montecarlo: Montecarlo, Argentina

Fajvan M, Wood JM (1996) Stand structure and development after gypsy moth defoliation in the Appalachian Plateau. For Ecol Manag 89:79–88

FAO (2005) Global Forest resources assessment–2005: Progress towards sustainable forest management, Forestry Paper 147:320. Food and Agriculture Organization of the United Nations, Rome

Fettig CJ, Klepzig KD, Billings RF, Munson AS, Nebeker TE, Negrón JF, Nowak JT (2007) The effectiveness of vegetation management practices for prevention and control of bark beetle infestations in coniferous forests of the western and southern United States. For Ecol Manag 238(1–3):24–53

Frankenhuyzen KV, Beardmore T (2004) Current status and environmental impact of transgenic forest trees. Can J For Res 34:1163–1180

Fry HRC, Quiring DT, Ryall KL et al (2008) Relationships between elm spanworm, *Ennomos subsignaria*, juvenile density and defoliation on mature sycamore maple in an urban environment. For Ecol Manag 255(7):2726–2732

Funes H, Griffo R, Zerba E, Gonzalez-Audino P (2011) Mating disruption of the ambrosia beetle *Megaplatypus mutatus* in poplar and hazelnut plantations using reservoir systems for pheromones. Entomol Exp Appl 139(3):226–234

Haack RA, Law KR, Mastro VC, Ossenbruggen HS, Raimo BJ (1997) New York's battle with the Asian long-horned beetle. J For 95:11–15

Haack RA, Britton KO, Brockerhoff EG, Cavey JF, Garrett LJ, Kimberley M, Lowenstein F, Nuding A, Olson LJ, Turner J, Vasilaky KN (2014) Effectiveness of the international phytosanitary standard ISPM No. 15 on reducing wood borer infestation rates in wood packaging material entering the United States. PLoS One 9(5):e96611

Hajek AE, Lund J, Smith MT (2008) Reduction in fitness of female Asian longhorned beetle (*Anoplophora glabripennis*) infected with *Metarhizium anisopliae*. J Invertebr Pathol 98:198–205

Herms DA, McCullough DG (2014) Emerald ash borer invasion of North America: history, biology, ecology, impacts and management. Annu Rev Entomol 59:13–30

Holvoet B, Muys B (2004) Sustainable Forest management worldwide: a comparative assessment of standards. Int For Rev 6:99–122

Hunter AF, Dwyer G (1997) Outbreaks and interacting factors: insect population explosions synthesized and dissected. Integr Biol Issues News Rev 1(5):166–177

Jacobi WR, Hardin JG, Goodrich BA, Cleaver CM (2012) Retail firewood can transport live tree pests. J Econ Entomol 105(5):1645–1658

Jeger M, Bragard C, Caffier D, Candresse T, Chatzivassiliou E, Dehnen-Schmutz K, Gilioli G, Jaques-Miret JA, MacLeod A, Navajas-Navarro M (2018) Pest categorisation of the *Gonipterus scutellatus* species complex. EFSA J 16:5107

Johnson EW, Wittwer D (2008) Aerial detection surveys in the United States. Aust For 71:212–215

Kenis M, Hurley BP, Hajek AE, Cock MJW (2017) Classical biological control of insect pests of trees: facts and figures. Biol Invasions 19(11):3401–3417

Knight KS, Brown JP, Long RP (2013) Factors affecting the survival of ash trees (*Fraxinus* spp.) infested by emerald ash borer (*Agrilus planipennis*). Biol Invasions 15:371–383

Lesk C, Coffel E, D'Amato AW, Dodds K, Horton R (2017) Threats to north American forests from southern pine beetle with warming winters. Nat Clim Chang 7:713–717

Liu H, Bauer LS (2008) Microbial control of emerald ash borer, *Agrilus planipennis* (Coleoptera: Buprestidae) with *Beauveria bassiana* strain GHA: greenhouse and field trials. Biol Conserv 45:124–132

Loureiro E, Oliveira N, Wilcken C, Filho A (2004) Patogenicidade de *Verticillium lecanii* ao pulgao do pinus. Rev Árvore 28:765–770

Lyons DB, Lavallée R, Kyei-Poku G, Van Frankenhuyzen K, Johny S, Guertin C, Francese JA, Jones GC, Blais M (2012) Towards the development of an auto-contamination trap system to manage populations of emerald ash borer (Coleoptera: Buprestidae) with the native entomopathogenic fungus, *Beauveria bassiana*. J Econ Entomol 105(6):1929–1939

Martini A, Baldassari N, Baronio P, Anderbrant O, Hedenstrom E, Hogberg HE, Rocchetta G (2002) Mating disruption of the pine sawfly *Neodiprion sertifer* (Hymenoptera: Diprionidae) in isolated pine stands. Agric For Entomol 4(3):195–201

Meng PS, Hoover K, Keena MA (2015) Asian longhorned beetle (Coleoptera: Cerambycidae), an introduced pest of maple and other hardwood trees in North America and Europe. J Integr Pest Manag 6:4

Montalva C, Rojas E, Ruiz C, Lanfranc D (2010) El pulgón del ciprés en Chile: Una revisión de la situación actual y antecedentes del control biológico. Bosque 31:81–88

Montalva C, Valenzuela E, Barta M, Rojas E, Arismendi N, Rodrigues J, Humber R (2017) *Lecanicillium attenuatum* isolates affecting the invasive cypress aphid (Cinara cupressi) in Chile. BioControl 62:625–637

Mutun S, Ceyha Z, Sözen C (2009) Invasion by the oak lace bug, *Corythucha arcuata* (Say) (Heteroptera: Tingidae), in Turkey. Turk J Zool 33:263–268

Nair KSS (2007) Tropical Forest insect species- ecology, impact and management. Cambridge University Press, Cambridge

Nair KSS, Varma RV (1981) Termite control in eucalypt plantations, KFRI research report no. 6. Kerala Forest Research Institute, Peechi

Paini DR, Mwebaze P, Kuhnert PM, Kriticos DJ (2018) Global establishment threat from a major forest pest via international shipping: *Lymantria dispar*. Sci Rep 8:13723

Pan HY (2005) Review of the Asian Longhorned Beetle: research, biology, distribution and management in China, Forest Department Working Paper. FBS/6E. FAO, Rome

Penteado S, Trentini R, Tadeo E, Reis W (2000) Ocorrencia, distribucicao, dano e controle de Pulgoes do genero Cinara em Pinus spp. no Brasil. Floresta 30:55–64

Powell JA, Bentz BJ (2014) Phenology and density-dependent dispersal predict patterns of mountain pine beetle (*Dendroctonus ponderosae*) impact. Ecol Model 273:173–185

Price PW, Denno RF, Eubanks MD, Finke DL, Kaplan I (2011) Insect ecology: behavior, populations and communities. Cambridge University Press, Cambridge

Rhainds M, Kettela EG, Silk PJ (2012) Thirty-five years of pheromone-based mating disruption studies with *Choristoneura fumiferana* (Clemens) (Lepidoptera: Tortricidae). Can Entomol 144(3):379–395

Ric J, Groot P, Gasman B, Orr M, Doyle J, Smith MT, Dumouchel L, Scarr T, Turgeon JJ (2007) Detecting signs and symptoms of Asian longhorned beetle injury: a training guide. Agriculture and Agri-Food Canada, Ottawa

Ross DW, Daterman GE (1997) Using pheromone-baited traps to control the amount and distribution of tree mortality during outbreaks of the Douglas-fir beetle. For Sci 43:65–70

Ross DW, Daterman GE (1998) Pheromone-baited traps for *Dendroctonus pseudotsugae* (Coleoptera: Scolytidae): influence of selected release rates and trap designs. J Econ Entomol 91:500–506

Ruiu L, Mannu R, Olivieri M, Lentini A (2021) Gypsy moth management with LdMNPV Baculovirus in Cork Oak Forest. Forests 12:495

Seybold SJ, Bentz BJ, Fettig CJ, Lundquist JE, Progar RA, Gillette NE (2018) Management of western North American bark beetles with semiochemicals. Annu Rev Entomol 63:407–432

Simov N, Grozeva S, Langourov M, Georgieva M, Mirchev P, Georgiev G (2018) Rapid expansion of the Oak lace bug *Corythucha arcuata* (Say, 1832) (Hemiptera: Tingidae) in Bulgaria. Hist Nat Bulg 27:51–55

Somers B, Asner GP (2012) Hyperspectral time series analysis of native and invasive species in Hawaiian rainforests. Remote Sens 4(9):2510

Sonmez E, Demirba Z, Demir I (2016) Pathogenicity of selected entomopathogenic fungal isolates against the oak lace bug, *Corythucha arcuata* Say (Hemiptera: Tingidae), under controlled conditions. Turk J Agric For 40:715–722

Sparks B, Braman SK, Nair S (2015) Control of lace bugs on ornamental plants, The University of Georgia, UGA Extension, Bulletin No. 1102. The University of Georgia, Athens

Stone C, Mohammed C (2017) Application of remote sensing technologies for assessing planted forests damaged by insect pests and fungal pathogens: a review. Curr For Rep 3:75–92

Strauss SH, Howe GT, Goldfarb B (1991) Prospects for genetic engineering of insect resistance in forest trees. Forest Ecol Manag 43:181–209

Sweeney J, Silk PJ, Rhainds M, MacKay W, Hughes C, Van Rooyen K, MacKinnon W, LeClair G, Kettela EG (2017) First report of mating disruption with an aggregation pheromone: a case study with *Tetropium fuscum* (Coleoptera: Cerambycidae). J Econ Entomol 110(3):1078–1086

Thompson I, Mackey B, McNulty S, Mosseler (2009) A Forest resilience, biodiversity, and climate change: a synthesis of the biodiversity/resilience/stability relationship in Forest ecosystems, Technical series no. 43. Secretariat of the Convention on Biological Diversity, Montreal

Tobin PC, Blackburn LM (2007) Slow the spread: a National Program to manage the gypsy moth, USDA forest service general technical report NRS-6. USDA Forest Service, Northern Research Station, Newtown Square

Vannatta AR, Hauer RH, Schuettpelz NM (2012) Economic analysis of emerald ash borer (Coleoptera: Buprestidae) management options. J Econ Entomol 105:196–206

Vettori C, Gallardo F, Häggman H, Kazana V, Migliacci F, Pilate G, Fladung M (2016) Biosafety of forest transgenic trees: improving the scientific basis for safe tree development and implementation of EU policy directives, vol 82. Springer

Watson G, Voegtlin D, Murphy S, Fottit R (1999) Biogeography of the *Cinara cupressi* complex (Hemiptera: Aphididae) on Cupressaceae, with description of a pest species introduced into Africa. Bull Entomol Res 89:271–283

Yemshanov D, Koch FH, McKenney DW, Downing MC, Sapio F (2009) Mapping invasive species risks with stochastic models: a cross-border United States-Canada application for *Sirex noctilio* Fabricius. Risk Anal 29(6):868–884

Zondag R (1969) A nematode infection of *Sirex noctilio* (F.) in New Zealand. N Z J Sci 12(4):732–747

Eco-tech Strategies: Revolutionizing Forest Insect Pest Control Through Biological and Biotechnological Innovations

15

Gadigavarahalli Subbareddy Uma, Shruti Godara, Athulya Ramakrishnan, J. Nandini, and Abhishek D. Kattappanavar

Abstract

Forest ecosystems are a cornerstone of global environmental health, supporting a vast array of biodiversity, regulating climate through carbon sequestration, and providing numerous economic and cultural benefits. However, these ecosystems face significant threats from forest insect pests, which can lead to tree mortality, alter forest structure and function, and trigger cascading ecological effects. Traditional pest control methods, primarily reliant on chemical pesticides, have raised concerns due to their environmental impacts, non-target effects, and the development of resistant pest populations. In response, there is a growing emphasis on sustainable, eco-friendly pest management strategies. This chapter delves into the realm of eco-tech strategies, focusing on the intersection of biological and biotechnological innovations in forest insect pest control. These strategies represent a transformative approach, moving away from conventional chemical treatments towards methods that harness the power of nature and cutting-edge science.

Keywords

Pest control · Forest management · Biological control

G. S. Uma · A. Ramakrishnan (✉) · J. Nandini · A. D. Kattappanavar
Forest Protection Division, Institute of Wood Science and Technology, Bengaluru, Karnataka, India

S. Godara
Genetics and Tree Improvement Division, Forest Research Institute, Dehradun, Uttarakhand, India

R. K. Kalia, R. Pathak (eds.), *Tree Biology and Biotechnology*,
https://doi.org/10.1007/978-981-96-0002-1_15

15.1 Introduction

Forests, often referred to as the lungs of the earth, are indispensable for maintaining ecological balance, supporting biodiversity, and providing myriad ecosystem services. They act as significant carbon sinks, mitigate climate change, and offer economic resources, recreational opportunities, and cultural values. The forest has a dynamic ecosystem teeming with diverse biodiversity. Alongside beneficial arthropods, various pests such as defoliators, gall formers, root damagers, shoot borers, sap suckers, and bark borers are also present. Pests like bark beetles, emerald ash borers, spruce budworms, leaf miners, pine sawflies, and weevils can significantly impact tree growth, vigour, and survival. They also deteriorate the standard of forest products with a significant decrease in the yield. Under favourable conditions, these pests can cause severe outbreaks, leading to substantial economic and ecological damage.

In recent times, invasive pests have become more prevalent in many regions due to climatic changes, globalization and developmental activities. These non-native species possess remarkable dispersal and adaptation capabilities, and the damage they cause can degrade wildlife habitats by reducing local biodiversity. The destructive activities of certain pests might result in the destruction of vast tracts of forest, disrupt ecological processes, incur substantial economic losses and hinder reforestation efforts (Guegan et al. 2023).

The management of forest insect pests is of utmost importance in maintaining the well-being, efficiency, and long-term viability of these invaluable ecosystems (Samways et al. 2020). Forest insect pests encompass a wide array of insect species that have undergone co-evolutionary processes with forest ecosystems for an extensive period of time, spanning millions of years (Allio et al. 2021). Certain insect species, referred to as indigenous pests, inherently inhabit forest ecosystems and typically sustain a harmonious equilibrium with their host trees and surrounding organisms. Nevertheless, the equilibrium of ecosystems can be disrupted by various factors such as human activities, climate change, and the introduction of invasive species, which can subsequently result in insect pest outbreaks (Harvey et al. 2023). These occurrences can lead to extensive leaf loss, death of trees, and substantial financial damages. Throughout history, forest managers have employed diverse strategies to manage insect pests, with chemical pesticides emerging as a predominant method (Sweeney et al. 2023). Although initially successful, pesticide utilisation has generated apprehension regarding its potential for environmental pollution, adverse effects on unintended organisms, and the emergence of insecticide resistance within pest communities (Paradva and Kalla 2023). In response to these challenges, eco-tech strategies have emerged at the forefront of forest pest management. These strategies encompass a range of biological and biotechnological innovations designed to control pest populations while minimizing ecological disruption. The integration of these advanced techniques holds the promise of revolutionizing pest management, ensuring the health and resilience of forest ecosystems. In an era where environmental sustainability is paramount, eco-tech strategies represent a significant leap forward in forest pest control. They offer the dual benefits of effective pest management and environmental conservation, paving the way for healthier, more resilient forest ecosystems.

This chapter aims to offer a comprehensive review of eco-tech approaches for forest insect pest control, highlighting their potential contribution to sustainable forest management practices. It explores the potential of natural pests, viz., entomopathogenic fungi, predators and parasitoids, highlighting their role in regulating pest populations naturally. It also examines biotechnological innovations, including genetic engineering and sterile insect techniques, which offer targeted and species-specific solutions to pest problems.

15.2 Biological Control

Protecting forests from pests through biological approaches is an integral part of sustainable forest management. The use of chemical insecticides is often impractical and unacceptable in many natural vegetation landscapes. Interestingly, forest species often exhibit higher tolerance to pest infestations than agricultural crops. Therefore, to reduce insect pest populations without harming the ecosystem, fungi, bacteria, viruses, nematodes etc., predators, and parasitoids can be advantageous. Based on the organisms used for biological pest control, it can be termed microbial control and pest control using natural enemies (Fig. 15.1). In nature, several microbes act as pathogens, while many reptiles, birds, and insects serve as predators. Additionally, certain insects function as parasitoids, playing a crucial role in maintaining pest populations at manageable levels.

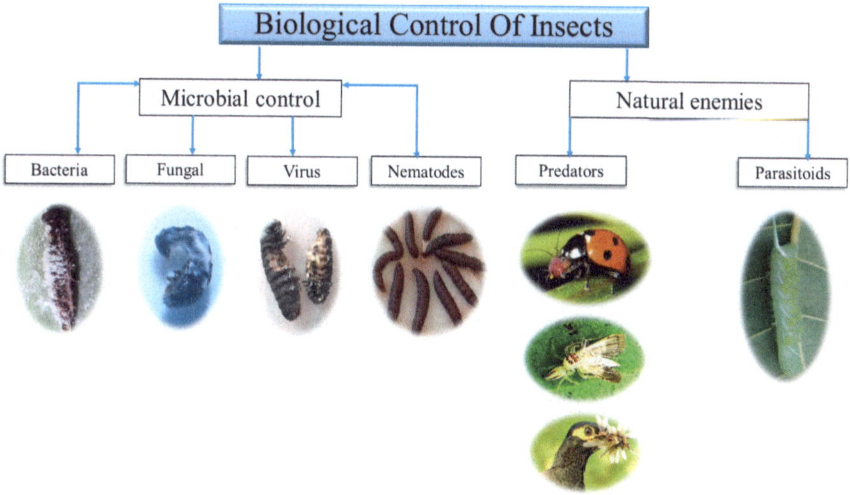

Fig. 15.1 Biological control of insects through various organisms in forest ecosystem

15.2.1 Microbial Control

Microbial pesticides are eco-friendly as they control pest populations with target-specific pathogenesis. They include bacteria, fungi, nematodes, and viruses that are commonly used for pest management. Based on the dieting habits of pests, entomopathogenic bacteria and entomopathogenic viruses are preferentially used against defoliators and borers. In contrast, entomopathogenic fungi generally opt against various pests, mainly of the lepidopteran group, since they possess various modes of entry into the host. Important entomopathogens and their mode of infection process are further dealt with.

15.2.1.1 Entomopathogenic Viruses

Entomopathogenic viruses (EPVs) are obligate intracellular parasites with either DNA or RNA as genetic material encapsulated within their protein coat (known as capsid) and cause insect disease. In general, these viruses are effectively used to manage lepidopteran and hymenopteran pests. As EPVs are highly specific in nature and safe for humans and other non-target species, they can be considered for sustainable pest management (Raj et al. 2022). A brief information on the major EPVs is given below:

Baculovirus It belongs to the family Baculoviridae and is the most prominently used EPV (Sosa-Gómez et al. 2020) characterized by infective particles called 'occlusion bodies' (OBs) that are either polyhedral or granular. Nuclear Polyhedroviruses (NPVs) and Granular Viruses (GVs) are the most often used baculovirus to manage various forest pests. Several assessment studies have analysed management of pest outbreaks through aerial sprayings of NPV against pine tree lappet moth (*Dendrolimus pini*) infesting Pine forests (Maaß and Kehlenbeck 2024), spruce budworm (*Choristoneura fumiferana*) infesting *Picea* spp., (Chang et al. 2012), teak defoliator (*Hyblaea puera*) on *Tectona grandis* (Bindu et al. 2022) etc. The GVs, *Hyphantria cunea* granulovirus (HycuGV) was found to infect forest lepidopteran pests *Malacosoma neustria* of Lasiocampidae family, *Lymantria dispar* and *Hyphantria cunea* of Erebidae family, *Helicoverpa armigera* from Noctuidae family, and *Spodoptera exigua* from Noctuidae family (Bayramoğlu et al. 2021; Cuartas-Otálora et al. 2019). Baculoviruses cause systemic infection and mainly target the fat bodies of host insects. Severe infection of baculoviruses depletes the oxygen-absorbing capacity in infected insects, and this makes them crawl apically for oxygen and is found dead in an inverted position later.

Entomopoxvirus It belongs to the family Poxviridae. The sub-family Entomopoxvirinae are insect pathogenic viruses and comprises Alphaentomopoxvirus, which infects coleopteran (e.g., *Melolontha melolontha* entomopoxvirus), Betaentomopoxvirus infects Lepidoptera and Oorthoptera (e.g., *Amsacta moorei* entomopoxvirus), Gammaentomopoxvirus infects only Dipteran (e.g., *Chironomus luridus* entomopoxvirus) and the unassigned fourth group attacks hymenopterans.

During the period 2003 and 2011, an entomopox virus, ItEPV was observed in spruce bark beetles, *Ips typographus* in Beglika and Borino State Forests (Takov et al. 2006, 2007). It was also recorded from Turkey against the same host infesting *Picea orientalis* (Yaman and Baki 2011). White coloration of fat bodies confined to the posterior part of the abdomen is considered a significant feature of entomopox-viral infection. In some coleopterans, the rectal sac is observed swollen due to the accumulation of spheroids and infective particles. Compared to baculoviruses, the infection process of entomopoxvirus is very slow.

Cypovirus Cytoplasmic polyhedroviruses (CPVs) belong to the family Reoviridae. Their host range includes lepidoptera, coleoptera and dipteran insects. CPVs are generally spread through oral ingestion via infected insect's food sources. On their ingestion, the virions attack midgut epithelial cells and their infection is fatal for the host insect. Some key symptoms are reduced egg-laying capacity, flightlessness, and body discoloration, but they are intact. Isolation of virulent strains of CPVs is reported from lepidopteran pests, namely, *Dendrolimus sibiricus*, infesting boreal forests (Martemyanov et al. 2023) and *Thaumetopoea pityocampa* infesting Pine forests (Ince et al. 2007).

Iridoviruses They belong to the family Iridoviridae, which infect several insects, including Diptera, Hemiptera, Coleoptera, Lepidoptera and Hymenoptera orders. All host species of Iridovirus are associated with aquatic environments (Del Rincón-Castro and Ibarra 2011). They are generally transmitted through co-habitation. They are multiplied in cytoplasma of host cells and are characterized by iridescent spots on severely infected tissues, mainly on appendages and thorax. The natural population possesses low infectivity and chronic infection rates (Harish et al. 2021).

15.2.1.2 Entomopathogenic Bacteria (EPB)

Bacteria are omnipresent and found in a variety of interactions with biotic and abiotic factors in nature. Entomopathogenic bacteria include both spore-forming genera like *Bacillus* spp., *Paenibacillus* spp., and *Clostridium* spp., and non-spore-forming genera *Pseudomonas*, *Serratia* and *Yersinia*. The most studied bacterium is *Bacillus thuringiensis* (Bt) of the family Bacillaceae used majorly to control Lepidopteran and Coleopterans. Bt produces a toxic protein, Bt delta endotoxin. On ingestion, the protoxin gets activated in an alkaline condition at the mid-gut of the insect to produce the toxin. The activated toxin binds to specific membrane receptors of epithelia cells and induces the pore formation. Increased cell permeability disrupts gut integrity and finally results in peat mortality either from starvation or septicaemia (Deka et al. 2021). *B. sphaericus, P.aenibacillus popilliae* were reported as potent biocontrol agents against Japanese beetles (Glare et al. 2017); apart from them *S. entomophila, Brevibacillus laterosporus, Chromobacterium subtsugae,* and *Yersinia entomophaga* are also found as effective EPBs. *C. subtsugae* and *Burkholderia rinojensis* are reported as potent biological controls against mites (Dara et al. 2019; Ruiu 2018). Host insects of these bacteria are summarized in Table 15.1.

15.2.1.3 Entomopathogenic Nematodes (EPNs)

Nematodes that induce pathogenesis in insects through symbiotic bacteria are called entomopathogenic nematodes (EPNs). They are microscopic, soil-dwelling insect parasites and belong to families Steinemernatidae and Heterorhabditidae of phylum Nematoda. EPNs are used as biological control agents to kill insects through a symbiotic association with bacteria like *Xenorhabdus* spp., and *Photorhabdus* spp. (Boemare 2002). Infective juvenile EPNs enter the host through natural openings (anus, mouth, spiracles) or directly through the cuticle and release their symbiotic bacteria into the host's hemocoel. These bacteria produce insecticidal proteins, cause septicemia, and digest the host's tissues, causing mortality in the pest population (Ciche and Ensign 2003). EPNs, namely, *H. indica* and *S. glaseri* were reported to infect 17 significant forest pests belonging to orders, viz., Lepidoptera, Coleoptera, and Hemiptera (Sharmila et al. 2018). EPNs are effective against a wide range of insect hosts across all insect orders with rapid mortality rates. They are considered a promising option for large-scale use and the selection of the right EPN species or strain to determine their efficacy against a particular pest.

15.2.1.4 Entomopathogenic Fungi (EPF)

Entomopathogenic fungi (EPF) are one of the significant biological control entities in pest management. They are a heterogeneous group with nearly 700 recognized species. Features like vast adaptability, low nutrition requirement, broad host range, and minimal or no ecological risks make them an ideal approach instead of synthetic chemicals. Most of them belong to the orders Hypocreales and Entomophtharales (Alfina and Haneda 2022). EPFs infect insects by penetrating the insect's cuticle or through body openings. After penetration, their progression eventually invades vital organs, leading to insect death. EPFs have evolved specialized mechanisms to overcome insect defense compounds and break down the insect's integument. They are also found to synthesize secondary metabolites broadly termed 'mycotoxins' (e.g., beauvericin, cyclodepsipeptide, destruxin and desmethyldestruxin) which are toxic against host insects. They are cost-effective and environmentally friendly alternatives to chemical pesticides.

EPFs are found on the forest grounds with leaf litter and also below the soil surface. The prevalence and diversity of EPF varied depending on the type of forest or vegetation. In tropical forests, a high diversity of EPFs was observed, whereas temperate forests exhibited relatively low EPF diversity (Aung et al. 2008; Prischepa et al. 2011). Environmental niche plays a major role in the prevalence of EPFs. Hypocreales dominated the forest floor, including litter and soil surface at Białowieia (Poland), whereas Entomophthorales was reported majorly from canopy and vegetation between forest canopy and forest floor (Majchrowska-Safaryan and Tkaczuk 2021). The existence and diversity of EPFs vary based on climate and host insects. Humid tropical forests favour diverse EPF, majorly Hypocreales. Members of genus *Cordyceps* (Ascomycota: Hypocreales) are found more frequently as 'zombie

Table 15.1 Various entomopathogens/microbes reported from forest trees to manage insect pests

Forest pest	Tree host/type of forest	Epizootic microbe
Cinara pinea	Pine forest	*Erynia canadenis*
Dendrolimus pini	*Pinus silvestris*	*Isaria farinosa, Cordyceps militaris, Beauveria bassiana, Lecanicillium lecanii*
D. punctatus	Masson pine	*B. basiiana*
Fiorinia externa	*Tsuga grandis*	*Colletotrichum* sp., *L. lecanii, B. bassiana*
Hyblaea puera	*Tectona grandis*	Parasites, Baculovirus, HpNPV, *B. Bassiana*
Leucoma salicis	*Populus nigra*	NPV, *B. Bassiana*
Lymantria dispar	*Larix leptolepis*	*Entomophthora aulicae, PaecilomycesCanadensis*
–	Mixed hardwood forests	*Entomophaga maimaiga,* NPV
–	*Larix leptolepis* and *Alnus japonica*	NPV, *E. maimaiga, Isaria javanica*
Malacosoma disstria	–	*Furia gastropachae*
Lepidopterans and sawflies	Different coniferous deciduous trees	*Cordyceos militaris, Cordyceps* sp.
Orgyia leucostigma	*Abies balsamea*	*Entomophaga aulicae,* NPV
Panolis flammea	*Pinus contorta*	*I. farinosa, E. aulicae, Nomuraea rileyi,* *B. basssiana* and NPV
Syntypistis punctatella	*Fasus crenata*	*Cordyceps militaris, I. farinosa, B. Bassiana*
Helicoverpa armigera	–	*Helicovervap armigera nucleopolyhedrovirus* (HearNPV)
Spodoptera litura	*Sesbania bispinosa*	*Spodoptera litura nucleopolyhedrovirus* (SpliNPV)
Douglas-fir tussock moth	*Pseudotsuga menziesiivar, Abies* spp.	*Orgyia pseudotsugata* NPV
Rhynchophorus ferrugiueus	*Palm trees*	*Bacillus thuringiensis* *Metarhizium anisopliae* ZZ-A1 and *Fusarium oxysporum* ZZ-L1
Popillia japanica	–	*Paenibacillus popilliae*
Earias vitella, Helicoverpa armigera and *Plutella xylostella*	–	*Bacillus thuringiensis*subsp. galleria
Lepidoptera Diptera Lepidoptera Coleoptera	–	*Bacillus thuringiensis* subsp. aizawal *B. thuringiensis* subsp. israelensis *B. thuringiensis* subsp. kurstaki *B. thuringiensis* subsp. tenebrionis
Several orders of soil borne pests	–	*Heterorhabditis bacteriophora* *H. heliothidis,* *Steinernema carpocapsae, S. feltiae*

Source: Augustyniuk-Kram and Kram (2012), Kalha et al. (2014), Pu et al. (2017, 2023), Raj et al. (2022), Irsad et al. (2023)

fungus' infecting ants or spiders (Aung et al. 2008). Other members like *Beauveria, Metarhizium* and *Isaria* are usually soil-borne and found infecting soil-dwelling insects (Keller et al. 1989). Entomophthorales, highly prevalent in temperate forests (Burges 1988), were reported with the fewest numbers in tropical forests (Evans 1982). The existence of EPFs, even in extreme conditions, is marked by the report of *Tolypocladium cylindrosporum, B. bassiana* and *Metarhizium anisopliae* from the Arctic Circle of Norway (Klingen et al. 2015) and endemic *Paecilomyces antarctica* from Antarctic region (Bridge et al. 2005).

15.2.2 Predators and Parasitoids

Traditionally, insect pest management depends on synthetic chemicals, which have an array of adverse effects on the environment and other living organisms. Organic contaminants in the air can linger and enter the soil domain. In support of this, pesticide organochlorine was eventually reported from soil, and its original source was found to be airborne (Neves et al. 2018). Pesticides not only endanger the environment and ecology but also seriously jeopardize human health. They have the potential to be potent carcinogens that affect the immune and endocrine systems as well as potentially harm the body over time.

Researchers are looking more and more to biological control approaches for solutions because of the serious threat of pests and the shortcomings of other methods. Biological control has been used to manage pests since the 1800s. In 1889, Riley made attempts to control invasive scale insect *Icerya purchasi* infesting citrus trees using the predatory ladybug *Rodolia cardinalis*. It was the first documented case of this type of strategy. The strategy dramatically decreased the number of pests and damaged them within just 2 years. Since then, a great deal of effective uses of biological control agents have been reported, suggesting biological control's position as an essential pest management technique.

Effective management of invasive forest pests, like the emerald ash borer (EAB), through sustainable methods such as biological control is crucial for restoring damaged ecosystems. From 2015 to 2017, parasitoids from Northeast Asia were released in the northeastern states of the U.S.A. to control EAB populations. A study by Duan et al. (2022) found that introducing two parasitoids (*Spathius galinae* and *Tetrastichus planipennisi*) led to 76% reduction in EAB larvae density from 2015 to 2020. This decline was mainly due to increased parasitism by *S. galinae* and some predation by local natural enemies like woodpeckers. *S. galinae* alone reduced the pest population growth rate by 31–57%. These results highlight that the timely introduction of suitable natural enemies can significantly suppress major pest activity and aid in forest recovery.

In northern Minnesota *Sarcophaga aldrichi*, a sarcophagid fly, was reported to cause 98% pupal parasitism in forest tent caterpillars, which was a serious annoyance (Schowalter 2017). In Canada, early instars of tent caterpillar were found parasitized by braconid wasp *Aleiodes malacosomatos* and its later instars were infected with two tachinid flies, namely, *Leschenaultia exul* and *Patelloa pachypyga* (Mondor and Roland 1997, 1998). Compared to other host trees, aspen is the tree that attracts these two tachinid fly species the most to the feeding damage caused by the forest tent caterpillar (Mondor and Roland 1997, 1998). When caterpillars eat the leaves that contain the eggs, they become parasitized. Female parasitoids deposit their eggs on the injured foliage.

In Canadian forests of aspen and mixed-hardwood vegetation, predation and parasitism of forest tent caterpillars were studied by Nixon and Roland (2012) using exclusion procedures. Accordingly, early instars of the pest were predated by general arthropods and later instars of the pest were fed by birds on aspen strands, constituting a significant pest reduction of around 40%. Parasitoids including *Itoplectis quandricingulata, I. conquisitor*, and *Theronia atalantae fulvescens* were effective in lowering pest population up to 30%. A few examples of utilizing predators and parasitoids for forest insect pest management are listed in Table 15.2.

15.3 Biotechnological Approaches for Forest Insect Pest Management

Biotechnological approaches for forest insect pest management offer innovative solutions that can be more sustainable and targeted as compared to conventional methods of chemical pesticides. These approaches leverage advances in genetics, microbiology, and other biological sciences. The major biotechnological approaches for forest insect pest management are depicted in Fig. 15.2.

15.3.1 Genetically Engineered Entomopathogens

Genetically engineered entomopathogens have been explored as innovative solutions for controlling forest pests, which can cause significant ecological and economic damage. For instance, Bt, a naturally occurring bacterium known for producing toxins lethal to certain insects, has been genetically engineered to enhance its potency against the gypsy moth (*Lymantria dispar*). These modified Bt strains are more virulent and have been successfully used in forest management programs to curb the populations of gypsy moth. Researchers have transferred genes encoding chitinases from various Bt strains, which significantly boost its insecticidal effectiveness against pests like *Tuta absoluta*, *Galleria mellonella*, and *Spodoptera frugiperda*. Furthermore, the incorporation of kinase genes and the fusion of cytolytic proteins with accessory proteins have further increased the potency of Bt against specific insect pests. For example, the expression of the *ppk* gene in *Bt* subsp. *israelensis* enhanced its toxicity against *Culex quinquefasciatus* by 7.7 times, while the co-expression of Cyt1A98 and P20 proteins improved insecticidal activity against lepidopteran pests. These genetic enhancements demonstrate the potential of engineered Bt strains to be more effective in controlling a wider range of insect pests (Azizoglu et al. 2020).

Table 15.2 Examples of forest insect biocontrol using parasitoids and predators

Name of the pest	Biocontrol agent	Place	Reference
European pine sawfly, *Neodiprion sertifer* Diprionidae	*Lophyroplectus luteator* *Exenterus abruptorius* *Lamachus eques* Ichneumonidae Larval parasite	Europe	Pschorn-Walcher (1977)
	Exenterus Canadensis Diprionidae	Canada	
Larch sawfly, *Pristiphora erichsonii* Tenthridinidae	*Mesoleius tenthredinis*, Ichneumonidae (parasitized over 80% of the population)	Manitoba	Drooz et al. (1985)
Larch case bearer, *Coleophora laricella* Hübner Coleophoridae	*Agathis pumila* Braconidae *Chrysocharis laricinellae* Eulopidae	Millbridge, Ontario	Ryan et al. (1987)
Winter moth, (Oak) *Operophtera brumata* Geometridae	*Cyzenis albicans* Tachinidae *Agrypon flaveolatum* Ichneumonidae	Canada	Hassell (1980)
Colombian defoliator, (Pine and Cypress plantations) *Oxydia trychiata* Geometridae	*Telenomus alsophilae* Scelionidae (99% Parasitized)	Colombia	Drooz et al. (1977)
European pine shoot moth, *Rhyacionia buoliana* Tortricidae	*Orgilus obscurator* Braconidae *Eulimneria rufifemur* Ichneumonidae	Southern Ontario	Syme (1977)
Gypsy moth, *Lymantria dispar* Lymantriideae	*Ooencyrtus kuwanae* Chalcididae *Anastatus disparis* Eupelmidae	Melrose Highlands, Massachusetts	Weseloh et al. (1983)
European spruce sawfly, *Gilpinia hercyniae* Diprionidae	*Dahlbominus fuscipennis* Eulophidae *Exenterus confusus* Kerr. Ichneumonidae	United States	Coppel and Mertins (2012)
European wood wasp, *Sirex noctilio* (F.) Siricidae	*Ibalia leucospoides* Ibalidae *Rhyssa hoferi* Ichneumonidae *Deladenus siricidicola* (Parasitic Nematode)	Tasmania	Taylor (1981)
Greater European spruce beetle, *Dendroctonus* Scolytidae	*Rhizophagus grandis* Monotomidae	Kobakhidze	Kobakhidze et al. (1968)
Teak leaf defoliator*Hyblaea puera* Hyblaeidae Teak leaf skeletonizer *Eutectona machaeralis* Pyralidae	*Trichogramma raoi* Nagaraja Trichogrammatidae	India	Roychoudhury (2015)

(continued)

Table 15.2 (continued)

Name of the pest	Biocontrol agent	Place	Reference
Emerald ash borer, *Agrilus planipennis* Buprestidae	*Oobius agrili* *Encyrtidae* *Tetrastichus planipennisi* Eulophidae	United States	Duan et al. (2018)
European spruce bark beetle *Ips typographus* (Spruce-fir forest) Curculionidae	*Coeloides bostrichorum* Braconidae	Bulgaria	Doychev et al. (2016)
Pine processionary moth *Th Aumetopoea pityocampa* Notodontidae	*Ooencyrtus pityocampae* Encyrtidae *Exorista segregata* Tachinidae	Europe	De Boer and Harvey (2020)
Gypsy moth *Lymantria dispar* Lymantridae	*Exorista larvarum* Tachinidae *Lymantrichneumon disparis* Ichneumonidae	Iran	Babaei et al. (2009)
Forest tent caterpillar *Malacosoma disstria* Hübner Lasiocampidae	*Sarcophaga aldrichi* Parker (30% of pupal parasitism) Sarcophagidae *Itoplectis quandricingulata* Provancher Ichneumonidae	Aspen, Colorado	Nixon and Roland (2012)
Forest tent caterpillar *Malacosoma disstria* Lasiocampidae	*Icterus galbula* (Predatory bird) Icteridae	Northern Oriole	Parry et al. (1997)
Pine processionary moth *Thaumetopoea pityocampa* Notodontidae	*Clamator glandarius* Cuculidae *Cuculus canorus* Cuculidae	Central Spain	Barbaro and Battisti (2011)

Another example is *Beauveria bassiana*, a fungus that naturally infects and kills a variety of insect species. Through genetic modifications, its effectiveness against the emerald ash borer (*Agrilus planipennis*) and the Asian longhorned beetle (*Anoplophora glabripennis*) has been significantly improved (Fan et al. 2007). Similarly, *Metarhizium anisopliae*, a fungus used as a biological control agent, has been genetically engineered to increase its virulence and specificity toward the mountain pine beetle (*Dendroctonus ponderosae*), a major threat to pine forests in North America (Zhao et al. 2016). This could offer a powerful tool for protecting large forest areas from beetle infestations.

In addition, *Nucleopolyhedrovirus* (NPV), a group of viruses that infect and kill insects, has been genetically engineered to enhance its pathogenicity or expand its host range to target specific pests like the spruce budworm (*Choristoneura fumiferana*), a significant defoliator of spruce and fir trees (Szewczyk et al. 2006). These

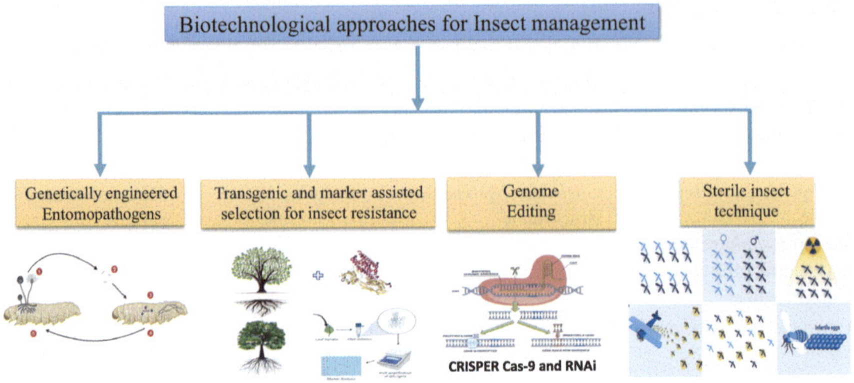

Fig. 15.2 Major biotechnological approaches for forest insect pest management

engineered NPV strains have been developed to reduce the requirement for chemical insecticides in forest management. These genetically engineered entomopathogens represent promising tools for sustainable and environmentally friendly forest pest management, though their deployment is carefully regulated and subjected to rigorous environmental assessments.

15.3.2 Transgenic and Marker-Assisted Selection for Insect Resistance

15.3.2.1 Transgenic Trees

Transgenic trees, also known as genetically modified (GM) trees, are trees altered through genetic engineering to express desirable traits, conservation, or commercial purposes. These modifications can provide resistance to specific pests, reducing the need for chemical pesticides and promoting sustainable forest management. A prominent example of genetic modification in trees is the incorporation of toxins from Bt. Researchers have also isolated numerous other insect-resistant genes, including protease inhibitors (PI), AaIT, α-amylase inhibitors, lectins, and chitinases. Each of these genes targets different insect pests and varies in its effectiveness. *Bacillus thuringiensis* (Bt) produce toxic proteins which binds to the receptors in the insect gut, causing cell lysis and death of the insect when ingested by the insects. Genes encoding Bt toxins can be inserted into the genome of trees to provide resistance against specific insect pests. The Bt family includes various toxins targeting different insects (Tang et al. 2004). Commonly used toxins are CryIA and CryIIIA, targeting Lepidoptera and Coleoptera, respectively. Combining these endotoxins can broaden insect resistance, as shown in several studies (Ying et al. 2012; Jiang et al. 2016).

Many transgenic trees have been developed to enhance resistance to specific insect pests by inserting various insecticidal genes. For example, *Populus nigra* has been modified with the gene CryIA(c) to develop resistance against *Orthosia incerta* and *Apocheima cinerarius*. These pests are known for their larval stages, which feed on the leaves and cause extensive damage to the trees. Introducing the CryIA(c) gene enables the trees to produce toxic proteins for these pests (Hu et al. 2001). *Pinus radiate* and *P. taeda* have been transformed with the CryIA(c) gene to target *Teia anartoides* and *Dendrolimus punctatus,* which are known for causing severe defoliation in pine forests (Tang and Tian 2003; Grace et al. 2005). In the hybrid species *P. tremula* × *P. tremuloides*, the gene CryIIIA(a) has been introduced to fight against *Chrysomela tremulae*, a pest that feeds on the foliage of these trees, leading to reduced growth and productivity (Génissel et al. 2003). Similarly, *Eucalyptus camaldulensis* has been engineered with the gene CryIIIA to target *Chrysophtharta* spp., a group of beetles that are significant pests of eucalyptus plantations. These beetles can cause severe defoliation, affecting the vitality and yield of the trees. The incorporation of the gene CryIIIA helps maintain the health and productivity of these hybrid trees by providing effective pest resistance (Harcourt et al. 2000).

Many other transgenic events targeting insecticidal genes have been attempted. For instance, *Populus euramericana* and *P. trichocarpa* × *P. deltoides* have been engineered to express the gene CryIIIA that provides resistance against *Chrysomela scripta*, a beetle known to defoliate poplar trees. This genetic modification enhances the ability of trees to withstand damage caused by these pests (Meilan et al. 2000). *Juglans regia*, commonly known as walnut, has been transformed with the gene CryIA(c) to control *Laspeyresia pomonella*. This genetic enhancement helps reduce the infestation levels and thus improves the yield and quality of walnuts (Dandekar et al. 1998). *P. nigra* has been genetically modified with the gene CryIA(c) to resist two pests, *Apocheima cinerarius* and *Lymantria dispar*. These insects are notorious for causing severe defoliation in poplar forests, and the introduction of CryIA(c) enhances the resistance of trees (Wang et al. 1996). *Larix decidua*, a species of larch, has been engineered to express the Btδ endotoxin gene, which targets *Coleophora varicella*, a moth species that feeds on larch foliage (Shin et al. 1994). Furthermore, *P. nigra* transformed with the gene CryIAc646 exhibited resistance to *L. dispar* and *Hyphantria cunea*. These pests are known for their devastating impact on poplar trees, and the incorporation of CryIAc646 enhances the ability of trees to defend against these insects (Confalonieri et al. 1994). In addition to these examples, *P. alba* × *P. grandidentata* has been genetically modified with the CryIA(a) gene to target *Malacosoma disstria*. This pest damages the foliage of poplar trees (McCown et al. 1991). These genetic modifications effectively protect trees such as poplar, walnut, larch, etc., from damage caused by pests, ensuring their sustainable growth and health.

Transgenic trees encode proteins toxic to certain insect pests, significantly increasing the trees' resistance to these pests. This enhanced resistance leads to healthier and more productive trees, reducing the need for chemical insecticides. Consequently, this minimizes environmental contamination and protects beneficial insect populations and other non-target species, contributing to greater biodiversity and ecological balance. Additionally, the reduced reliance on chemical pesticides results in cost savings for forest managers and farmers, potentially increasing timber yield and quality, thus offering economic benefits. However, potential concerns include the risk of non-target effects on other organisms and the development of pest resistance to these toxins. These factors must be carefully assessed and managed through ecological monitoring and appropriate regulatory measures. Combining multiple Bt genes with diverse insecticidal properties specificities can expand the spectrum of insect resistance in trees. However, the number of tree varieties possessing such multi-gene resistance traits remains limited, and further research is necessary to fully comprehend this strategy's potential advantages and drawbacks. However, the number of tree varieties possessing such multi-gene resistance traits remains limited, and further research is necessary to fully comprehend this strategy's potential advantages and drawbacks. One significant concern associated with the use of Bt trees is the potential impact on non-target organisms. While Bt toxins are designed to target specific insect pests, there is a possibility that they may also affect beneficial insects or other non-target species that are ecologically important. Careful assessment of the environmental impact is crucial before deploying Bt trees on a large scale. Another critical issue is the potential development of resistance in pest populations. Insects may evolve resistance to Bt toxins over time, rendering the resistance traits ineffective. To mitigate this risk, strategies such as the deployment of refuge areas (non-Bt host plants) and the rotation of different Bt genes can be implemented. Monitoring for the development of resistance in pest populations is also essential for the long-term success of Bt tree plantations. While combining multiple Bt genes can enhance insect resistance in trees, further research is needed to fully understand the associated benefits and potential issues. Careful consideration of non-target effects and resistance management strategies is crucial for the sustainable deployment of Bt trees in forestry and agroforestry systems.

15.3.3 Marker-Assisted Selection (MAS)

Marker-Assisted Selection (MAS) is a cutting-edge technique that leverages genetic markers linked to desirable traits to enhance breeding programs. This method is particularly advantageous in breeding pest-resistant tree varieties, offering a more efficient and precise alternative to traditional breeding methods. MAS involves the utilisation of genetic markers linked to desirable traits to accelerate the breeding of pest-resistant tree varieties. The main goal of tree breeding is to improve the quantity and quality of wood products from plantations. Significant progress has been achieved through progressive selection in genetically diverse breeding populations to harness additive genetic variation. However, the lengthy generation times of trees and the weak correlations between early-stage and late-stage traits have spurred interest in MAS. MAS involves identifying DNA markers that account for a substantial portion of phenotypic variation. Genetic linkage maps for most commercial tree species have enabled the identification of chromosomal regions where DNA markers are associated with quantitative traits loci (QTL). MAS based on QTL is most effective for within-family selection in a few elite families that can be clonally propagated. Challenges include the low resolution of marker-trait associations, the small proportion of phenotypic variation explained by QTL, and the difficulty in validating QTL across different genetic backgrounds and environments. Consequently, research has shifted towards association mapping to identify DNA sequence variations in genes directly responsible for phenotypic traits (e.g., gene-assisted selection, GAS). GAS offers advantages such as high resolution of marker-trait associations and the ability to transfer markers across families and species. Association studies are exploring variations in genes related to wood formation and quality, pathogen resistance, cold tolerance, and drought tolerance. Single nucleotide polymorphisms (SNPs) that are significantly associated with trait variation can be used for early selection. SNPs can be applied across individuals regardless of pedigree, broadening their use in tree-breeding programs in both developing and developed countries. Reductions in genotyping costs and advancements in gene discovery will further augment these opportunities.

15.3.3.1 Benefits of MAS

Accelerated Development: MAS significantly reduced the time required to develop pest-resistant tree varieties. Traditional breeding methods could take decades to achieve similar results, whereas MAS streamlined the process to a few years.

Resource Efficiency: By selecting resistant seedlings early, MAS reduced the need for extensive field trials and mature tree assessments, conserving valuable resources and reducing costs.

Reduced Chemical Use: The development of pest-resistant trees through MAS decreased the reliance on chemical pesticides. This not only lowered environmental impact but also promoted sustainable forest management practices.

Improved Forest Health: The introduction of pest-resistant varieties contributed to healthier and more resilient forests, capable of withstanding pest pressures and maintaining ecological balance.

15.3.4 RNA Interference (RNAi)

RNA interference (RNAi) is a biological process where RNA molecules inhibit gene expression by neutralizing targeted mRNA molecules. This technique can be utilised to silence essential genes in insect pests, leading to their death or reduced reproductive capacity. RNAi has emerged as a promising technique for managing various insect pests by targeting specific genes and inducing significant mortality or developmental abnormalities. RNAi involves the introduction of double-stranded RNA (dsRNA) corresponding to a specific gene in the pest. Once ingested, the dsRNA is processed into small interfering RNAs (siRNAs) that degrade the target mRNA, effectively silencing the gene. This approach provides a highly specific and environmentally friendly method for pest management.

In the case of the spruce budworm (*Choristoneura fumiferana*), targeting the *V-ATPase* gene through feeding with dsRNA resulted in a 48% mortality rate, demonstrating the potential of RNAi in controlling this pest (Martinez et al. 2021). Additionally, injection of dsRNA targeting the Chitin deacetylase (*CDA*) gene led to abnormal phenotypes and higher mortality, further confirming the efficacy of RNAi-based interventions in spruce budworms (Quan et al. 2013). Similarly, RNAi has shown remarkable results in the mountain pine beetle (*Dendroctonus ponderosae*). Feeding dsRNA targeting heat shock protein (*hsp*), Shibire (*shi*), and inhibitor of apoptosis protein (*iap*) genes caused a mortality rate ranging from 75% to 85% (Kyre et al. 2020). This high mortality indicates the significant impact of RNAi on this destructive pest. For the Asian longhorn beetle (*Anoplophora glabripennis*), RNAi targeting 28 different genes through injection resulted in mortality rates between 40% and 100% (Dhandapani et al. 2020a). Additionally, feeding dsRNA targeting *IAP*, *SNF7*, and *SSK* genes led to 75–90% mortality (Dhandapani et al. 2020b). The emerald ash borer (*Agrilus planipennis*) also showed high susceptibility to RNAi. Feeding dsRNA targeting *shi* and *hsp* genes resulted in mortality rates of 69.44% and 46.66%, respectively (Leelesh and Rieske 2020). Further studies indicated mortality rates ranging from 30% to 90% when targeting *hsp* and *shi* genes, and 24–78% when targeting *cop* and *iap* genes through feeding (Rodrigues et al. 2017, 2018). Injection of dsRNA targeting the *AplaScrB-2* gene also led to decreased expression of the targeted gene, highlighting the gene-specific effects of RNAi (Zhao et al. 2015).

The use of RNAi as a pest management strategy has also given promising results in controlling the European gypsy moth (*Lymantria dispar*) and other pests. In the European gypsy moth, injection of dsRNA targeting the ecdysone receptor and ultraspiracle (*usp*) genes resulted in shortened development, indicating a disruption in the insect's hormonal regulation and development process (Wen et al. 2020). Another study targeting the ocular albinism type 1 (*OA1*) gene via injection resulted in a 4.8-fold increase in mortality (Sun et al. 2019). Additionally, oral delivery of dsRNA targeting juvenile hormone epoxide hydrolase (jheh) led to delayed larval development, further demonstrating the potential of RNAi to disrupt critical developmental stages in the gypsy moth (Wen et al. 2018). Feeding dsRNA targeting specific loci, such as Locus 365 and Locus 28,365, resulted in a 60% reduction in body mass (Ghosh and Gundersen-Rindal 2017). For the willow leaf beetle (*Plagiodera versicolora*), a bacteria-mediated delivery system of dsRNA targeting genes such as signal recognition particle protein 54 k (*srp54*), *hsp*, Cactus (*cact*), *shi* and Actin (*act*) achieved mortality rates between 26% and 82% (Zhang et al. 2019). In the case of the Chinese white pine beetle (*Dendroctonus armandi*), injection of dsRNA targeting the *DaAqp12L* gene resulted in increased mortality after cold stress, demonstrating the potential of RNAi to enhance susceptibility to environmental stresses (Fu et al. 2019). Additionally, targeting the *csp2* gene through injection led to reduced development, further confirming the impact of RNAi on the growth and survival of this beetle (Li et al. 2018). These findings underscore the broad applicability and effectiveness of RNAi in pest management. RNAi can significantly reduce pest populations and improve the efficacy of integrated pest management strategies by targeting critical genes involved in development, growth, and stress responses.

15.3.5 Genome Editing

Genome editing is an advanced biotechnological approach with great promise for enhancing tree insect resistance. This method involves precise modifications to the DNA of tree species to introduce traits that confer resistance to targeted insect pests. The first step in genome editing for insect resistance is the identification of genes associated with natural pest resistance. The genomes of various tree species are analysed to pinpoint specific genes that play a role in defending against insect attacks. After identifying the target genes, new genes are inserted or existing ones are modified to improve insect resistance. For instance, genes from other organisms that produce natural insect-repelling compounds can be introduced into the genome of the tree. Edited cells are then regenerated into whole plants through tissue culture techniques, and these genetically edited trees are tested under controlled environments to evaluate their resistance to insect pests. Successful modifications will result in trees that exhibit strong resistance without adverse effects on growth or health. Trees that show promising results in the laboratory are subsequently planted in field trials to assess their performance in natural conditions. These trials help to determine the effectiveness of genome editing in providing long-term resistance to insect pests. Before genetically edited trees can be widely planted, they must undergo rigorous regulatory review to ensure their environmental and human health safety. Once approved, these trees can be deployed in forests and commercial plantations. Genome editing offers numerous benefits for insect resistance in trees. It allows for precise modifications, minimizing unintended effects on the tree's genome and ensuring specific enhancement of desired traits. Compared to traditional breeding methods, genome editing can significantly accelerate the development of insect-resistant tree varieties.

One of the primary advantages of genome editing technologies, such as CRISPR-Cas9, is their ability to allow for highly precise modifications at specific locations in the genome. Researchers can target endogenous genes directly, either by enhancing their expression or by disrupting genes that confer susceptibility to pests or diseases. This precision minimizes off-target effects and reduces the risk of unintended genetic changes, which is a common concern with transgenic methods. Unlike transgenic, which involves introducing foreign genes from other species into an organism's genome, genome editing focuses on the organism's own genetic material. By modifying endogenous genes, scientists can harness and enhance natural traits already present in the organism. By making precise edits to existing genes, researchers can achieve desired traits in a manner that is similar to natural mutations. This can be particularly useful for breeding programs, as it allows for the improvement of traits without altering the genetic integrity of the species. The use of genome editing to modify endogenous genes is often viewed more favourably by regulatory bodies compared to transgenic methods. Additionally, public acceptance of genome-edited products is generally higher, as they are perceived as less artificial and more aligned with natural breeding processes. Genome editing reduces the risk of gene silencing by targeting endogenous genes, ensuring more consistent and stable expression of the desired traits.

One of the most commonly used tools for genome editing is CRISPR-Cas9 (Clustered Regularly Interspaced Short Palindromic Repeats-CRISPR associated protein 9), which allows precise cuts in the DNA at specific locations. By using CRISPR-Cas9, researchers can either enhance the expression of resistance genes or disrupt genes that make trees susceptible to insect damage. The CRISPR-Cas9 system consists of a guide RNA (gRNA), which provides the specificity needed for CRISPR-Cas9 to target a precise DNA sequence consisting of a 20-nucleotide sequence complementary to the target DNA. This ensures that Cas9 cuts the DNA only at the intended location, minimizing off-target effects, and is critical for accurate gene editing, whether for knocking out genes, inserting new sequences, or making specific corrections. This mechanism allows for the insertion, deletion, or modification of genes to confer desired traits such as pest resistance. CRISPR-Cas9 enables precise edits at specific genomic locations, making it a versatile tool for genetic improvement.

Gene editing in trees using CRISPR has been gaining interest, but several challenges and complexities have limited the progress compared to other organisms. Trees have long lifecycles, meaning experiments can take many years to yield results, slowing down research and making it difficult to quickly iterate on experiments. Additionally, many tree species have complex and large genomes, which make genetic manipulation more challenging. Identifying specific genes and their functions within these large genomes is a daunting task. The methods used to introduce CRISPR components into tree cells are less developed and efficient compared to those for model organisms and crop plants. Trees have woody tissues and other structural characteristics that make transformation more difficult, and after gene editing, regenerating a whole tree from edited cells can be challenging due to their specific regeneration conditions. Furthermore, gene editing in trees raises additional regulatory and ethical concerns, given their long-lived nature and the potential ecological impacts of releasing genetically edited trees into the environment. Economic factors also play a role, as research funding and incentives for tree gene editing might be lower than crops, providing immediate economic benefits through improved yields and disease resistance. Much genetic research has traditionally focused on model organisms and staple crops due to their immediate applicability to food security and agriculture, leading to less emphasis on trees. However, despite these challenges, progress is being made in the field, with advancements in CRISPR technology, improved transformation techniques, and a growing understanding of tree genetics gradually overcoming these barriers.

CRISPR-Cas technology offers two primary strategies for plant protection. The first involves genome editing in insects to modify gene expressions that aid in pest management. For instance, the codling moth, a significant pest of pome fruits globally, was targeted using CRISPR-Cas to edit the *CpomOR1* gene, leading to altered egg viability and production (Garczynski et al. 2017). The second strategy focuses on editing plant genomes, specifically targeting genes involved in plant-insect interactions to enhance resistance. For example, the deletion of the *CYP71A1* gene, which encodes tryptamine 5-hydroxylase responsible for converting tryptamine to serotonin in plants, resulted in reduced growth of plant hoppers. This approach was successfully applied to rice, making it resistant to pests like the striped stem borer (*Chilo suppressalis*) and the brown plant hopper (*Nilaparvata lugens*) using CRISPR-Cas9 technology (Lu et al. 2018).

15.3.6 Sterile Insect Technique

The sterile insect technique (SIT) is an eco-friendly, autocidal strategy designed to control insect pests on a large scale. It involves mass-producing sexually sterile insects of a specific species and then releasing them into the wild (Hendrichs et al. 2021). Although these sterile males compete with wild males for females, their mating does not result in viable offspring, leading to a gradual decline in the pest population as the ratio of fertile to sterile insects decreases over time. With consistent releases, this can eventually lead to local eradication. SIT is considered ecologically sound because it is highly target-specific and has gained popularity as a sustainable method of insect control. Originally, SIT was implemented in government and public area-wide integrated pest management programs to eradicate targeted insect populations. However, due to the requirement of specific landscape isolation and higher costs associated with the programs, most current SIT efforts focus on population suppression rather than eradication (Enkerlin 2021). SIT has been explored for managing various agricultural pests, including fruit flies, lepidopteran moths, and beetles, as well as pathogen-carrying insects that affect humans and livestock.

However, its application against forest insect pests has been less common, with only a few examples of successful implementation. For instance, SIT was tested on the spruce budworm (*Choristoneura fumiferana*), a major defoliator in North America. In laboratory conditions, sterile male moths were mass-produced by irradiation and released, resulting in a 90% reduction in male fertility (Ennis 1982). Similar experiments were conducted on the eight-toothed bark beetle (*Ips typographus*), where irradiated males showed increased competition for females compared to control males. However, females mated with irradiated males laid significantly fewer eggs, and the egg hatch rate was considerably lower than in the control group (Turčáni and Vakula 2007). Another significant application of SIT has been against the gypsy moth (*Lymantria dispar*), which produces one generation per year, with females mating only once and laying egg masses containing 200–1600 eggs (Reardon and Mastro 1993). Mastro and Schwalbe (1988) reviewed various SIT approaches for the gypsy moth, including (1) the release of fully sterile male pupae, (2) the release of male pupae given a sub-sterilizing dose of radiation, and (3) the dissemination of F1 sterile eggs. The potential of sub-sterilizing doses was especially appealing, as they resulted in infertile offspring. However, the use of F1 sterile eggs was not successful in practice, and the direct release of sterilized or sub-sterilized males proved to be the most effective method. Despite this, operational use of SIT against the gypsy moth has been limited due to unpredictable efficacy, high costs, and low moth production (Reardon and Mastro 1993). The application of SIT in forest ecosystems requires careful consideration of factors such as the target species' biology, environmental suitability, and potential impacts on non-target species.

15.4 Combining Biotechnological Methods

A combination of biotechnological methods, such as genetic engineering and MAS, holds significant promise for developing pest-resistant trees. Genetic engineering, particularly through CRISPR/Cas9, allows for precise modifications of specific genes associated with desired traits. MAS, on the other hand, facilitates the identification and selection of these traits by linking them to genetic markers. By integrating these two approaches, researchers can efficiently enhance the genetic resistance of trees to pests, leading to more robust and resilient varieties. For example, researchers can use CRISPR/Cas9 to knock out genes that make trees susceptible to pests or to enhance the expression of genes that produce defensive chemicals against insects. This combination of MAS and CRISPR/Cas9 ensures that the selected traits are not only present but are also optimally expressed in the resulting tree varieties. Combining MAS and genetic engineering techniques represents a synergistic approach to tree breeding.

The amalgamation of RNAi and microbial control presents a robust strategy for simultaneously managing multiple pest species in trees. RNAi targets specific genes within pest organisms, disrupting their normal function and leading to their death or reduced fitness. When integrated with microbial control methods, such as the application of entomopathogenic fungi, bacteria, or viruses, this approach provides a comprehensive pest management system that can tackle a broad range of pests effectively. A broad-spectrum pest management strategy is achieved by integrating Bt sprays with RNAi-based gene silencing. While RNAi targets specific pests like the emerald ash borer, Bt sprays manage caterpillar pests, ensuring that multiple pest species are controlled simultaneously. This integrated pest management system not only protects trees from a range of pests but also reduces the likelihood of pest resistance development, as it employs multiple modes of action.

15.5 Challenges and Considerations

15.5.1 Environmental Impact

Assessing the long-term environmental impacts of genetically modified trees and microbial control agents is crucial to ensure ecological balance and biodiversity. Potential risks include unintended effects on non-target species, changes in ecosystem dynamics, and gene flow to wild relatives. For example, genetically modified trees might cross-pollinate with wild relatives, potentially altering their natural characteristics and interactions within the ecosystem. Similarly, microbial control agents like Bt, while effective against target pests, might also affect beneficial insects, disrupting ecological balances. Long-term monitoring and comprehensive environmental impact assessments are essential to mitigate these risks.

15.5.2 Regulatory Approval

Navigating the regulatory landscape is a significant challenge for biotechnological products, which must undergo rigorous evaluation to ensure safety and efficacy before approval. This process involves detailed risk assessment, environmental impact studies, and public consultations, which can be time-consuming and costly. Regulatory bodies require comprehensive data on potential impacts and benefits, as well as mitigation strategies for identified risks. This thorough vetting process is essential to protect public health and the environment.

15.5.3 Public Perception

Public acceptance of genetically modified organisms (GMOs) in forests is influenced by concerns about safety, environmental impact, and ethics. Addressing public concerns is vital for successful implementation. Transparent communication, public engagement, and education are essential to address misconceptions and build trust. Informing the public about the benefits, risks, and rigorous regulatory processes of GMOs can help achieve acceptance and support.

15.5.4 Cost and Accessibility

Ensuring the cost-effectiveness and accessibility of biotechnological solutions is crucial for widespread adoption by forest managers. The development and deployment of these technologies can be expensive, and economic barriers may limit their practical implementation. Strategies to reduce costs and improve scalability include optimizing production processes, securing funding for research, and providing financial support or incentives for adoption. Balancing regulation with economic viability is essential for practical implementation.

15.6 Future Prospects

The future of pest management in trees will be greatly influenced by advancements in genomics and biotechnology. Tools such as CRISPR/Cas9, RNAi, and high-throughput sequencing are expected to offer new and more precise methods for developing pest-resistant tree varieties. These innovations promise to revolutionize pest management with more accurate and efficient solutions. The focus on sustainability will drive the development of environmentally friendly pest management solutions. Combining biotechnological methods with traditional practices will create integrated systems that minimize environmental impact while maintaining forest health. Approaches such as using biopesticides, conserving beneficial organisms, and employing integrated pest management will balance effective pest control with ecosystem preservation. Effective implementation of biotechnological pest management will require collaboration among researchers, forest managers, policymakers, and the public. Building interdisciplinary teams and fostering partnerships will be essential for developing and adopting successful strategies. Such collaboration ensures consideration of diverse perspectives, the establishment of appropriate regulatory frameworks etc.

15.7 Conclusion

Integrating biotechnological methods with traditional pest management strategies offers a comprehensive approach to managing forest insect pests. Success stories like Bt cottonwood and RNAi-based products demonstrate the potential of these technologies. Addressing challenges such as environmental impact, regulatory approval, public perception, and cost-effectiveness is essential for widespread adoption. Future prospects in genomics, sustainable practices, and collaborative efforts will continue to advance pest management solutions, ensuring healthy and resilient forest ecosystems.

References

Alfina T, Haneda NF (2022) Entomopathogeni€c fungi as biological agents in forest plant pest control: a systematic review. IOP Conf Ser Earth Environ Sci 959:012013. IOP Publishing

Allio R, Nabholz B, Wanke S et al (2021) Genome-wide macro-evolutionary signatures of key innovations in butterflies colonizing new host plants. Nat Commun 12:354

Augustyniuk-Kram A, Kram KJ (2012) Entomopathogenic fungi as an important natural regulator of insect outbreaks in forests. Forest ecosystems-more than just trees. pp 265-294, www.intechopen.com

Aung OM, Soytong K, Hyde KD (2008) Diversity of entomopathogenic fungi in rainforests of Chiang Mai Province, Thailand. Fungal Divers 30:15–22

Azizoglu U, Salehi Jouzani G, Yilmaz N, Baz E, Ozkok D (2020) Genetically modified entomopathogenic bacteria, recent developments, benefits and impacts: a review. Sci Total Environ 734:139169

Babaei MR, Barari H, Kara K (2009) Weight differences of male and female pupae of gypsy moth (*Lymantria dispar*) and host-sex preference by two parasitoid species *Lymantrichneumon disparis* and *Exorista larvarum*. Pak J Biol Sci 12(5):443–446

Barbaro L, Battisti A (2011) Birds as predators of the pine processionary moth (Lepidoptera: Notodontidae). Biol Control 56(2):107–114

Bayramoğlu Z, Gençer D, Demir İ (2021) Bioactivity of a betabaculovirus, *Hyphantria cunea* granulovirus, in six lepidopteran insects as potential hosts. Turk J Entomol 45(4):417–424

Bindu TN, Balakrishnan P, Sajeev TV, Sudheendrakumar VV (2022) Seeding baculovirus HpNPV in the epicentre populations of teak defoliator, *Hyblaea puera* to prevent large-scale outbreaks. Curr Sci 122(7):812

Boemare N (2002) Interactions between the partners of the entomopathogenic bacterium nematode complexes,*Steinernema-Xenorhabdus* and *Heterorhabditis-Photorhabdus*. Nematology 4(5):601–603

Bridge ES, Jones AW, Baker AJ (2005) A phylogenetic framework for the terns (Sternini) inferred from mtDNA sequences: implications for taxonomy and plumage evolution. Mol Phylogenet Evol 35(2):459–469

Burges MN (1988) Fungi in biological control systems. Manchester University Press, Manchester

Chang WY, Lantz VA, Hennigar CR, MacLean DA (2012) Benefit-cost analysis of spruce budworm (*Choristoneura fumiferana* Clem.) control: incorporating market and non-market values. J Environ Manag 93(1):104–112

Ciche TA, Ensign JC (2003) For the insect pathogen *Photorhabdus luminescens*, which end of a nematode is out? Appl Environ Microbiol 69(4):1890–1897

Confalonieri M, Balestrazzi A, Bisoffi S (1994) Genetic transformation of *Populus nigra* by agrobacterium tumefaciens. Plant Cell Rep 13:256–261

Coppel HC, Mertins JW (2012) Biological insect pest suppression (4). Springer Science & Business Media

Cuartas-Otálora PE, Gómez-Valderrama JA, Ramos AE, Barrera-Cubillos GP, Villamizar-Rivero LF (2019) Bio-insecticidal potential of nucleopolyhedrovirus and granulovirus mixtures to control the fall armyworm *Spodoptera frugiperda* (JE Smith, 1797) (Lepidoptera: Noctuidae). Viruses 11(8):684

Dandekar AM, McGranahan GH, Vail PV, Uratsu SL, Leslie CA, Tebbets JS (1998) High levels of expression of full-length cryIA (c) gene from *Bacillus thuringiensis* in transgenic somatic walnut embryos. Plant Sci 131(2):181–193

Dara SK, Montalva C, Barta M (2019) Microbial control of invasive forest pests with entomopathogenic fungi: a review of the current situation. Insects 10(10):341

De Boer JG, Harvey JA (2020) Range-expansion in processionary moths and biological control. Insects 11(5):267

Deka B, Baruah C, Babu A (2021) Entomopathogenic microorganisms: their role in insect pest management. Egypt J Biol Pest Control 31:1–8

Del Rincón-Castro MC, Ibarra JE (2011) Entomopathogenic viruses. Biological controlof insect pests, 1st edn. Studium Press, pp 29–64

Dhandapani RK, Duan JJ, Palli SR (2020a) Orally delivered dsRNA induces knockdown of target genes and mortality in the Asian long-horned beetle, *Anoplophora glabripennis*. Arch Insect Biochem Physiol 104(4):e21679

Dhandapani RK, Gurusamy D, Duan JJ, Palli SR (2020b) RNAi for management of Asian long-horned beetle, *Anoplophora glabripennis*: identification of target genes. J Pest Sci 93:823–832

Doychev D, Kechev M, Todorov I, Mirchev P, Bencheva S, Georgiev G (2016) New Entomophagous enemies of *Ips typographus* (Linnaeus, 1758) (Coleoptera: Curculionidae) from Bulgaria. Acta Zool Bulg 68(1):131–134

Drooz AT, Bustillo AE, Fedde GF, Fedde VH (1977) North American egg parasite successfully controls a different host genus in South America. Science 197(4301):390–391

Drooz AT, Quimby JW, Thompson LC, Kulman HM (1985) Introduction and establishment of *Olesicampe benefactor* Hinz (Hymenoptera: Ichneumonidae), a parasite of the larch sawfly, *Pristiphora erichsonii* (Hartig) (Hymenoptera: Tenthredinidae), in Pennsylvania. Environ Entomol 14(4):420–423

Duan JJ, Bauer LS, Van Driesche RG, Gould JR (2018) Progress and challenges of protecting North American ash trees from the emerald ash borer using biological control. Forests 9(3):142

Duan JJ, Van Driesche RG, Schmude J, Crandall R, Rutlege C, Quinn N, Elkinton JS (2022) Significant suppression of invasive emerald ash borer by introduced parasitoids: potential for North American ash recovery. J Pest Sci 95(3):1081–1090

Enkerlin WR (2021) Impact of fruit fly control programmes using the sterile insect technique. In: Sterile insect technique. CRC Press, pp 979–1006

Ennis TJ (1982) The sterile male approach to the control of the spruce budworm. *Choristoneura fumiferana*. In: Sterile insect technique and radiation in insect control. IAEA, Vienna, pp 389–393

Evans HC (1982) Entomogenous fungi in tropical forest ecosystems: an appraisal. Ecol Entomol 7(1):47–60

Fan Y, Fang W, Guo S, Pei X, Zhang Y, Xiao Y, Li D, Jin K, Bidochka MJ, Pei Y (2007) Increased insect virulence in *Beauveria bassiana* strains overexpressing an engineered chitinase. Appl Environ Microbiol 73(1):295–302

Fu D, Dai L, Gao H, Sun Y, Liu B, Chen H (2019) Identification, expression patterns and RNA interference of aquaporins in *Dendroctonus armandi* (Coleoptera: Scolytinae) larvae during overwintering. Front Physiol 10:967

Garczynski SF, Martin JA, Griset M, Willett LS, Cooper WR, Swisher KD, Unruh TR (2017) CRISPR/Cas9 editing of the codling moth (Lepidoptera: Tortricidae) CpomOR1 gene affects egg production and viability. J Econ Entomol 110(5):1847–1855

Génissel A, Leplé JC, Millet N, Augustin S, Jouanin L, Pilate G (2003) High tolerance against *Chrysomela tremulae* of transgenic poplar plants expressing a synthetic cry3Aa gene from *Bacillus thuringiensisssp tenebrionis*. Mol Breed 11:103–110

Ghosh SKB, Gundersen-Rindal DE (2017) Double strand RNA-mediated RNA interference through feeding in larval gypsy moth, *Lymantria dispar* (Lepidoptera: Erebidae). Eur J Entomol 114:170–178

Glare TR, Jurat-Fuentes JL, O'callaghan M (2017) Basic and applied research: entomopathogenic bacteria. In: Microbial control of insect and mite pests. Academic, pp 47–67

Grace LJ, Charity JA, Gresham B, Kay N, Walter C (2005) Insect-resistant transgenic *Pinus radiata*. Plant Cell Rep 24:103–111

Guegan JF, de Thoisy B, Gómez-Gallego M, Jactel H (2023) World forests, global change, and emerging pests and pathogens. Curr Opin Environ Sustain 61:101266

Harcourt RL, Kyozuka J, Floyd RB, Bateman KS, Tanaka H, Decroocq VLDJ, Dennis ES (2000) Insect-and herbicide-resistant transgenic eucalypts. Mol Breed 6:307–315

Harish S, Murugan M, Kannan M, Parthasarathy S, Prabhukarthikeyan SR, Elango K (2021) Entomopathogenic viruses. In: Microbial approaches for insect pest management, pp 1–57

Harvey JA, Tougeron K, Gols R et al (2023) Scientists' warning on climate change and insects. Ecol Monogr 93:1–37

Hassell MP (1980) Foraging strategies, population models and biological control: a case study. J Anim Ecol 307:603–628

Hendrichs J, Vreysen MJB, Enkerlin WR, Cayol JP (2021) Strategic options in using sterile insects for area-wide integrated pest management. In: Sterile insect technique. CRC Press, pp 841–884

Hu JJ, Tian YC, Han YF, Li L, Zhang BE (2001) Field evaluation of insect-resistant transgenic *Populus nigra* trees. Euphytica 121:123–127

Ince IA, Demir I, Demirbag Z, Nalcacioglu R (2007) A cytoplasmic polyhedrosis virus isolated from the pine processionary caterpillar, *Thaumetopoea pityocampa*. J Microbiol Biotechnol 17(4):632

Irsad, Shahid M, Haq E, Mohamed A, Rizvi PQ, Kolanthasamy E (2023) Entomopathogen-based biopesticides: insights into unraveling their potential in insect pest management. Front Microbiol 14:1208237

Jiang F, Zhang T, Bai S, Wang Z, He K (2016) Evaluation of Bt corn with pyramided genes on efficacy and insect resistance management for the Asian corn borer in China. PLoS One 11(12):e0168442

Kalha CS, Singh PP, Kang SS, Hunjan MS, Gupta V, Sharma R (2014) Entomopathogenic viruses and bacteria for insect-pest control. In: Integrated pest management. Academic, pp 225–244

Keller S, Zimmermann G, Wilding N, Collins NM, Hammond PM, Webber JF (1989) Mycopathogens of soil insects. In: Insect-fungus interactions, pp 239–270

Klingen I, Westrum K, Meyling NV (2015) Effect of Norwegian entomopathogenic fungal isolates against *Otiorhynchus sulcatus* larvae at low temperatures and persistence in strawberry rhizospheres. Biol Control 81:1–7

Kobakhidze DN, Tvaradze MS, Yashvill GV, Kraveishvili IK (1968) Artificial rearing of *Rhizophagus grandis* Gyll. for the control of *Dendroctonus micans* KUG. Georgia, Soobshch. Akad Nauk GruzSSR 51(2):435–440

Kyre BR, Bentz BJ, Rieske LK (2020) Susceptibility of mountain pine beetle (*Dendroctonus ponderosae* Hopkins) to gene silencing through RNAi provides potential as a novel management tool. For Ecol Manag 473:118322

Leelesh RS, Rieske LK (2020) Oral ingestion of bacterially expressed dsRNA can silence genes and cause mortality in a highly invasive, tree-killing pest, the emerald ash borer. Insects 11(7):440

Li Z, Dai L, Chu H, Fu D, Sun Y, Chen H (2018) Identification, expression patterns, and functional characterization of chemosensory proteins in *Dendroctonus armandi* (Coleoptera: Curculionidae: Scolytinae). Front Physiol 9:291

Lu HP, Luo T, Fu HW, Wang L, Tan YY, Huang JZ, Wang Q, Ye GY, Gatehouse AMR, Lou YG et al (2018) Resistance of rice to insect pests mediated by suppression of serotonin biosynthesis. Nat Plants 4(6):338–344

Maaß O, Kehlenbeck H (2024) Cost–benefit analysis of monitoring insect pests and aerial spraying of insecticides: the case of protecting pine forests against *Dendrolimus pini* in Brandenburg (Germany). Forests 15(1):104

Majchrowska-Safaryan A, Tkaczuk C (2021) Abundance of entomopathogenic fungi in leaf litter and soil layers in forested habitats in Poland. Insects 12(2):134

Martemyanov VV, Akhanaev YB, Belousova IA, Pavlusin SV, Yakimova ME, Kharlamova DD, Tokarev YS (2023) A new cypovirus-1 strain as a promising agent for Lepidopteran pest control. Microbiol Spectrum 11(3):e03855–e03822

Martinez Z, De Schutter K, Van Damme EJ, Vogel E, Wynant N, Broeck JV, Smagghe G (2021) Accelerated delivery of dsRNA in lepidopteran midgut cells by a *Galanthus nivalis* lectin (GNA)-dsRNA-binding domain fusion protein. Pestic Biochem Physiol 175:104853

Mastro VC, Schwalbe CP (1988) Status and potential of F, sterility for the control of noxious Lepidoptera. In: Proceedings of the international symposium on modern insect control: nuclear techniques and biotechnology, pp 15–40

McCown BH, McCabe DE, Russell DR, Robison DJ, Barton KA, Raffa KF (1991) Stable transformation of Populus and incorporation of pest resistance by electric discharge particle acceleration. Plant Cell Rep 9:590–594

Meilan R, Ma C, Cheng S, Eaton JA, Miller LK, Crockett RP, Strauss SH (2000) High levels of roundup® and leaf-beetle resistance in genetically engineered hybrid cottonwoods. In: Hybrid poplars in the Pacific Northwest: culture, commerce and capability, pp 29–38

Mondor EB, Roland J (1997) Host locating behaviour of *Leschenaultia exul* and *Patelloa pachypyga*: two tachinid parasitoids of the forest tent caterpillar, Malacosoma disstria. Entomol Exp Appl 85(2):161–168

Mondor EB, Roland J (1998) Host searching and oviposition by *Leschenaultia exul*, a tachinid parasitoid of the forest tent caterpillar, Malacosoma disstria. J Insect Behav 11(4):583–592

Neves PA, Colabuono FI, Ferreira PA, Kawakami SK, Taniguchi S, Figueira RC, Bícego MC (2018) Depositional history of polychlorinated biphenyls (PCBs), organochlorine pesticides (OCPs) and polycyclic aromatic hydrocarbons (PAHs) in an Amazon estuary during the last century. Sci Total Environ 615:1262–1270

Nixon AE, Roland J (2012) Generalist predation on forest tent caterpillar varies with forest stand composition: an experimental study across multiple life stages. Ecol Entomol 37(1):13–23

Paradva KC, Kalla S (2023) Nanopesticides: a review on current research and future perspective. Chem Select 8(26):e202300756

Parry D, Spence J, Volney W (1997) Responses of natural enemies to experimentally increased populations of the forest tent caterpillar, *Malacosoma disstria*. Ecol Entomol 22(1):97–108

Prischepa L, Mikulskaya N, Gerasimovich M, Sosnowska D, Balazy S (2011) Diversity of entomopathogenic microorganisms in pest populations of Bialowieza forest stands. Vytauto Didžiojo universiteto Botanikos sodo raštai 15:72–81

Pschorn-Walcher H (1977) Biological control of forest insects. Annu Rev Entomol 22(1):1–22

Pu YC, Ma TL, Hou YM, Sun M (2017) An entomopathogenic bacterium strain, *Bacillus thuringiensis*, as a biological control agent against the red palm weevil, *Rhynchophorus ferrugineus* (Coleoptera: Curculionidae). Pest Manag Sci 73(7):1494–1502

Pu YC, Zheng ZW, Ding CH, Chen XD (2023) Development of potential microbial agents with two new entomopathogenic fungal strains to control the red palm weevil *Rhynchophorus ferrugineus*(Olivier)(Coleoptera: Curculionidae). Egypt J Biol Pest Control 33(1):107

Quan G, Ladd T, Duan J, Wen F, Doucet D, Cusson M, Krell PJ (2013) Characterization of a spruce budworm chitin deacetylase gene: stage-and tissue-specific expression, and inhibition using RNA interference. Insect Biochem Mol Biol 43(8):683–691

Raj MN, Samal I, Paschapur A, Subbanna ARNS (2022) Entomopathogenic viruses and their potential role in sustainable pest management. In: New and future developments in microbial biotechnology and bioengineering. Elsevier, pp 47–72

Reardon RC, Mastro VC (1993) Development and status of the sterile insect technique for managing gypsy moth. NA-TP, pp 13–93

Rodrigues TB, Rieske LK, Duan J, Mogilicherla K, Palli SR (2017) Development of RNAi method for screening candidate genes to control emerald ash borer, *Agrilus planipennis*. Sci Rep 7(1):7379

Rodrigues TB, Duan JJ, Palli SR et al (2018) Identification of highly effective target genes for RNAi-mediated control of emerald ash borer, *Agrilus planipennis*. Sci Rep 8:5020

Roychoudhury N (2015) Biological control of teak defoliator and leaf skeletonizer. TFRI, pp 1–100

Ruiu L (2018) Microbial biopesticides in agroecosystems. Agronomy 8(11):235

Ryan RB, Tunnock S, Ebel FW (1987) The larch casebearer in North America. J For 85(7):33–39

Samways MJ, Barton PS, Birkhofer K et al (2020) Solutions for humanity on how to conserve insects. Biol Conserv 242:108427

Schowalter TD (2017) Biology and management of the forest tent caterpillar (Lepidoptera: Lasiocampidae). J Integr Pest Manag 8(1):24

Sharmila R, Subramanian S, Poornima K (2018) Host range of entomopathogenic nematodes. J Entomol Zool Stud 6(3):1310–1312

Shin DI, Podila GK, Huang Y, Karnosky DF (1994) Transgenic larch expressing genes for herbicide and insect resistance. Can J For Res 24(10):2059–2067

Sosa-Gómez DR, Morgado FS, Corrêa RFT, Silva LA, Ardisson-Araújo DMP, Rodrigues BMP, Ribeiro BM (2020) Entomopathogenic viruses in the neotropics: current status and recently discovered species. Neotrop Entomol 49:315–331

Sun L, Liu P, Zhang C, Du H, Wang Z, Moural TW, Cao C (2019) Ocular albinism type 1 regulates deltamethrin tolerance in *Lymantria dispar* and *Drosophila melanogaster*. Front Physiol 10:766

Sweeney J, Dodds KJ, Fettig CJ, Carnegie AJ (2023) IPM: the Forest context. In: Allison DJ, Paine TD, Slippers B, Wingfield MJ (eds) Forest entomology and pathology. Springer International Publishing, Cham, pp 581–646

Syme PD (1977) Observations on the longevity and fecundity of *Orgilus obscurator* (Hymenoptera: Braconidae) and the effects of certain foods on longevity. Can Entomol 109(7):995–1000

Szewczyk B, Hoyos-Carvajal L, Paluszek M, Skrzecz I, Lobo de Souza M (2006) Baculoviruses-Re-emerging biopesticides. Biotechnol Adv 24(2):143–160

Takov D, Pilarska D, Wegensteiner R (2006) Entomopathogens in *Ips typographus* (Coleoptera: Scolytidae) from several spruce stands in Bulgaria. Acta Zool Bulg 58(3):409–420

Takov D, Doychev D, Wegensteiner R, Pilarska D (2007) Study of bark beetle (Coleoptera, Scolytidae) pathogens from coniferous stands in Bulgaria. Acta Zool Bulg 59(1):87–96

Tang W, Tian Y (2003) Transgenic loblolly pine (*Pinus taeda* L.) plants expressing a modified δ-endotoxin gene of *Bacillus thuringiensis* with enhanced resistance to *Dendrolimus punctatus* Walker and *Crypyothelea formosicola* Staud. J Exp Bot 54(383):835–844

Tang W, Harris L, Newton RJ (2004) Plant biotechnology: a case study of *Bacillus thuringiensis* (BI) and its application to the future of genetic engineered trees. J For Res 15:1–10

Taylor KL (1981) The Sirex wood wasp: ecology and control of an introduced forest insect (pine pest). CSIRO

Turčáni M, Vakula J (2007) The influence of irradiation on the behaviour and reproduction success of eight toothed bark beetle *Ips typographus* (Coleoptera: Scolytidae). J For Sci 53:31–37
Wang GeJiao WG, Castiglione S, Chen Ying CY, Li Ling LL, Han YiFan HY, Tian YingChuan TY, Sala F (1996) Poplar (*Populus nigra* L.) plants transformed with a *Bacillus thuringiensis* toxin gene: insecticidal activity and genomic analysis. Transgenic Res 5:289–301
Wen R, Wang B, Wang B, Ma L (2018) Characterization and expression profiles of juvenile hormone epoxide hydrolase from *Lymantria dispar* (Lepidoptera: Lymantridae) and RNA interference by ingestion. J Insect Sci 18(1):13
Wen R, Zhao Q, Wang B, Ma Y, Ma L (2020) Molecular characterization and functional analysis of USP-1 by RNA interference in the Asian gypsy moth *Lymantria dispar*. J For Res 31:1923–1931
Weseloh RM, Andreadis TG, Moore REB, Anderson JF, Dubois NR, Lewis FB (1983) Field confirmation of a mechanism causing synergism between *Bacillus thuringiensis* and the gypsy moth parasitoid, *Apanteles melanoscelus*. J Invertebr Pathol 41(1):99–103
Yaman M, Baki H (2011) First record of Entomopoxvirus of *Ips typographus* (Linnaeus) (Coleoptera: Curculionidae, Scolytinae) for Turkey. Acta Zool Bulg 63:199–202
Ying WG, Sheng YM, Mei HX, Jie LX (2012) Comparison of exogenous gene expression and insect-resistance ability of transgenic 741 poplars with single and double Bt genes. Acta Entomol Sin 55:798–803
Zhang Y, Xu L, Li S, Zhang J (2019) Bacteria-mediated RNA interference for management of *Plagiodera versicolora* (Coleoptera: Chrysomelidae). Insects 10(12):415
Zhao C, Gonzales MAA, Poland TM, Mittapalli O (2015) Core RNAi machinery and gene knockdown in the emerald ash borer (*Agrilus planipennis*). J Insect Physiol 72:70–78
Zhao H, Lovett B, Fang W (2016) Genetically engineering in entomopathogenic fungi. Adv Genet 94:137–163

Biotechnological Approaches for Combatting Tree Diseases to Enhance the Forest Ecosystem Sustainability

16

K. Darshan, Aditi Tailor, Komal Rani, Amoghavarsha Chittaragi, Diksha Loona, E. Santhoshinii, Sonali Nigam, and S. Shreedevasena

Abstract

Forest tree diseases pose a significant threat to the health and sustainability of forests due to climate change, human-driven deforestation, and global trade, which can introduce invasive pathogens. As forests' environmental and economic importance becomes increasingly obvious, developing effective strategies to combat these diseases is imperative. Biotechnology offers innovative and sustainable solutions for the management and mitigation of tree diseases. The multi-

K. Darshan (✉)
Forest Protection Division, ICFRE-Tropical Forest Research Institute, Jabalpur, Madhya Pradesh, India

A. Tailor
Genetics and Tree Improvement Division, ICFRE-Arid Forest Research Institute, Jodhpur, Rajasthan, India

K. Rani
Genetics and Tree Improvement Division, ICFRE-Tropical Forest Research Institute, Jabalpur, Madhya Pradesh, India

A. Chittaragi
Department of Plant Pathology, CCSHAU, Hisar, Haryana, India

D. Loona
Department of Plant Pathology, Punjab Agricultural University, Ludhiana, Punjab, India

E. Santhoshinii
University of Agricultural Sciences, Dharwad, Karnataka, India

S. Nigam
Department of Botany and Microbiology, St. Aloysius College, Jabalpur, Madhya Pradesh, India

S. Shreedevasena
ICAR-Directorate of Medicinal and Aromatic Plants Research, Anand, Gujarat, India

R. K. Kalia, R. Pathak (eds.), *Tree Biology and Biotechnology*, https://doi.org/10.1007/978-981-96-0002-1_16

omics approach involves integrating data through different omics fields, including genomics, transcriptomics, proteomics, and metabolomics, to realize a wider insight into biological systems. By analysing plants' genetic, transcriptomic, proteomic, and metabolomic profiles and associated microbes, we can gain insights into the mechanisms underlying plant-microbial interactions and develop predictive models for managing plant diseases under changing environmental conditions. Marker-assisted breeding is a powerful method for enhancing breeding programs' efficiency by using DNA markers associated with loci or QTLs of interest. Marker-assisted selection has been widely implemented to introduce disease resistance genes into susceptible lines of several commercial crop species. It also allows for gene pyramiding to broaden the disease tolerance spectrum. This chapter comprehensively explores genetic resistance and marker-assisted selection (MAS), genome editing, RNA interference (RNAi), plant immune system activation and omics approaches in forest disease management. These approaches not only contribute to disease control but also promote the preservation of biodiversity and the overall health of forest ecosystems.

Keywords

Tree diseases · Biotechnological approaches · Multi-omics approach · Marker-Assisted Selection · Forest disease management

16.1 Introduction

Forests are a diverse ecosystem, and trees have a wide genetic base, which makes them susceptible to various diseases. Due to increasing food demand and modifications in climatic trends like altered precipitation, rising temperatures impacted the distribution and behaviour of pathogens and pests. Forest trees are long-term plants with extended growth periods and complex genomes. They are highly lignified woody plants appearing in different shapes, shortly known as macrophanerophytes, bushes or subshrubs. They play a major role in surface morphology, biodiversity, and combatting climatic changes. Due to global warming, the need of the hour is to protect the forest trees to prevent the extinction of rare woody plants. Besides these, abiotic stress such as drought and salinity as well as biotic stress such as pathogen and pest attacks are also major bottlenecks. The sustainable forest area in India is about 24.62% (Anonymous 2023). Therefore, forest trees need to be more climate-resilient to resist any biotic and abiotic stress and ensure ecosystem sustainability and long-term health. The trait improvement such as regulating the production of secondary metabolites, namely, carotenoids, anthocyanin, and salicylate, regulating the plant hormones for early flowering, and enriching the timber properties like high lignin content is crucial for forest afforestation as well as contributing to forest economy. Conventional methods of disease control, such as the use of pesticides, often have drawbacks due to their potential for environmental pollution and the development of pest resistance. Therefore, the development of sustainable and environmental friendly disease management strategies is essential for the protection of forest trees.

Plant diseases are global threats that adversely affect both plant growth and productivity and thus have great economic implications for any country. Disease management, therefore, is an important objective that needs considerable attention. This includes strategies to mitigate damages by eliminating or suppressing the disease-causing element or pathogen and improving plants' resistance to the pathogen. Since important tree species are grown in plantations with a narrow genetic base, rendering them more vulnerable to diseases, it becomes crucial to identify newer genotypes by undertaking disease resistance assessment across populations to prevent economic losses. Utilization of advanced molecular tools has the potential to serve as an excellent alternative or complementary strategy to traditional breeding and could greatly accelerate this process of selection for resistance in forest trees as compared to the traditional phenotypic evaluation. Genetic engineering approaches like double-stranded RNA mediated interference (RNAi) have proved to be a significant tool in managing forest tree disease. RNAi is a cellular process degrading target-specific mRNA, thereby suppressing transcription (Gouthu et al. 2022). It has been widely explored in the various forest trees in insect-pest management (Li et al. 2019; Kyre et al. 2019; Yuan et al. 2023) and fungal control (Bhagta et al. 2023), but bacterial control studies are very scanty.

The forest trees are prone to infection of several fungi viz., *Ceratocystis fimbriata, Fusarium circinatum, Hymenoscyphus fraxineus, Melampsora columbiana, Neonectria faginata, Phyllosticta citricarpa, Phytophthora pluvialis, P. agathidicida, Ophiostoma novo-ulmi, Sphaerulina musiva* which are the recently reported emerging pathogens. Some of the commonly infecting pathogens which cause damage to greater extent are *Phytophthora ramorum, Fomes annosus, Poria weirii, Nectria galligena, Botryodiplodia theobromae, Fomes pini, Ganoderma lucidum, Phanerochaete salmonicolor, Elytroderma deformans* and *Ravenelia hobsoni* (Parthasarathy et al. 2021). Plant disease ecology relies on various fields like microbial ecology, plant physiology, epidemiology and genetics to apprise thoughtful findings. Maintaining plant health is vital for good yields in managed environments like forests or farms. Recent technological progress has revolutionized plant disease studies, allowing a comprehensive understanding of pathogenesis. Tools like omics aid in thoroughly examining plant and microbial traits across genetics and appearances. This in-depth understanding has led to the discovery of defence mechanisms and better management strategies, from refining breeding to on-field practices that limit pathogen spread. These omics-driven advancements are crucial given the ever-changing environment influencing plant health. Omics-driven methods have enabled the creation of genome-scale tools and advanced translational research by merging insights to grasp the dynamics of Plant-Pathogen Interactions (PPI). It includes understanding pathogen ecology, molecular mechanisms, and the core principles of disease development. To delve into these interactions, this field explores genomics, transcriptomics, proteomics, metagenomics, metabolomics, volatilomics, and spectranomics. To overcome the barriers associated with conventional genetic engineering for developing resistance via plant breeding, the genome editing tool CRISPR/

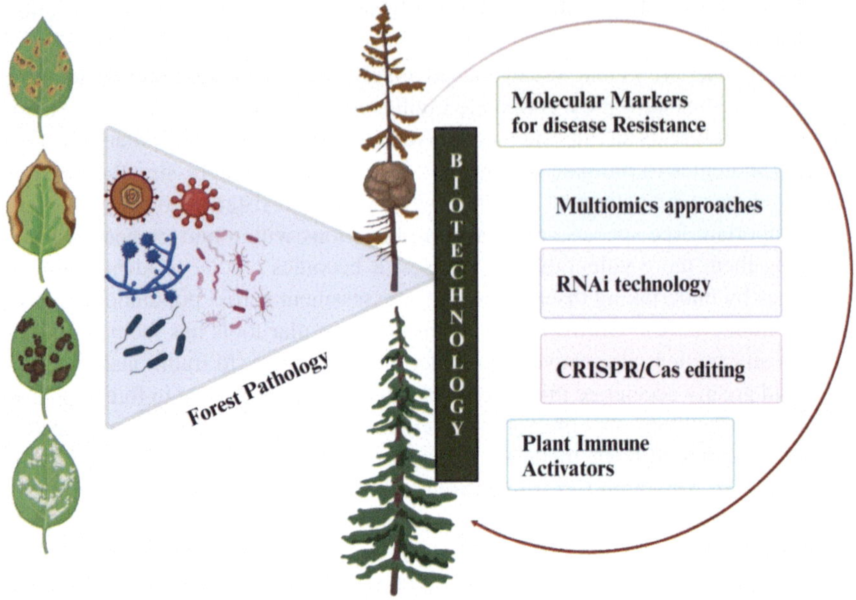

Fig. 16.1 Illustration of different biotechnological techniques in forest tree disease management

Cas9 is used to precisely edit complex genomes with higher transformation efficiency, developing an improved woody plant in a short time. CRISPR/Cas9 technology is one of the potential tools for precise genome editing in woody plants. Biotechnological approaches offer promising alternatives to conventional disease control methods by utilizing Plant Immune Activators (PIAs) to enhance the natural defence mechanisms of trees against pathogens. PIAs are elicitors that trigger plant defence responses, enabling them to recognize and effectively combat invading pathogens. In brief, this chapter highlights the different biotechnological approaches like genetic resistance and marker-assisted selection (MAS), genome editing, RNA interference (RNAi), plant immune system activation and omics approaches for combatting tree diseases to enhance the forest ecosystem sustainability (Fig. 16.1).

16.2 Molecular Markers for Disease Tolerance

Marker-Assisted Selection (MAS) is an advanced molecular approach that speeds up the selection process. It takes advantage of molecular markers, which, in the most basic sense, are DNA sequences that serve as molecular tags for particular genomic regions. MAS principally makes use of the linkage disequilibrium (LD) among DNA markers and their neighbouring QTLs or genes associated with the desirable trait (Muranty et al. 2014). Molecular markers comprise an important category of molecular tools and have profoundly impacted plant breeding by helping to increase genetic gain, shorten breeding programs aimed at improvement in relation to various important traits, improve the selection precision and efficiency of evaluation, and reduce reliance on conventional screening methods (Muranty et al. 2014; Nadeem et al. 2018). Some of the early-developed markers were based on hybridization (e.g., RFLP, Restriction Fragment Length Polymorphism), which gradually evolved into PCR-based marker systems (e.g., AFLP, Amplified Fragment Length Polymorphism; RAPD, Random Amplified Polymorphic DNA; ISSR, Inter Simple Sequence Repeat; SSR, Simple Sequence Repeat etc.), and more recently sequencing-based (Next generation sequencing or NGS) systems, genotyping-by-sequencing (GBS), and association mapping (Table 16.1). With advancements in genomics and sequencing technologies, high-throughput markers are being developed and made available to fulfil specific requirements and targeted use by the end-users. In addition to being an indispensable part of modern-day breeding programs, the advantages of molecular markers have been utilized for other applications, such as genetic map construction, genetic diversity assessment, as well as phylogenetic relationship analysis (Porth and El-Kassaby 2014; Zhang et al. 2020; Younessi-Hamzekhanlu and Gailing 2022; Rakída et al. 2023). The main factors attributable to the widespread acceptance of molecular markers include their genetic stability, efficiency, cost-effectiveness, and ease of use.

The popularization of MAS is on account of the ease with which the number of resistance genes may be determined through the linked DNA markers, which would have been difficult to assess with phenotypic evaluation given that they exert individually non-distinguishable effects on the traits (Goswami et al. 2023). With MAS, it has even been possible to achieve single gene introgressions and gene pyramiding or integration of multiple gene determinants of disease resistance into a single genotype to develop a more durable and broader spectrum resistance (Muranty et al. 2014). These determinants include the qualitative resistance genes that can prove more effective against the ever-evolving pathogens when pyramided. Qualitative resistance is controlled by a single gene, referred to as the *R* gene, which imparts almost complete resistance to the disease (Jones and Dangl 2006). These are thus the major genes that can be identified by both QTL mapping and genome-wide association studies (GWAS) (Weiss et al. 2020). QTL mapping and GWAS are prerequisites for MAS application to breeding programs. Along with the development of markers, it is also crucial that genes or QTLs contributing to a trait are identified, in this case, those associated with a greater resistance or lower susceptibility. In general, methods like QTL mapping and association studies as well as the construction of genetic maps have been adopted to investigate disease resistance traits which, in turn, can accelerate the breeding of forest tree species. This QTL discovery can be done at a larger genome-wide scale using the high-throughput GWAS (Gangurde et al. 2022). Both have been used for discovering genomic regions associated with disease resistance, but QTL mapping suffers from linkage drag which makes it non-deployable many times, a limitation that is overcome by GWAS as it uses high-throughput markers and is based on LD (Shamari et al. 2018; Gangurde et al. 2022). Hence, GWAS has been more popularly used to detect and narrow down QTLs linked with disease resistance traits to single genes to achieve a greater precision and lower false discovery rate (Gangurde et al. 2022). The other category of

Table 16.1 Molecular markers linked with genes/loci associated with resistance to pathogens causing various diseases in tree species

Plant species		Disease	Causal organism/ pathogen	Locus/ gene involved	Markers utilized	Reference(s)
Timber/forest trees						
Pine	Sugar pine (*Pinus lambertiana*)	White pine blister rust	*Cronartium ribicola*	–	RAPD: 10 (OPD-19/1120, OPE-16/800, OPF-03/810, OPF-07/950, OPK-01/1110, OPT-15/650, OPAD-09/920, OPAG-05/610, OPAI-03/650, OPAN-10/590)	Devey et al. (1995)
				Cr1 locus	RAPD markers (BC_432_1110 and OPG_16_950) converted to SCAR markers	Jermstad et al. (2011)
				Cr1^{R} locus	CAPS marker based on 5 SNPs: *Cr1AM1*	Wright et al. (2022)
	Japanese black pine (*Pinus thunbergii*)	Pine needle gall	Pine needle gall midge (*Thecodiplosis japonensis*)	*R* locus	RAPD: *OPC06580*, *OPD01700*, and *OPAX192100*	Kondo et al. (2000)
	Western white pine (*Pinus monticola*)	White pine blister rust	*Cronartium ribicola*	*Cr2* locus	RAPD markers: 8 (Six of them (U181-420, U256-1385, U550-2500, U570-843, OA11-2500, and OI10-1250) in coupling, and 2 (U550-2200r and U561-690r) in repulsion	Liu et al. (2006)
					SNP: snp58688-82Y and snp41490-1778 M locus (SNP genotyping by TaqMan and PCR-KASP)	Liu et al. (2017)
				PmPR10-2	SSR marker: 5′-AATTATTT-3′	Liu et al. (2013)
	Loblolly pine (*Pinus taeda*)	Fusiform rust	*Cronartium quercuum* f.sp. *fusiforme*	*Fr1* locus	RAPD: J_7_485A	Kuhlman et al. (2007)
	Limber pine (*Pinus flexilis*)	White pine blister rust	*Cronartium ribicola*	*Cr4* locus	SNP markers: 32	Liu et al. (2016)

Elm	Chinese Elm (*Ulmus parvifolia* J)	Black leaf spot	*Stegophora ulmea* (	–	RAPD: $OPP17_{1590}$, $OPC20_{900}$, $OPP17_{1630}$	Benet et al. (1995)
Poplar	*Populus* sp.	Leaf rust	*Melampsora larici-populina*	*Mer* locus	AFLP: 3 tightly linked markers: E39. G01.270, E40.G37.320, and E44. G09.100	Cervera et al. (1996)
	Hybrid poplar [*Populus trichocarpa* × *P. deltoides*]	Leaf rust	*Melampsora columbiana* (pathotype 3)	*MXC3* locus	AFLP: 7 co-segregated with locus; CCG. GCT_01 converted to STS marker (*BVS1*)	Stirling et al. (2001)
	Populus deltoides	Marssonina leaf spot	–	–	RAPD markers: OPAI17-1550 and OPAI13-900	Zhang et al. (2002)
	Populus deltoides	Black spot disease	–	–	RAPD markers: OPAI17-1550 and OPAI13-900 converted into SCAR markers SAI17-1570 and SAI13-292, respectively	Zhang et al. (2007)
Eucalyptus	*Eucalyptus grandis*	Myrtle rust	*Puccinia psidii* Winter	*Ppr1*	RAPD: AT9/917	Junghans et al. (2003)
					RAPD marker AT9/917 converted into SCAR marker	Laia et al. (2015)
Hazelnut	*Corylus avellana*	Eastern filbert blight	*Anisogramma anomala*	–	RAPD: 7 ($AA12_{850}$, 73_{500}, 152_{800}, 122_{825}, 275_{1130}, $H19_{650}$ and $O16_{1250}$. Marker 268_{580})	Mehlenbacher et al. (2004)
Sea buckthorn	*Hippophae*	Dried-shrink disease	–	–	ISSR markers:887_{190}, 835_{700}, 809_{290}, and 811_{280}	Ruan et al. (2009)
					SRAP markers: 11 (e1-$Em3_{600}$, Me1-$Em1_{680}$, Me2-$Em1_{650}$, Me2-$Em1_{950}$, Me3-$Em6_{1300}$, Me2-$Em6_{320}$, Me2-$Em6_{400}$, Me1-$Em2_{600}$, Me1-$Em1_{1200}$, Me1-$Em1_{1700}$, Me2-$Em2_{250}$)	Li et al. (2010)

(continued)

Table 16.1 (continued)

Plant species		Disease	Causal organism/ pathogen	Locus/ gene involved	Markers utilized	Reference(s)
Casuarina	*Casuarina equisetifolia*	Bacterial wilt	*Ralstonia solanacearum*	–	AFLP	Sun et al. (2014)
Cocoa	*Theobroma cacao* L.	Vascular streak dieback	*Ceratobasidium theobromae* [now *Rhizoctonia theobromae*	–	SSR: mTcCIR42	Chandrakant (2014)
					ISSR: UBS857	Tulshiram (2016)
Ash	European ash (*Fraxinus excelsior*)	Dieback disease	*Hymenoscyphus fraxineus*	G allele	cDNA-based SNP or cSNP using Associative transcriptomics	Harper et al. (2016)
Oak	Persian oak (*Quercus brantii*)	Dieback	–	–	ISSR markers	Shamari et al. (2018)
Albizzia	Sengon or Moluccan Albizzia (*Falcatria moluccana*)	Gall rust disease	*Uromycladium tepperianum*	–	SSR markers: Pafa05 allele 2	Siregar et al. (2019)
Fruit trees						
Apple	*Malus floribunda*	Apple scab	*Venturia inaequalis*	*Vf* gene locus	RAPD: $OPAM19_{2200}$ and $OPAL07_{580}$	Tartarini (1996)
	Malus sieboldii	Crown gall	*Agrobacterium tumefaciens*	*Cg*	SSR marker: NZmsEB119405	Moriya et al. (2010)
	Malus × *domestica*	Apple scab	*Venturia inaequalis*	*Rvi15*	RAPD: OPK14	Kamyak et al. (2016)
	Malus × *domestica*	Scab	–	*MfVf* *MdAlt* locus	SSR marker: MgVf-GS	Kikuchi et al. (2017)
	Malus niedzwetzkyana	Fire blight	*Erwinia amylovora*	FBF7 QTL	SCAR markers: AE10-375 and GE80-19	Kolchenko et al. (2023)
	Malus sp.	Apple scab	*Venturia inaequalis*	–	RAPD, SSR, and SCAR	Uzun et al. (2023)

Pear	Japanese pear [*Pyrus pyrifolia*	Black spot	*Alternaria alternata*	*A*	RAPD markers: A4-A, AY7-a	Iketani et al. (2001)
				Aki	SSR marker	Terakami et al. (2016)
		Scab	*Venturia nashicola*	*Vn*	RAPD markers: AW13-C, N17-a, U17-b, W2-A, O9-A, AQ11-B	Iketani et al. (2001)
	Japanese and European pear [*Pyrus pyrifolia*) and *Pyrus communis*)	Scab	*Venturia nashicola*	*Vnlf*	SSR: LFA02a09	Takeuchi et al. (2023
Peach	Greek peach [*Prunus persica*]	Bacterial spot	*Xanthomonas arboricola* pv. *pruni*	*Xap1* and *Xap6*	SSR marker: XapF1-SSR, Ppe-XapF6-SSR	Frett (2016)
	Prunus persica	Shoot blight	*Cylindrocarpon destructans* (now *Nectria radicicola*)	–	ISSR markers: 8	Drogoudi et al. (2023)
Fig	Wild fig. (*Ficus erecta*)	Canker	*Ceratocystis* sp.	*C* allele	CAPS marker based on SNP: 7 416 805 of FER_r1.1chr02	Shirasawa et al. (2020)
Apricot	*Prunus armeniaca*	Plum pox	Plum pox virus	*PPVres*	SSR: 3	Rakída et al. (2023)

resistance that can be used for improving disease resistance includes quantitative resistance governed by a variable number of QTLs or multiple genes (Weiss et al. 2020). The QTLs associated with a trait can be variable with regard to both the number and their individual effect or contribution to the trait. While some can be major contributors to the trait, other QTLs may only have a little effect. These QTLs can be more easily evaluated using molecular markers (Muranty et al. 2014) and can also be identified through mapping.

MAS is more effective for simple traits governed by fewer QTLs, each contributing largely to the trait, as opposed to complex quantitative traits with multiple QTLs, like the wood properties (Lebedev et al. 2020). Since disease resistance is a simple trait governed by only a few QTLs, each with large effects, MAS retains its effectiveness in the genotypic evaluation of disease resistance (Lebedev et al. 2020). Hence, some success has been achieved in this respect and genotypes have been screened for disease resistance using MAS. However, predictive precision is more for traits controlled by few, large-effect QTLs (Ahmar et al. 2021).

MAS has witnessed greater success and adoption in the case of agricultural crops, annuals, and perennials with short rotation (Muranty et al. 2014; Shamari et al. 2018; Ahmar et al. 2021). Molecular markers have been developed in perennial, woody plants but have only witnessed limited application and have not been as successful as in agricultural crops. They have been exploited, majorly for genetic diversity studies, population assessment, and for discovering new genotypes for a desirable trait. Many factors present hindrances to the successful use of MAS in forest trees, including long rotation cycles, delayed attainment of sexual maturity, large genomes that pose a limitation to the availability of markers, along with other added levels of complexity in the case of trees, particularly forest tree species, are majorly undomesticated and showed lower levels of LD (Shamari et al. 2018; Lebedev et al. 2020). Additionally, forest trees tend to exhibit strong QTL-environment and QTL-genetic background interactions and exhibited greater allele frequency fluctuation over generations (Lebedev et al. 2020).

QTL mapping in trees is extremely complicated as long generation times account for a considerable delay in producing mapping populations to be used for mapping studies (Harper et al. 2016). This situation is further aggravated by the scanty investigations that have been carried out on tree species to unravel their genetics and genome structure. This hinders the investigators interested in working out genetic components of variations observed in a given trait in a bind, having to resort to de novo genome assembly which, in turn, is a time-consuming process, requiring lengthy periods before the final assembly can be furnished. All these setbacks call for newer molecular tools to be developed so that when a new disease emerges, these challenges can be overcome and the disease can be controlled promptly. NGS-based procedures provide promising alternatives and overcome some of these limitations, allowing the generation of enormous amounts of data and enabling the detection of a huge number of markers, like single nucleotide polymorphisms (SNPs), from a population, small or large, even from species with large genome sizes that can serve as valuable genomic resources (Liu et al. 2017; Gangurde et al. 2022). These can help assess the resistance mechanism to different diseases and utilize the information to increase the efficiency of selecting genotypes exhibiting higher disease resistance.

Nevertheless, some significant advancements have been made by the incorporation of MAS in breeding strategy, allowing more precise selection of forestry species. Markers and their derived markers, like sequence-characterized amplified regions (SCAR), have been utilized to identify linked QTLs and predict disease resistance in many forest trees (Table 16.1). Using markers that co-segregate with the gene governing resistance to a disease, it is possible to introduce resistance from the resistant to susceptible cultivars or varieties. For instance, SSR markers closely linked to *Vlnf*, a resistance gene from *Pyrus communis, were developed to enable the successful detection of Vlnf integration* into *P. pyrifolia* (Takeuchi et al. 2023). In the case of *Populus*, several markers including RFLP, AFLP, RAPD, STS, SSR and SNP, have been used for the identification of Linkage Groups (LGs) and QTLs therein for resistance to pathogens like *Septoria populicola*, *Melampsora larici-populina*, *Melamspora columbiana,* and *Melampsora medusa* (Biselli et al. 2022). Similarly in *Pinus radiate,* LGs with one or more QTLs linked to pitch canker disease resistance were discovered by AFLP, ISSR, Selective Amplification of Microsatellite Polymorphic Loci (SAMPL), along with SSR markers to generate the linkage map (Moraga-Suazo et al. 2014). QTLs were identified for resistance to ink disease pathogen *Phytophthora cinnamomi* in Japanese chestnut (*Castanea crenata*) using microsatellite and SNP markers and resistance to *Ceratocystis manginecans* in *Eucalyptus* species using microsatellite markers (Indrayadi et al. 2023). In *Picea abies*, four SNP markers were identified to be linked with QTLs for resistance to *Heterobasidion parviporum* infection, including a leucoanthocyanidine reductase, and three factors involved in the lignification process, 4-coumarate CoA ligase, and R2R3-MYB transcription factor (Lind et al. 2014).

Using QTL mapping in *Eucalyptus globulus,* two major QTLs were documented for resistance against leaf disease caused by *Mycosphaerella cryptica* (Freeman et al. 2008). More recently, population-level association mapping using GWAS has helped reveal alleles with SNPs linked to variable disease resistance (Calic et al. 2017; Chaudhary et al. 2020). For example, GWAS in Persian oak (*Quercus brantii*) for tolerance to dieback using ISSR markers revealed the trait tolerance associated with trunk size or diameter at breast height (Shamari et al. 2018). The same approach was employed for the identification of *Mt* gene in *Fagus grandifolia*, an important resistance gene for beech bark disease caused by the woolly beech scale insect (*Cryptococcus fagisuga*), and the fungus *Neonectria faginata* or *N. ditissima*, which probably encodes a metallothionein-like protein (metal ion binding) (Calic et al. 2017). Two SNPs were discovered in *Fraxinus excelsior* associated with resistance to fungus *Hymenoscyphus fraxineus*, with one being present in a gene that encode a subtilisin-related peptidase S8/S53 domain protein and the other being associated with an *LRR* gene (Chaudhary et al. 2020). Association mapping in *Pinus taeda* identified 10 SNP markers associated significantly with pitch canker resistance caused by *Fusarium circinatum* and SNP2374 for fusiform rust caused by *Cronartium quercuum* f.sp. *fusiforme* (Cumbie et al. 2020). QTL mapping and GWAS may also be combined to enhance QTL identification discovery as exemplified in sugar pine wherein 14 SNPs were discovered that were involved in quantitative disease resistance controlled by NBS-LRR using both QTL mapping and GWAS (Weiss et al. 2020). Markers may also be utilized for the identification of pathogen genetic diversity as done in *Hevea brasilliensis* for *Rigidoporus microporus* triggering white root disease of rubber trees through ISSR markers or to evaluate the disease extent or aggressiveness; for example, RAPD markers were used for assessing aggressiveness of *Sphaeropsis sapinea* (now *Diplodia sapinea*) isolated from conifers, and SNPs for low versus high damage caused due to *Hymenoscyphus fraxineus* in European ash (*Fraxinus excelsior*) (Kaewchai et al. 2009; Stocks et al. 2019).

High-density genetic maps have been generated to identify QTLs for resistance to various diseases like blight and *Phytophthora* root rot (caused by *P. cinnamomi*, also known as ink disease) in *Castanea*, Witches' broom disease (caused by *Moniliophthora perniciosa*), in *Theobroma cacao*, white pine blister rust (caused by *Cronartium ribicola)* in sugar pine (*Pinus lambertiana*) (Royaert et al. 2016; Santos et al. 2017; Weiss et al. 2020). SNP genotyping was done to decipher markers co-localizing with loci governing disease resistance, for example, in loblolly pine and limber pines (Eckert et al. 2008; Liu et al. 2016). In limber pine (*Pinus flexilis*), 32 SNP markers were localized on the *Cr4* linkage group, a locus associated with resistance to white pine blister rust caused by the fungal pathogen *Cronartium ribicola* (Liu et al. 2016).

Genomic Selection (GS) has surfaced as another powerful approach of MAS, whereby genome-wide markers (often SNPs) are employed to construct a genomic predictive model that predicts the performance with respect to a particular trait (Grattapaglia 2022). GS offers enhanced prediction accuracy and thus can help shorten the breeding cycle and increase selection intensity (Grattapaglia 2022).

16.2.1 Multi-omics Approaches in Forest Disease Management

The field of omics explores genomics, transcriptomics, proteomics, metagenomics, metabolomics, volatilomics, and spectranomics to delve into these interactions. Moreover, it has paved the way for enhancing crop resistance, offering immense potential for crop breeding and enhancement programs (Sundin et al. 2016).

Genomics flourished since the groundbreaking publication of the human genome draft (Lander et al. 2001). Microbial genomics examines genomes of microscopic organisms like bacteria, fungi, and viruses, focusing on structure, function, evolution, mapping, and editing. Integrating genomics tools can assist ecological inquiries, especially regarding environmental changes. In studying plant disease ecology and health, two major branches, i.e., structural and functional genomics of microbe, are essential. Structural genomics, forming a physical genome map, plots the genes to chromosomes. Functional genomics is associated with genome sequences to an organism's transcripts and proteins to explain the gene functions and interactions (Hieter and Boguski 1997; Barone et al. 2008). Nobrega and Pennacchio (2004) suggested that comparative genomics identifies shared genomic elements within or across species.

Transcriptomics, also termed expression genomics, focuses on the transcribed part of the genome, specifically mRNA molecules, in varying environments and times. It enables the study of gene expression in diverse tissues and conditions. Comparing gene expression helps pinpoint genes linked to specific functions. For instance, Ralph et al. (2006) identified over 1000 up-regulated genes in hybrid poplar responding to insect attacks. In the *Populus-Melampsora* pathosystem, transcript profiling unveiled differential regulation of pathogenesis-related genes in susceptible versus incompatible interactions (Rinaldi et al. 2007). Proteomics is an omics approach that involves the study of the complete set of proteins produced by an organism. Several methods, viz., mass spectrometry, electrospray ionization, matrix-assisted laser desorption/ionization (MALDI), gel- (2D and 3D) and non-gel-based methods have amplified the volume of data and assisted in realizing the associations at proteomic level (Mehta et al. 2008; Quirino et al. 2010). Proteomics has helped researchers to identify the proteins that are differentially expressed during plant-microbe interactions such as pathogenesis-related (PR) proteins, and to understand the functions of these proteins. Metabolomics profiles all the minor molecules and represents the endpoint of omics, capturing outcomes from the genome, transcriptome, and proteome (Liu and Locasale 2017). Cutting-edge technology like ultra-HPLC, high-resolution mass spectrometry, and nuclear magnetic resonance spectroscopy allows detailed analysis of thousands of compounds in study organisms (Allwood et al. 2008). Targeted metabolomics focuses on a known set of compounds useful for specific processes and method optimization (Griffiths et al. 2010), while untargeted approaches explore unknown compounds, aiming for a comprehensive view of vast compound profiles in biological samples (Schrimpe-Rutledge et al. 2016).

Comparative genomics seeks shared genomic features within or across species, unveiling the diverse and dynamic genome structures of fungal plant pathogens, including promptly evolving regions from chromosomes rich in transposons to whole accessory chromosomes (Moller and Stukenbrock 2017). Metagenomics transcends traditional genomic techniques, delving into the genetic diversity of unculturable microbes (Solden et al. 2016). Metagenomic processes involve collecting environmental DNA, bioinformatic analysis, taxonomic profiling, and revealing functional elements crucial for understanding plant growth through endophytic microbes and soil microbes' role in suppressing plant pathogens (Fadiji and Babalola 2020; Larkin 2015). Volatilomics, while technically a subset of metabolomics, comprises much more than a simple category within it (Cumeras and Correig 2018). It focuses on profiling high vapor pressure compounds released by organisms, specifically volatile organic compounds (VOCs) (Rosenkranz and Schnitzler 2016). These VOCs, detectable in the air and environmental samples, play fundamental roles in plant-microbe interactions, aiding in pathogen defense and attracting beneficial microbes. Volatilomics identifies these emitted VOCs from plants and microbes, examining their production, regulation, and reception by other organisms in the environment. A real-time volatilomics approach enables continuous monitoring of living organisms' VOC emissions, offering an understanding of the temporal dynamics of plant-microbe interactions.

Combining spectroscopy with chemistry, taxonomy, and ecology forms spectranomics (Asner and Martin 2009, 2016). This approach relies on the idea that plants have unique chemical fingerprints that are more distinct with added constituents and that spectroscopic signatures reveal the plant's chemical portfolio (Jacquemoud et al. 1995; Ustin et al. 2004). In-situ and imaging spectroscopy, like hyperspectral imaging, have the potential for early, nondestructive identification of biotic stress in natural and agro-ecosystems (Fallon et al. 2020). Spectral analysis of plant traits helps to sense the beneficial and parasitic plant-microbe interactions, aiding in modelling biochemical and physiological pathosystem processes (Fallon et al. 2020; Gold et al. 2020a, b). Reflectance indices like NDVI have been used in broadband and multispectral methods since the 1980s for late-stage plant disease detection (Nagarajan et al. 1984; Jackson 1986; Hatfield and Pinter 1993; Nilsson 1995).

16.2.2 RNAi in Forest Tree Disease

RNAi has been utilized to confer resistance to many perennial plants as they are susceptible to virus attacks, which causes a substantial loss of yield and quality. Honeysweet, a virus-resistant variety of plum developed resistance against plum pox virus (PPV), has been prepared using RNAi technology (Scorza et al. 1994), and it is still effective against PPV and widely cultivated across the globe (Singh et al. 2022). PPV coat protein gene construct has been cloned to give resistance against viruses in other woody perennial plants like *Prunus domestica* (Hily et al. 2007; Ravelonandro et al. 2019).

Apart from the application of RNAi in virus infection, it holds a very significant role in combating fungal infections. Exogenous application of dsRNA has been done to control *Venturia inaequalis*, a fungal pathogen of apple trees. Spraying of dsRNA is a novel approach to combat the pathogens on trees as the use of fungicide is not appreciable to maintain the biodiversity and health of natural forests (Bhagta et al. 2023). Zhang et al. (2014) identified and characterised RNA-dependent polymerases that induce post-transcriptional gene silencing (PTGS) in *Cryphonectria parasitica,* the devastating fungal pathogen of chestnut trees. Carneiro et al. (2010) performed RNAi-mediated gene knockdown of gene encoding pectinolytic polygalacturonase enzymes in *Ophiostomo novo-ulmi*, a fungal pathogen of *Ulmus americana*. These enzymes are mainly responsible for the virulence of the fungi. Similarly, Chahed (2016) found two genes, GIP and NPP1, upon the infection of *Phytophthora cinnamomi* into the *Castanea sativa,* which were crucial for maintaining the virulence of this highly aggressive fungus in the host. The pathogen's molecular mechanism of pathogenesis is crucial to dissecting many ways of transforming the dsRNA into the cell, including Agrobacterium-mediated transformation (AMT). Pareek and Rajam (2017) performed AMT of genes involved in the signalling pathway of MAP kinase of phytopathogenic fungi, *Fusarium oxysporum,* to decipher the mechanism of pathogenesis used by the fungi. Through functional genomics, Shanmugam et al.

(2017) identified two genes, *Fow2* and *chsV*, involved in transcription activation and chitin synthesis, respectively. These genes are associated with pathogenicity in *F. oxysporum* and *F. solani* infections. Spada et al. (2021) prepared dsRNA-based fungicides against *Botrytis cinerea*, a fungal pathogen of various woody perennials that are effective against fungal progression and spread under in-vitro and in-vivo conditions. The study knocked down *BcBmp1* and *BcPls1* genes involved in the fungal invasion and penetration into the host. In another study Spada et al. (2023) knocked down MAP kinase *Bmp3* gene and successfully prepared dsRNA-based fungicide to control grey mould infections.

16.2.3 CRISPR/Cas Technology in Forest Trees

The clustered regularly interspaced short palindromic repeats (CRISPR) and CRISPR-associated protein (Cas) systems are located in 90% of archaea and 40% of bacteria. In some cases, it is part of bacterial adaptive immunity against bacteriophages, plasmids, and RNA. When phage viruses infect bacteria, the specific spacer sequences in the CRISPR array are transcribed to crRNA, which mediates the Cas9 protein, cleaving the complementary DNA or RNA. This CRISPR/Cas9 technology can be employed with an efficient delivery mechanism for genetic engineering research on forest trees. Better qualitative research on forest trees will help in understanding their complex genomes, which aids in the development of disease resistance and improved traits (Table 16.2).

Dothideomycetes fungus, namely, *Dothistroma septosporum,* is one of the important pathogens of forest trees causing Dothistroma needle blight in pine trees. This filamentous fungus has *AflR* gene responsible for the biosynthesis of the dothistromin virulence factor. The genes associated with the biosynthesis of anthraquinone toxin dothistromin and secreted protein *Ds74283* having virulence factors associated with the pathogenicity may be deleted with the help of CRISPR/Cas9 technology. The functions of other virulence factors may also be explored through CRISPR/Cas9 to understand the host-pathogen interactions (McCarthy et al. 2022). Papermaking, pulping, and wood processing all depend on the intricate process of cellulose synthesis occurring in plant cells. CRISPR/Cas is utilized to break the *LFY* gene function in *Eucalyptus* and appears to be a helpful method for sexual containment without having a large or statistically significant negative impact on the morphology of the leaves or juvenile vegetative growth (Elorriaga et al. 2021). *Populus* tree species is one of the typical species for molecular studies of forest trees. The application of this technology is useful in understanding the *Melampsora-Populus* pathosystems and also for exploring genes in *Melampsora* resistance by developing knockout mutants to enhance productivity and forest health. The pathogenic dimorphic fungus *Ophiostoma novo ulmi* is capable of adjusting between yeast and mycelial form for pathogenicity. The candidate genes involved in transition were identified earlier. Using CRISPR/Cas9 to trigger the NHEJ repair mechanism, candidate genes involved in the transition can be knocked out, creating mutants. These knockout mutants cannot switch to the yeast phase, which impedes the pathogen

Table 16.2 CRISPR/Cas applications on forest pathogen and pest management

Target species	Target genes	CRISPR findings	References
Dothideomycetes	*AflR*	Pathogenicity factor	McCarthy et al. (2022)
Poplar trichocharpa	*PtrWRKY18* and *PtrWRKY35*	Resistance to *Melampsora*	Jiang et al. (2017)
Populus tomentosa	*PtSGT1* and *PtSGT4*	Cellulose synthesis	Xue et al. (2021)
Poplar spp.	*ROCK3/4*	Cytokinin biosynthesis gene	Riefler et al. (2022)
Poplar spp.	*CSE*	Increased lignin synthesis	Jang et al. (2021)
Ophiostoma novo ulmi	Introgressed genes	Virulence factor	Soltis et al. (2013)
Dendrolimus punctatus	*DpWnt-1*	aberrant growth and malformed organ development	Singh et al. (2022)
Theobroma cacao	*N*PR3	Resistant to *Phytophthora tropicalis*	Fister et al. (2018)
Vitis vinefera	*WRKY52*	Tolerance to *Botrytis cinerea*	Min et al. (2022)
Citrus spp.	*CsLOB1*	Varying levels of resistance to *xanthomonas* sp. causing citrus canker	Jia and Wang (2020)
Malus domestica	*MdDIPM1* and *MdDIPM4*	Susceptibility to fire blight	Dalla et al. (2020)
Malus domestica	*MdCNGC2*	Increased resistance to *Botrytis dothidea*	Zhou et al. (2020)
Eucalyptus	*LFY* gene	Sexual containment	Elorriaga et al. (2021)

virulence, preventing entry into the elm tree. Pathogenicity of *Ophiostoma novo ulmi* depends upon its introgressed genes and secondary metabolites involved in virulence. The introgressed genes are involved in transferring genetic material between species via hybridization and backcrossing to the parental species (Soltis et al. 2013). The function of introgressed genes can be explored by this technology to further understand the pathogen strategies used to infect a host. An oomycete hemibiotroph, *Phytophthora ramorum,* causing sudden oak or Larch death, is an invasive forest pathogen. The effector plays a significant role in transforming the pathogen from an asymptomatic biotrophic phase to an aggressive necrotrophic phase, contributing to virulence in the host. Hence, CRISPR can be used as a tool to explore the function of effectors in the lifestyle transition of the pathogen. A basidiomycetes rust fungus, *Cronartium ribicola* initiating white pine blister rust is one of the important diseases infecting pine trees. The dominant *Cr* gene provides complete resistance, where the resistance can be broken by pathogen effector mutation. A partial resistance strategy can be employed by targeting multiple resistance genes to overcome this hindrance. To provide long-term stable resistance, immune receptors can be mutated to exhibit resistance against phylogenetically diverse pathogens. This multiplex gene editing can be developed by using CRISPR/Cas successfully. CRISPR-engineered genetic sexing strains, gene drive, sex ratio distortion, etc., are a few of the CRISPR-based strategies against insect pests (Moon et al. 2022). In the pine caterpillar moth, *Dendrolimus punctatus*, a mutation in the *DpWnt-1* gene mediated by CRISPR/Cas9 significantly changed the gene expression, resulting in aberrant growth and malformed organ development (Singh et al. 2022). Some of the targeted traits using CRISPR in woody species are lignin engineering, enhanced photosynthetic efficiency, improved CO_2 sequestration, sterility induction, insect-pest resistance, improved nitrate processing efficiency, reduced branching, improved wood density, augmentation of growth rate, modulation of microfibrillar angle.

16.3 Application of Plant Immune Activators (PIAs) in Tree Defense

Plant immune activators (PIAs) are elicitors that trigger plant defense responses, enabling them to recognize and effectively combat invading pathogens. Several biotechnological approaches are being employed to develop and apply PIAs for disease management in forest trees. Natural PIAs can be isolated from various sources, such as microorganisms, plants, and marine organisms. These compounds can be isolated and purified as PIAs or as inspiration for the development of synthetic analogues. The development of synthetic PIAs is designed to mimic the structure and function of natural PIAs, often with improved stability and efficacy. These synthetic compounds can be tailored to target specific pathogens or groups of pathogens. Genetic engineering techniques can be used to introduce genes encoding PIAs into plants, resulting in transgenic plants with enhanced disease resistance. This approach offers the potential for long-term protection against various diseases. Beyond naturally occurring PIAs, exciting advancements in biotechnology offer the potential to engineer even more potent and targeted defense solutions for the forests.

Biotechnological approaches offer promising alternatives to conventional disease control methods by utilizing PIAs to enhance the natural defense mechanisms of trees against pathogens. Researchers have engineered PIAs by using Β-1,3-glucans with increased branching patterns, leading to significantly higher activation of plant defense genes and improved resistance against fungal pathogens in pines (Hao et al. 2022). Scientists also fused chitosan with antimicrobial peptides, creating chimeric PIA molecules with broader spectrum activity against both fungal and bacterial pathogens in poplar trees (Liu et al. 2020). Another way is by silencing negative regulators by gene editing using CRISPR-Cas9 technology. Researchers have successfully silenced genes that suppress plant defense responses in eucalyptus, leading to enhanced resistance against fungal diseases (Wang et al. 2021). Scientists have engineered trees to express defense genes only upon pathogen attack, minimizing resource expenditure and potential negative side effects (Bi et al. 2019) and developed nanocapsules coated with plant-specific recognition molecules, allowing for targeted delivery of PIAs to diseased cells, minimizing off-target effects and maximizing efficacy (Prabhu et al. 2023). Nanopore-based delivery systems can be designed to release PIAs in response to specific environmental cues or pathogen signals, providing precise and efficient disease control (Chen et al. 2022). The application of PIAs in forest tree disease management has shown promising results in various studies. For instance, the application of the β-glucan elicitor laminarin has enhanced the resistance of oak trees against fungal pathogens (Korte and Pretorius 2019). Similarly, applying the jasmonic acid analogue has improved the resistance of pine trees against fungi and insects (Zhu et al. 2019). The synergistic effects of combining different PIAs with contrasting mechanisms of action are also being explored. For example, mixing B-1,3-glucans with chitosan has shown enhanced protection against fungal and bacterial pathogens in oak trees, potentially due to their complementary action on different defense pathways (Martinez-Medina et al. 2018). Research suggests that applying specific bacterial strains can trigger systemic resistance against fungal pathogens in poplar trees (Plett et al. 2014). In some studies, combining PIAs with beneficial microbes like *Trichoderma* fungi can further amplify defense responses. Studies have shown that co-application of B-1,3-glucans with *Trichoderma harzianum* led to significantly reduced fungal disease severity in pine seedlings compared to either treatment alone (Li et al. 2019). RNAi technology can be used to silence essential genes in the pathogen, rendering it less virulent and preventing disease development. Initial studies using RNAi against fungal cell wall biosynthesis genes have shown promising results in reducing fungal infections in trees (Seidl et al. 2018). Certain PIAs can trigger plants' production of volatile organic compounds (VOCs). These VOCs can attract natural enemies of pathogens and activate defense responses in neighbouring trees, creating a collective defense system. Studies have shown that chitosan application in citrus trees can induce the release of VOCs with antifungal properties (Aziz et al. 2017).

16.4 Scopes and Challenges

Only a few whole genome sequences are available to design specific gRNA and have less off-target effects. For example, conifer species, which are dominant in forest ecosystems across the Northern Hemisphere, possess notably larger and more complex genomes compared to those of most agricultural angiosperms. The problems associated with transformation protocols in forest tree species are due to the requirement of transforming their DNA and regenerating the whole plant afterward. Controversy over Genetically Engineered Organisms (GMOs) is greater for woody forest species than agricultural sector because of their perennial nature and presence in semi-natural ecosystems (Dort et al. 2020).

Biotechnological approaches offer promising avenues for the development and utilization of PIAs for disease management in forest trees. These approaches have the potential to provide sustainable and environmentally friendly alternatives to conventional disease control methods. However, further research is needed to address the challenges associated with PIAs, such as variability in plant responses, long-term efficacy, and environmental considerations, to ensure their effective and widespread adoption in forest tree protection. Despite the potential benefits of PIAs, several challenges need to be addressed for their widespread adoption in forest tree disease management. The effectiveness of PIAs mainly depends upon plant species, targeted pathogens, and environmental conditions. Therefore, careful selection and optimization of PIAs are necessary for each specific application. The long-term efficacy of PIAs needs to be evaluated to ensure their sustainability as a disease management strategy. The potential environmental impacts of PIAs need to be thoroughly assessed to ensure their safe and sustainable use.

16.5 Conclusion

The revolutionary potential of CRISPR-based gene editing technology holds the promise to transform the forestry sector. By this technology, trees can be made more resilient to both biotic and abiotic stresses, while also increasing their yield potential by incorporating multiple desired traits. Trees with medicinal significance possess substantial potential in the pharmaceutical industry, where they can play a crucial role. By manipulating genes responsible for the biosynthesis of alkaloids and flavonoids, it becomes possible to significantly increase the yield of bioactive compounds from the trees. This advancement could lead to the development of more potent and effective medicines. Understanding the host pathogen interactions and improved disease resistance strategies are essential for maintaining long-term forest sustainability. By harnessing the power of plant immunity, we can protect our forests and ensure their continued contribution to environmental health and human well-being.

References

Ahmar S, Ballesta P, Ali M, Mora-Poblete F (2021) Achievements and challenges of genomics-assisted breeding in forest trees: from marker-assisted selection to genome editing. Int J Mol Sci 22:10583

Allwood JW, Ellis DI, Goodacre R (2008) Metabolomic technologies and their application to the study of plants and plant–host interactions. Physiol Plant 132(2):117–135

Anonymous (2023). https://indianexpress.com/about/deforestation-inc/

Asner GP, Martin RE (2009) Airborne spectranomics: mapping canopy chemical and taxonomic diversity in tropical forests. Front Ecol Environ 7(5):269–276

Asner GP, Martin RE (2016) Spectranomics: emerging science and conservation opportunities at the interface of biodiversity and remote sensing. Glob Ecol Conserv 8:212–219

Aziz A, Sedaghat AR, Khalizadeh R et al (2017) Chitosan elicits volatile organic compounds with antifungal properties from citrus plants. App Microbiol Biotechnol 101(12):4855–4864

Barone A, Chiusano ML, Ercolano MR, Giuliano G, Grandillo S, Frusciante L (2008) Structural and functional genomics of tomato. Int J Plant Genomics 2008:820274

Benet H, Guries RP, Boury S, Smalley EB (1995) Identification of RAPD markers linked to a black leaf spot resistance gene in Chinese elm. Theor Appl Genet 90:1068–1073

Bhagta S, Bhardwaj V, Kant A (2023) Exogenous dsRNA trigger RNAi in *Venturia inaequalis* resulting in down regulation of target genes and growth reduction. Mol Biol Rep 50(10):8421–8429

Bi Y, Liu Y, Wang Y et al (2019) Inducible overexpression of a defense gene OsWRKY47 enhances rice resistance to blast and bacterial blight. Plant Mol Biol 99(1–2):141–155

Biselli C, Vietto L, Rosso L et al (2022) Advanced breeding for biotic stress resistance in Poplar. Plan Theory 11:2032

Calic I, Koch J, Carey D et al (2017) Genome-wide association study identifies a major gene for beech bark disease resistance in American beech (*Fagus grandifolia* Ehrh.). BMC Genomics 18:547

Carneiro JS, de la Bastide PY, Chabot M, Lerch L, Hintz WE (2010) Suppression of polygalacturonase gene expression in the phytopathogenic fungus *Ophiostoma novo-ulmi* by RNA interference. Fungal Genet Biol 47(5):399–405

Cervera MT, Gusmão J, Steenackers M et al (1996) Identification of AFLP molecular markers for resistance against *Melampsora larici-populina* in Populus. Theor Appl Genet 93:733–737

Chahed A (2016) Silencing the gip gene of *Phytophthora cinnamomi* by iRNA and subcellular localization of GIP and NPP1 proteins. Dissertation submitted to Escola Superior Agrária de Bragança to obtain the Degree of Master in Biotechnological Engineering, p 83

Chandrakant ES (2014) Identification of molecular marker linked to the resistance for vascular streak dieback disease in cocoa (*Theobroma cacao* L.). M.Sc. thesis, Kerala Agricultural University, Thrissur

Chaudhary R, Rönneburg T, Stein Åslund M et al (2020) Marker-trait associations for tolerance to ash dieback in common ash (*Fraxinus excelsior* L.). Forests 11:1083

Chen Y, Zhang X, Jiang G et al (2022) Smart nanocomposite carriers for controlled release of plant disease control agents. ACS Appl Nano Mater 5(2):1397–1406

Cumbie WP, Huber DA, Steel VC et al (2020) Marker associations for fusiform rust resistance in a clonal population of loblolly pine (*Pinus taeda* L.). Tree Genet Genomes 16:86

Cumeras R, Correig X (2018) The Volatilome in metabolomics. In: Volatile organic compound analysis in biomedical diagnosis applications. Apple Academic Press, In, pp 23–50

Dalla CL, Piazza S, Pompili V et al (2020) Strategies to produce T-DNA free CRISPRed fruit trees via *Agrobacterium tumefaciens* stable gene transfer. Sci Rep 10(1):20155

Devey ME, Delfino-Mix A, Kinloch BB, Neale DB (1995) Random amplified polymorphic DNA markers tightly linked to a gene for resistance to white pine blister rust in sugar pine. Proc Natl Acad Sci USA 92:2066–2070

Dort EN, Tanguay P, Hamelin RC (2020) CRISPR/Cas9 gene editing: an unexplored frontier for forest pathology. Front Plant Sci 11:1126

Drogoudi P, Pantelidis G, Karapetsi L et al (2023) Exploring the genetic and morphological variation and disease resistance in local and foreign *Prunus persica* (L.) Batsch cultivars. Agriculture 13:800

Eckert CG, Samis KE, Lougheed SC (2008) Genetic variation across species' geographical ranges: the central–marginal hypothesis and beyond. Mol Ecol 17(5):1170–1188

Elorriaga E, Klocko AL, Ma C et al (2021) Genetic containment in vegetatively propagated forest trees: CRISPR disruption of LEAFY function in Eucalyptus gives sterile indeterminate inflorescences and normal juvenile development. Plant Biotechnol J 19(9):1743–1755

Fadiji AE, Babalola OO (2020) Elucidating mechanisms of endophytes used in plant protection and other bioactivities with multifunctional prospects. Front Bioengg Biotechnol 8:467

Fallon B, Yang A, Lapadat C, Armour I, Juzwik J, Montgomery RA, Cavender-Bares J (2020) Spectral differentiation of oak wilt from foliar fungal disease and drought is correlated with physiological changes. Tree Physiol 40(3):377–390

Fister AS, Landherr L, Maximova SN (2018) Transient expression of CRISPR/Cas9 machinery targeting TcNPR3 enhances defense response in *Theobroma cacao*. Front Plant Sci 9:268

Freeman JS, Potts BM, Vaillancourt RE (2008) Few mendelian genes underlie the quantitative response of a forest tree, *Eucalyptus globulus*, to a natural fungal epidemic. Genetics 178:563–571

Frett T J (2016) Genetic determinism of Xanthomonas arboricola pv. pruni (Xap) resistance, fruit quality, and phenological traits in peach and incorporation of marker-assisted selection (MAS)

in the University of Arkansas Peach and Nectarine Breeding Program. Ph.D. thesis in Plant Science, University of Arkansas, p 735

Gangurde SS, Xavier A, Naik YD, Jha UC, Rangari SK, Kumar R et al (2022) Two decades of association mapping: insights on disease resistance in major crops. Front Plant Sci 13:1064059

Gold KM, Townsend PA, Chlus A, Herrmann I, Couture JJ et al (2020a) Hyperspectral measurements enable pre-symptomatic detection and differentiation of contrasting physiological effects of late blight and early blight in potato. Remote Sens 12(2):286

Gold KM, Townsend PA, Larson ER, Herrmann I, Gevens AJ (2020b) Contact reflectance spectroscopy for rapid, accurate, and nondestructive *Phytophthora infestans* clonal lineage discrimination. Phytopathology 110(4):851–862

Goswami M, Attri K, Goswami I (2023) Applications of molecular markers in fruit crops: a review. Int J Econ Plants 9:121–126

Gouthu S, Mandelli C, Eubanks BA, Deluc LG (2022) Transgene-free genome editing and RNAi ectopic application in fruit trees: potential and limitations. Front Plant Sci 13:979742

Grattapaglia D (2022) Twelve years into genomic selection in forest trees: climbing the slope of enlightenment of marker assisted tree breeding. Forests 13(10):1554

Griffiths WJ, Koal T, Wang Y, Kohl M, Enot DP, Deigner HP (2010) Targeted metabolomics for biomarker discovery. Angew Chem Int Ed 49(32):5426–5445

Hao J, Yang X, Zhang Z et al (2022) Enhanced branching in B-1,3-glucans improves disease resistance in *Pinus sylvestris* seedlings. Int J Biol Macromol 207:117–124

Harper AL, McKinney LV, Nielsen LR et al (2016) Molecular markers for tolerance of European ash (*Fraxinus excelsior*) to dieback disease identified using associative transcriptomics. Sci Rep 6:19335

Hatfield P, Pinter P Jr (1993) Remote sensing for crop protection. Crop Prot 12(6):403–413

Hieter P, Boguski M (1997) Functional genomics: it's all how you read it. Science 278(5338):601–602

Hily JM, Ravelonandro M, Damsteegt V, Bassett C, Petri C, Liu Z et al (2007) Plum pox virus coat protein gene Intron-hairpin-RNA (ihpRNA) constructs provide resistance to plum pox virus in *Nicotiana benthamiana* and *Prunus domestica*. J Am Soc Hortic Sci 132(6):850–858

Iketani H, Abe K, Yamamoto T, Kotobuki K, Sato Y, Saito T et al (2001) Mapping of disease-related genes in Japanese pear using a molecular linkage map with RAPD markers. Breed Sci 51(3):179–184

Indrayadi H, Glen M, Kurniawan YR et al (2023) Using transferable eucalypt microsatellite markers to identify QTL for resistance to *Ceratocystis* wilt disease in *Eucalyptus pellita* F. Muel (Myrtales, Myrtaceae). Forests 14:1703

Jackson R (1986) Remote sensing of biotic and abiotic plant stress. Annu Rev Phytopathol 24(1):265–287

Jacquemoud S, Verdebout J, Schmuck G, Andreoli G, Hosgood B (1995) Investigation of leaf biochemistry by statistics. Remote Sens Environ 54(3):180–188

Jang HA, Bae EK, Kim MH (2021) CRISPR-knockout of CSE gene improves saccharification efficiency by reducing lignin content in hybrid poplar. Int J Mol Sci 22(18):9750

Jermstad KD, Eckert AJ, Wegrzyn JL et al (2011) Comparative mapping in *Pinus*: sugar pine (*Pinus lambertiana* Dougl.) and loblolly pine (*Pinus taeda* L.). Tree Genet Genomes 7:457–468

Jia H, Wang N (2020) Generation of homozygous canker-resistant citrus in the T0 generation using CRISPR-SpCas9p. Plant Biotechnol J 18(10):1990

Jiang Y, Guo L, Ma X et al (2017) The WRKY transcription factors PtrWRKY18 and PtrWRKY35 promote *Melampsora* resistance in *Populus*. Tree Physiol 37(5):665–675

Jones JDG, Dangl JL (2006) The plant immune system. Nature 444:323–329

Junghans DT, Alfenas AC, Brommonschenkel SH et al (2003) Resistance to rust (*Puccinia psidii* Winter) in Eucalyptus: mode of inheritance and mapping of a major gene with RAPD markers. Theor Appl Genet 108:175–180

Kaewchai S, Wang HK, Lin FC, Hyde KD, Soytong K (2009) Genetic variation among isolates of *Rigidoporus microporus* causing white root disease of rubber trees in southern Thailand revealed by ISSR markers and pathogenicity. Afr J Microbiol Res 3(10):641–648

Kamyak S, İşçi M, Özongun S, Özgönen H (2016) Determination of reaction levels of some apple genetic resources in Turkey to Apple scab (*Venturia inaequalis* (Cke.) Wint.). Bitki Koruma Bülteni 56:227–238

Kikuchi T, Kasajima I, Morita M, Yoshikawa N (2017) Practical DNA markers to estimate apple (*Malus* Ã,Â x *domestica* Borkh.) skin color, ethylene production and pathogen resistance. J Hortic 04:1000211

Kolchenko M, Nurtaza A, Pozharskiy A et al (2023) Wild *Malus niedzwetzkyana* Dieck ex Koehne as a genetic resource for fire blight resistance. Horticulturae 9:1066

Kondo T, Terada K, Hayashi E et al (2000) RAPD markers linked to a gene for resistance to pine needle gall midge in Japanese black pine (*Pinus thunbergii*). Theor Appl Genet 100:391–395

Korte K, Pretorius A (2019) Laminarin induces resistance against powdery mildew in oak trees (*Quercus robur* L.). Front. Plant Sci 10:1245

Kuhlman T, Zhang Z, Saier MH Jr, Hwa T (2007) Combinatorial transcriptional control of the lactose operon of *Escherichia coli*. Proc Nat Acad Sci U S A 104(14):6043–6048

Kyre BR, Rodrigues TB, Rieske LK (2019) RNA interference and validation of reference genes for gene expression analyses using qPCR in southern pine beetle, *Dendroctonus frontalis*. Sci Rep 9(1):5640

Laia M, Alfenas A, Brommonschenkel S et al (2015) Identification of a sequence characterized amplified region (SCAR) marker linked to the *Puccinia psidii* resistance gene 1 (Ppr1) in *Eucalyptus grandis*. Afr J Agric Res 10:1957–1964

Lander ES, Linton LM, Birren B, Nusbaum C, Zody MC et al (2001) Initial sequencing and analysis of the human genome. Nature 409(6822):860–921

Larkin RP (2015) Soil health paradigms and implications for disease management. Annu Rev Phytopathol 53:199–221

Lebedev VG, Lebedeva TN, Chernodubov AI, Shestibratov KA (2020) Genomic selection for forest tree improvement: methods, achievements and perspectives. Forests 11:1190

Li H, Ruan C-J, Teixeira Da Silva JA, Liu B-Q (2010) Associations of SRAP markers with dried-shrink disease resistance in a germplasm collection of sea buckthorn (*Hippophae* L.). Genome 53:447–457

Li G, Zhang T, Zhang N et al (2019) Combined application of *Trichoderma harzianum* and B-1,3-glucan elicitor for the biological control of *Botryosphaeria dothiorella* dieback in *Pinus tabuliformis* seedlings. Front Plant Sci 10:1303

Lind M, Källman T, Chen J et al (2014) A *Picea abies* linkage map based on SNP markers identifies QTLs for four aspects of resistance to *Heterobasidion parviporum* infection. PLoS One 9:e101049

Liu X, Locasale JW (2017) Metabolomics: a primer. Trends Biochem Sci 42(4):274–284

Liu J-J, Ekramoddoullah AKM, Hunt RS, Zamani A (2006) Identification and characterization of random amplified polymorphic DNA markers linked to a major gene (*Cr2*) for resistance to *Cronartium ribicola* in *Pinus monticola*. Phytopathology 96:395–399

Liu J-J, Hammett C, Sniezko RA (2013) *Pinus monticola* pathogenesis-related gene PmPR10-2 alleles as defense candidates for stem quantitative disease resistance against white pine blister rust (*Cronartium ribicola*). Tree Genet Genomes 9:397–408

Liu J-J, Schoettle AW, Sniezko RA et al (2016) Genetic mapping of *Pinus flexilis* major gene (*Cr4*) for resistance to white pine blister rust using transcriptome-based SNP genotyping. BMC Genomics 17:753

Liu J, Sniezko RA, Zamany A et al (2017) Saturated genic SNP mapping identified functional candidates and selection tools for the *Pinus monticola Cr2* locus controlling resistance to white pine blister rust. Plant Biotechnol J 15:1149–1162

Liu J, Wang Y, Xue M et al (2020) Chimeric chitosan-antimicrobial peptides: a promising antifungal and antibacterial agent for *Populus deltoides*. Carbohydr Polym 252:117256

Martinez-Medina A, Sanchez-Ballesteros J, Fernandez-Ortiz A et al (2018) Synergistic effect of chitosan and B-1,3-glucan against *Phytophthora cinnamomi* in *Quercus suber* L. seedlings. Int J Mol Sci 19(10):3102

McCarthy HM, Tarallo M, Mesarich CH et al (2022) Targeted gene mutations in the forest pathogen *Dothistroma septosporum* using CRISPR/Cas9. Plan Theory 11(8):1016

Mehlenbacher SA, Brown RN, Davis JW et al (2004) RAPD markers linked to eastern filbert blight resistance in *Corylus avellana*. Theor Appl Genet 108:651–656

Mehta A, Brasileiro AC, Souza DS, Romano E, Campos MA, Grossi-de-Sa MF, Rocha TL (2008) Plant–pathogen interactions: what is proteomics telling us? FEBS J 275(15):3731–3746

Min T, Hwarari D, Li DA et al (2022) CRISPR-based genome editing and its applications in woody plants. Int J Mol Sci 23(17):10175

Moller M, Stukenbrock EH (2017) Evolution and genome architecture in fungal plant pathogens. Nat Rev Microbiol 15(12):756

Moon TT, Maliha IJ, Khan AAM et al (2022) CRISPR-Cas genome editing for insect pest stress management in crop plants. Stress 2(4):493–514

Moraga-Suazo P, Orellana L, Quiroga P et al (2014) Development of a genetic linkage map for *Pinus radiata* and detection of pitch canker disease resistance associated QTLs. Trees 28:1823–1835

Moriya S, Iwanami H, Takahashi S et al (2010) Genetic mapping of the crown gall resistance gene of the wild apple *Malus sieboldii*. Tree Genet Genomes 6:195–203

Muranty H, Jorge V, Bastien C et al (2014) Potential for marker-assisted selection for forest tree breeding: lessons from 20 years of MAS in crops. Tree Genet Genomes 10:1491–1510

Nadeem MA, Nawaz MA, Shahid MQ et al (2018) DNA molecular markers in plant breeding: current status and recent advancements in genomic selection and genome editing. Biotechnol Biotechnol Equip 32:261–285

Nagarajan S, Seibold G, Kranza J, Saari E, Joshi L (1984) Monitoring wheat rust epidemics with the Landsat2 satellite. Phytopathology 74(5):585–587

Nilsson HE (1995) Remote sensing and image analysis in plant pathology. Can J Plant Pathol 17(2):154–166

Nobrega MA, Pennacchio LA (2004) Comparative genomic analysis as a tool for biological discovery. J Physiol 554(1):31–39

Pareek M, Rajam MV (2017) RNAi-mediated silencing of MAP kinase signalling genes (Fmk1, Hog1, and Pbs2) in *Fusarium oxysporum* reduces pathogenesis on tomato plants. Fungal Biol 121(9):775–784

Parthasarathy S, Thiribhuvanamala G, Muthulakshmi P, Angappan K (2021) Diseases of forest trees and their management. CRC Press

Plett JM, Bais H, Novitt N et al (2014) Beneficial microbial reprogramming of the rhizosphere: strategies for enhancing plant health and defense. Front Microbiol 5:393

Porth I, El-Kassaby Y (2014) Assessment of the genetic diversity in forest tree populations using molecular markers. Diversity 6:283–295

Prabhu S, Gnanasekaran D, Kumar V (2023) Nanoencapsulated *Trichoderma harzianum* conidial suspension for the biocontrol of fungal diseases in plants. J Appl Microbiol 134(4):988–1002

Quirino BF, Candido ES, Campos PF, Franco OL, Krüger RH (2010) Proteomic approaches to study plant–pathogen interactions. Phytochemistry 71(4):351–362

Rakida A, Çetínsağ N, Akparov Z et al (2023) Assessment of genetic diversity and search for Plum pox virus resistance alleles of apricot (*Prunus armeniaca* L.) genotypes spread in Azerbaijan using SSR markers. Turk J Agric For 47:529–540

Ralph S, Oddy C, Cooper D, Yueh H, Jancsik S, Kolosova N et al (2006) Genomics of hybrid poplar (*Populus trichocarpa* x *deltoides*) interacting with forest tent caterpillars (*Malacosoma disstria*): normalized and full-length cDNA libraries, expressed sequence tags, and a cDNA microarray for the study of insect-induced defences in poplar. Mol Ecol 15:1275–1297

Ravelonandro M, Scorza R, Briard P (2019) Innovative RNAi strategies and tactics to tackle plum pox virus (PPV) genome in *Prunus domestica-plum*. Plan Theory 8(12):565

Riefler M, Brügmann T, Fladung M et al (2022) A constitutively active cytokinin receptor variant increases cambial activity and stem growth in Poplar. Int J Mol Sci 23(15):8321

Rinaldi C, Kohler A, Frey P et al (2007) Transcript profiling of poplar leaves upon infection with compatible and incompatible strains of the foliar rust *Melampsora larici-populina*. Plant Physiol 144:347–366

Rosenkranz M, Schnitzler JP (2016) Plant volatiles. eLS, John Wiley & Sons, Ltd, Chichester

Royaert S, Jansen J, Da Silva DV et al (2016) Identification of candidate genes involved in Witches' broom disease resistance in a segregating mapping population of *Theobroma cacao* L. in Brazil. BMC Genomics 17:107

Ruan C-J, Li H, Mopper S (2009) Characterization and identification of ISSR markers associated with resistance to dried-shrink disease in sea buckthorn. Mol Breeding 24:255–268

Santos C, Nelson CD, Zhebentyayeva T et al (2017) First interspecific genetic linkage map for *Castanea sativa x Castanea crenata* revealed QTLs for resistance to *Phytophthora cinnamomi*. PLoS One 12:e0184381

Schrimpe-Rutledge AC, Codreanu SG, Sherrod SD, McLean JA (2016) Untargeted metabolomics strategies - challenges and emerging directions. J Am Soc Mass Spectrom 27(12):1897–1905

Scorza R, Ravelonandro M, Callahan AM, Cordts JM, Fuchs M, Dunez J et al (1994) Transgenic plums (*Prunus domestica* L.) express the plum pox virus coat protein gene. Plant Cell Rep 14:18–22

Seidl MF, Wandtke UC, Scheer E (2018) RNAi-mediated silencing of fungal cell wall biosynthesis genes: a potential strategy for plant disease control. Appl Microbiol Biotechnol 102(12):5271–5283

Shamari A-R, Mehrabi A-A, Maleki A, Rostami A (2018) Association analysis of tolerance to dieback phenomena and trunk form using ISSR markers in *Quercus brantii*. Cell Mol Biol (Noisy-le-Grand) 64:116–124

Shanmugam V, Sharma V, Bharti P, Jyoti P, Yadav SK, Aggarwal R et al (2017) RNAi induced silencing of pathogenicity genes of *Fusarium* spp. for vascular wilt management in tomato. Ann Microbiol 67:359–369

Shirasawa K, Yakushiji H, Nishimura R et al (2020) The *Ficus erecta* genome aids *Ceratocystis* canker resistance breeding in common fig (*F. carica*). Plant J 102:1313–1322

Singh S, Rahangdale S, Pandita S et al (2022) CRISPR/Cas9 for insect pests management: a comprehensive review of advances and applications. Agriculture 12(11):1896

Siregar UJ, Rahmawati D, Damayanti A (2019) Fingerprinting sengon (*Falcatria moluccana*) accessions resistant to boktor pest and gall rust disease using microsatellite markers. Biodiversitas 20:2698–2706

Solden L, Lloyd K, Wrighton K (2016) The bright side of microbial dark matter: lessons learned from the uncultivated majority. Curr Opin Microbiol 31:217–226

Soltis DE, Gitzendanner MA, Stull G et al (2013) The potential of genomics in plant systematics. Taxon 62(5):886–898

Spada M, Pugliesi C, Fambrini M, Pecchia S (2021) Silencing of the Slt2-type MAP kinase Bmp3 in *Botrytis cinerea* by application of exogenous dsRNA affects fungal growth and virulence on *Lactuca sativa*. Int J Mol Sci 22(10):5362

Spada M, Pugliesi C, Fambrini M, Palpacelli D, Pecchia S (2023) Knockdown of Bmp1 and Pls1 virulence genes by exogenous application of RNAi-inducing dsRNA in *Botrytis cinerea*. Int J Mol Sci 24(5): 4869.

Stirling B, Newcombe G, Vrebalov J et al (2001) Suppressed recombination around the MXC3 locus, a major gene for resistance to poplar leaf rust. Theor Appl Genet 103:1129–1137

Stocks JJ, Metheringham CL, Plumb WJ et al (2019) Genomic basis of European ash tree resistance to ash dieback fungus. Nat Ecol Evol 3:1686–1696

Sun S, Shu CW, Chen JL et al (2014) Screening for resistant clones of *Casuarina equisetifolia* to bacterial wilt and the analysis of AFLP markers in resistant clones. For Pathol 44:276–281

Sundin GW, Wang N, Charkowski AO, Castiblanco LF, Jia H, Zhao Y (2016) Perspectives on the transition from bacterial phytopathogen genomics studies to applications enhancing disease management: from promise to practice. Phytopathology 106(10):1071–1082

Takeuchi Y, Nishio S, Terakami S et al (2023) Genetic mapping of the pear scab resistance gene Vnlf using a pseudo-BC3 population derived from Japanese pear cultivars and European pear 'La France.'. Sci Hortic 321:112260

Tartarini S (1996) RAPD markers linked to the Vf gene for scab resistance in apple. Theor Appl Genet 92:803–810

Terakami S, Moriya S, Adachi Y et al (2016) Fine mapping of the gene for susceptibility to black spot disease in Japanese pear (*Pyrus pyrifolia* Nakai). Breed Sci 66:271–280

Tulshiram WS (2016) Characterization and validation of microsatellite markers for resistance to vascular streak dieback disease in cocoa (*Theobroma cacao* L.). M.Sc. thesis, Kerala Agricultural University, Thrissur

Ustin SL, Roberts DA, Gamon JA, Asner GP, Green RO (2004) Using imaging spectroscopy to study ecosystem processes and properties. Bioscience 54(6):523–534

Uzun A, Koçyiğit Ş, Pinar H, Turgunbaev K, Kaymak S (2023) Selection of Central Asian apple species for scab resistance genes using molecular markers. Zemdirbyste Agriculture 110(3):245–254

Wang Y, Yang X, Pan Y et al (2021) CRISPR/Cas9-mediated silencing of the MdWRKY1 transcription factor enhances disease resistance against *Phytophthora cinnamomi* in Eucalyptus globulus. Plant Biotechnol J 19(10):1265–1277

Weiss M, Sniezko RA, Puiu D et al (2020) Genomic basis of white pine blister rust quantitative disease resistance and its relationship with qualitative resistance. Plant J 104:365–376

Wright JW, Stevens KA, Hodgskiss P, Langley CH (2022) SNPs in a large genomic scaffold are strongly associated with $Cr1^R$, major gene for resistance to white pine blister rust in range-wide samples of sugar pine (*Pinus lambertiana*). Plant Dis 106:1639–1644

Xue Y, Li S, Miao D et al (2021) Investigation of PtSGT1 and PtSGT4 function in cellulose biosynthesis in *Populus tomentosa* using CRISPR/Cas9 technology. Int J Mol Sci 22(24):13200

Younessi-Hamzekhanlu M, Gailing O (2022) Genome-wide SNP markers accelerate perennial forest tree breeding rate for disease resistance through marker-assisted and genome-wide selection. Int J Mol Sci 23:12315

Yuan Y, Pan M, Shen L, Liu Y, Zhu Q, Hong J et al (2023) Effect of associated bacteria GD1 on the low-temperature adaptability of *Bursaphelenchus xylophilus* based on RNA-Seq and RNAi. Microorganisms 11(2):430

Zhang B, Huang MR, Zhuge Q et al (2002) Identification of markers linked to resistance locus of Marssonina leaf spot in poplars by bulked segregant analysis (BSA). Yi Chuan (Chinese) 24:543–547

Zhang B, Fei BH, Yu Y, Zhao RJ (2007) Microfibril angle variability in Masson pine (*Pinus massoniana* Lamb.) using X-ray diffraction. For Stud China 9:33–38

Zhang DX, Spiering MJ, Nuss DL (2014) Characterizing the roles of *Cryphonectria parasitica* RNA-dependent RNA polymerase-like genes in antiviral defense, viral recombination and transposon transcript accumulation. PLoS One 9(9):e108653

Zhang Y, Hu P, Zhong C et al (2020) Analyses of genetic diversity, differentiation and geographic origin of natural provenances and land races of *Casuarina equisetifolia* based on EST-SSR markers. Forests 11:432

Zhou H, Bai S, Wang N et al (2020) CRISPR/Cas9-mediated mutagenesis of MdCNGC2 in apple callus and VIGS-mediated silencing of MdCNGC2 in fruits improve resistance to *Botryosphaeria dothidea*. Front Plant Sci 11:575477

Zhu J, Zhou Y, Xu J, Zhao J, Hu J (2019) Methyl jasmonate-induced resistance against *Armillaria mellea* in Scots pine is associated with increased pathogenesis-related gene expression and enhanced defense-related enzyme activities. Front Plant Sci 10:213

Abiotic Stress Reduction with Molecular and Physiological Approaches in Trees

17

Anusree Sobhanan, Kavya, Rekha Meena, and G. P. Mishra

Abstract

Trees play a vital role in ecosystems as they preserve biodiversity, contribute to primary productivity, and support ecological functions. From an economic perspective, humans derive significant benefits from trees. Multiple factors, including biotic and abiotic stresses, adversely affect tree species, leading to extensive tree mortality and consequently disrupting ecological balance. Abiotic stresses such as cold, salinity, and drought are projected to escalate due to global climate change, posing adverse consequences for tree populations. These effects may include biodiversity loss, disturbances in the carbon cycle, and a reduction in the services provided by trees. Trees have evolved a range of mechanisms, including biochemical, physiological and molecular strategies, to adjust to these stressors. Understanding abiotic stress-tolerance processes will enable constructive pyramiding of some tolerances in a single tree via genetic engineering. Furthermore, developing phenotyping technology broadens the range of features that may be assessed, providing us with a less ill knowledge of stress tolerance. A comprehensive understanding of these tolerance mechanisms is essential for developing artificial techniques to mitigate environmental stress. This book chapter highlights the major abiotic stresses in trees, their impacts, response mechanisms and strategies to mitigate these issues in trees.

A. Sobhanan · Kavya · R. Meena (✉)
Department of Agriculture and Environmental Sciences, National Institute of Food Technology Entrepreneurship and Management, Kundli, Sonipat, Haryana, India

G. P. Mishra
ICAR-Indian Agricultural Research Institute, New Delhi, India

R. K. Kalia, R. Pathak (eds.), *Tree Biology and Biotechnology*,
https://doi.org/10.1007/978-981-96-0002-1_17

Keywords

Abiotic stress · Tolerance · Climate change · Genetic engineering · Phenotyping technologies

17.1 Introduction

Trees play an essential part in the ecosystem, providing various environmental, social, and economic benefits. They contribute to oxygen production, carbon sequestration, habitat provision, and aesthetic value. According to the Sustainable Development Report (2023), forest covers nearly 31% of the world and provide shelter to 80% of all terrestrial biodiversity. Trees assist in mitigating climate change and provide widespread products that suffice human needs, which consist of paper, wood, biomass, fuel, and biomaterials (Palmer 2021). From a functional perspective, stress is defined as the changed physiological state brought on by variables that tend to fluctuate the equilibrium (Kranner et al. 2010). Stress can be divided into biotic stress and abiotic stress. Biotic stress is due to living organisms, fungi, bacteria, pests, etc., while abiotic stress is essentially caused by the environmental condition that negatively affects the growth and general development of trees. The abiotic conditions are high-temperature waves due to increasing solar radiation (Teskey et al. 2015), prolonged drought (Cook et al. 2018), high saline soil and water (Corwin 2021), and abrupt flooding episodes (Tabari 2020). These abiotic stresses cause many changes in trees, including hormonal imbalance, a decrease in development and growth, and oxidative stress, and they respond to these by external and internal changes in different ways such as by decreasing the cell wall expansion, stomatal closure, etc. The present chapter discusses and integrates the most recent insight of abiotic stress and tree reactions in the form of biochemical, physiological, and molecular strategies to adjust these stressors.

17.2 Types of Abiotic Stress

17.2.1 Salinity Stress

Salinity is the primary global limitation on crop productivity (Majeed and Muhammad 2019). It denotes a state marked by elevated levels of soluble salts, predominantly calcium, sodium magnesium chlorides, and, to certain limits, sulphates and carbonates. Among these, sodium chloride (NaCl) reigns as a major prevalent and soluble in most saline environments. High levels of Cl^- and Na^+ ions in cell sap lead to a decreased osmotic gradient in the medium of nutrients, decreasing water and nutrient intake and impacting plant morphological traits (Sapre et al. 2018). This imbalance leads to the disruption of cell membranes, increased ion toxicity, and oxidative harm. Meanwhile, semi-arid and arid parts of the world are commonly associated with salinity. Soil affected by salts has been documented in almost every climatic region and at various elevations. Evapotranspiration of water leads to elevated concentrations of salt in the soil, which manifests as an overabundance of sodium ions within the rhizosphere and an accumulation of dissolved salts in the soil water. It is common in clay soils because interchangeable salt is bonded to the negative charges of the clay.

Salinity stress is a dual-temporal process. Firstly, there's a swift osmotic reaction triggered once the outside osmotic pressure surpasses a certain threshold, inducing water stress and hindering growth. This starting phase is succeeded by an ionic phase that expedites metabolic harm, leading to processes like senescence. Ionic stress progresses gradually, exerting a lesser impact compared to osmotic stress at lower to moderate salinity levels. When combined, these osmotic and ionic stressors result in water scarcity, ionic toxicity, nutrient imbalances, and oxidative pressure (Munns 2011). Salinity stress represents a significant abiotic challenge, profoundly impacting production across approximately 20% of globally irrigated areas (Majeed and Muhammad 2019). This environmental stress adversely affects cellular hyperosmolarity and ion equilibrium, necessitating coordinated processes within plants to alleviate these conditions. Salt-affected land is estimated to reach 50% of the total arable area by 2050 (Kumar and Sharma 2020), posing challenges to meeting the anticipated food production demands of the population (Tnay 2019).

17.2.2 Soil Acidity Stress

The soil could be classified as acidic when its pH falls between 5.0 and 5.5. The availability of plant nutrients in the soil declines when pH drops, making it more difficult for plants to absorb water and causing nutritional deficits and toxicities. Deficiencies in calcium (Ca), magnesium (Mg), potassium (K), phosphorus (P), and molybdenum (Mo) cause soils to become infertile. It was not realised for many years that trees are poisoned by high concentrations of H+ ions found in acid soil (Kidd and Proctor 2001). Toxic metal/metalloid ions like iron (Fe), zinc (Zn), copper (Cu), manganese (Mn), and aluminium (Al) become more soluble at low pH. Seed germination is slowed and eventually stops at extremely acidic pH levels (Deska et al. 2011; Kolodziejek and Patykowski 2015). Pollen viability was lowered by acidity stress, which consequently decreased the amount of seed available for germination. A low pH value was found to lower the chlorophyll level and increase membrane permeability. Low pH also reduced the transpiration rate and the water content in the root and shoot. Furthermore, low pH causes higher H+ concentrations, which lead to oxidative stress through the production of ROS in cells. Additionally, higher H_2O_2 and MDA levels in roots cause damage to cellular membranes. Antioxidant enzymes such as superoxide dismutase (SOD), catalase (CAT), ascorbate peroxidase (APX), monodehydroascorbate reductase (MDHAR), and dehydroascorbate reductase (DHAR) are similarly affected by lower pH, similar to abiotic stressors (Hasanuzzaman et al. 2012). Acidification may impact ectomycorrhizas directly by changing the pH of the soil, indirectly via altering the availability of nutrients and the solubility of metals.

17.2.3 Temperature Stress

Low temperatures beneath the freezing point of water can induce cold stress on plants that can affect physiological and metabolic processes, directly leading to plant death (Petruccelli et al. 2022). Cold stress encompasses exposure to temperatures ranging from 0 to 15 °C, including freezing at 0 °C, which detrimentally impacts tree growth and development. A comparison between freezing and chilling stress reveals that freezing stress significantly more adversely affects. The detrimental impact of freezing stress initiates with the nucleation of ice within cells, gradually expanding to form ice crystals, resulting in water leakage and cellular dehydration that leads to various physiological and metabolic damage to the plants (Wisniewski et al. 2018). Many morphological and biochemical consequences are observed in trees due to frost damage. The subtropical tree species, particularly, the olive tree, is highly susceptible to low temperature, and air temperature is the most important habitat factor restricting olive tree development and productivity (Petruccelli et al. 2022). Low temperature (−10 °C) in leaves can lead to cell death, thereby invasion of the pathogen. This leads to defoliation and affects respiration, reducing nutrient uptake and yield. Extensive cracks and discoloration are observed in shoots of olive trees due to frost damage. This ultimately triggers more water loss, dysfunction of vessels, and oxidation of phenolic compounds (Sanzani et al. 2012). Oxidative damage caused by freezing has accumulated higher concentrations of malondialdehyde content (Paliy and Paliy 2018).

Heat stress describes trees undergoing irreversible physical or metabolic harm after exposure to a particular temperature threshold, which varies across different species. Typically, ranging temperatures from 35° to 45 °C induce heat stress in numerous temperate and tropical species. Trees frequently experience high-temperature stress due to the release of greenhouse gases. High temperatures disrupt protein folding, leading to the denaturation of structure membrane and intracellular proteins. This stress induces the production of proteins with chaperone functions like heat-shock proteins (HSPs) and small HSPs. Oxidative damage to biomolecules is also attributed to heat stress. Furthermore, there was an observed heightened accumulation of certain eukaryotic translation initiation factors (eif4f and eif5a-3). This points toward significant cellular reorganization, potentially culminating in Programmed Cell Death (PCD) during heat stress. Increased heat events can reduce photosynthesis, leaf abscission, oxidative stress, etc. (Teskey et al. 2015). Tree mortality driven by heat stress is documented by Percival (2023), who reported massive tree mortality in pine, fir, and oak in North America.

17.2.4 Drought Stress

Unequal rainfall distribution and consequent dry spells due to climate change resulted in significant stress known as drought, which is a widely spread abiotic stress observed globally. It is a prolonged issue that recurs in nature, is difficult to monitor, and is not specific to a particular region or period (Salehi-Lisar and

Bakhshayeshan-Agdam 2016). It leads to cellular dehydration, decreased water potential, and poor plant growth. Low-water stress disrupts physiological and biochemical functions and slows growth. Drought leads to changes in trees that can be primary and secondary, physical and physiological. The effect of drought or dry soil conditions leads to direct damage to the roots and root death such as non-woody feeder roots, usually soil of top level, i.e., 15 inches. Effects of drought on roots as they are located in the above layer of soil, where the fastest drying occurs, seedlings and newly transplanted plants are more vulnerable. Furthermore, feeder roots are usually lost in recent transplants during the transplanting methods. Drought also impacts established trees and shrubs, particularly those in marginal sites with pavement over roots, pockets of soil on ledges or sandy soils, or improperly planted ones.

17.2.5 Heavy Metal Stress

Heavy metals, characterized by their high densities exceeding 5 g cm^{-3}, play a multifaceted role in living organisms. While certain heavy metals, including iron, copper, zinc, cobalt, and nickel, are necessary micronutrients indispensable for the proper functioning of numerous proteins, their excessive accumulation can inflict harm. Conversely, lead, cadmium, and mercury are non-essential metals and pose significant toxicity. Both underground and aboveground surfaces of plants can collect heavy metals (Patra et al. 2004). The concentration of essential and non-essential metals is a single important factor in the growth of trees, so their presence in excess can inhibit the growth. Trees require these elements in very small quantities for their growth, metabolism, and development (Zengin and Munzuroglu 2005). They have evolved intricate mechanisms to sustain the homeostasis of essential metal ions and detoxify non-essential ones, but exposure to high concentrations of heavy metals can disrupt these mechanisms and lead to oxidative stress, impaired biomolecule function, and displace the essential metal ions. Exposure to high redox-reactive metal concentrations has been shown to decrease the activity of antioxidant enzymes, limiting the ability to neutralize reactive oxygen species (ROS) generated by heavy metals.

17.3 Impact of Abiotic Stress on Trees and Their Responses

Abiotic stresses are expected to rise in the near future due to the growing impact of climate change. The longer growth cycle of trees makes them more prone to water stress, salinity, and low temperatures. These conditions will have detrimental effects on growth and productivity and cause extensive tree mortality. Stress-induced effects on trees will indirectly influence the services they offer to humanity. Trees are multifaceted in their importance, supplying food, medicine, fuel, paper, and industrial materials and playing a crucial role in maintaining ecological equilibrium and preserving wildlife (Liu et al. 2021). Global carbon cycling is greatly impacted by forest trees, as they store a significant share of the planet's carbon stock, making them a primary reservoir for atmospheric CO_2 (Hoover and Riddle 2020). The extreme abiotic environment is expected to cause considerable changes in primary productivity, biodiversity, and ecological functions (Sage 2020). According to McDowell et al. (2018), the massive mortality of trees is largely driven by biotic and abiotic stresses.

Abiotic stress in fruit trees and crops leads to a reduction in both yield and quality. The effects of salt stress can cascade into other responses, including reduced growth and yield, heightened toxicity, and the development of oxidative stress. The accumulation of Na^+ and Cl^- ions produced by salt stress in leaves reduces growth in citrus trees (Ziogas et al. 2021). The increased accumulation of these elements dehydrates the cells and adversely affects the enzymes involved in carbohydrate metabolism and the photosynthetic process, reducing photosynthesis along with leaf damage and alterations in chlorophyll content (Syvertsen and Garcia-Sanchez 2014).

Temperate forest trees are adapted to survive severe winter conditions by halting growth and building cold resistance in advance of winter arrival. As daylight diminishes and temperatures drop, their metabolism shifts focus to producing storage compounds like vegetative storage proteins, which triggers a series of biochemical and physical adjustments that enable plants to endure lower temperatures (Welling et al. 2002). Freezing injuries often target membranes, but plants fortify membrane arrangements through cold-acclimation processes to resist damage (Bhattacharyya 2022). In order to maintain the fluidity of the membrane for better freezing tolerance, the side chain alteration of fatty acid in the lipid membrane takes place, which is an essential step (De Palma et al. 2008). Similarly, to counter heat stress, various enzymatic and non-enzymatic antioxidants like dehydroascorbate reductase (DHAR), thioredoxin h-type (TRX H), glutathione-S-transferase (GST), and superoxide dismutase (SOD) that having the chloroplast as a precursor are upregulated, aiding in redox homeostasis. When faced with elevated temperatures, it generates a group of proteins called HSPs, which are thought to play a crucial role in aiding plants to withstand heat stress (Tian et al. 2021). The introduction of HSPs through genetic modification has demonstrated an enhancement in heat tolerance in numerous cases.

A study of drought stress in *Populus* species highlighted that stem growth exhibited the highest sensitivity and decreased the eco-physiological estimates due to water deficit (Yi et al. 2020). With increasing drought stress, the growth of shoot elongation ceased and the growth of roots carried on as far as a severe reduction in photosynthesis took place. Abscisic acid-mediated stomatal closure serves as the initial trigger for the decline in photosynthesis, with further reductions occurring as desiccation stress advances due to the degradation of the photosynthetic machinery, ultimately leading to leaf shedding.

The stress-induced mortality of trees in forests has far-reaching effects on both the organisms living there and their habitats, thereby destabilizing the ecosystem. These scenarios can potentially lead to species extinction, such as lichens that have thrived within these tree populations (Esseen et al. 2023). These stresses pose a significant threat to tree species incapable of adapting to climate shifts, potentially pushing them towards extinction (Fensham et al. 2020). Significant tree losses can be observed due to forest fires caused by prolonged drought conditions. For instance, in the United States, ~18% of tree loss is attributed to this phenomenon (Berner et al. 2017). Trees exposed to environmental stresses can release volatile organic compounds (VOCs), potentially contributing to air pollution (Faiola and Taipale 2020). The tree loss disrupts the carbon reservoir, causing carbon emissions to be released from dead trees. For example, the 2005 Amazon forest drought resulted in the loss of 1.21–1.60 Petagrams of inherent carbon storage due to tree loss. Similarly, severe drought in Texas, USA, led to a carbon loss of 24–30 teragrams (Yang et al. 2018). The decline in trees may negatively affect human life, along with a decrease in forest productivity and revenue loss (Aukema et al. 2011). Economic analysis should be conducted to find the impact of losses. The consequences of various abiotic stresses on trees and their responses have been presented in Table 17.1.

17.4 Approaches Towards Abiotic Stress

Multi-omics approaches includes genomics, proteomics, transcriptomics are being used to study the science of stress biology and to develope stress smart trees (Fig. 17.1). It is a fast-ongoing technique of the modern era that includes methodology, interdisciplinary, and trans-disciplinary approaches. These omics gives a new perspective to understand the reaction of stressed trees.

17.4.1 Proteomics for Stress Tolerance

Many techniques have been developed by proteomics to assess the prevalence of subcellular proteins, such as ROS scavengers under stressful conditions, transcriptional regulators, and water and ion transporters. One of the main strategies for producing tolerant crops is the overexpression of stress-responsive genes (Pandey and Dubey 2019). Numerous researchers have documented the amplification of subcellular proteins in the vacuole, plasma membrane, nucleus, mitochondria endoplasmic reticulum, and chloroplast. Of these, 42% were found in the nucleus, whereas 23%, 13%, and 10% were found in the chloroplast, plasma membrane, and endoplasmic reticulum, respectively. According to Nouri and Komatsu (2013), stress tolerance is increased in transgenic plants by overexpressing organelle proteins in plant life in a changing environment.

According to Kumar et al. (2018), the plasma membrane is involved in signal perception, cellular homeostasis, and the expression of many plasma membrane proteins that aid in plants' ability to withstand cold stress. It has been observed that during cold acclimation, phospholipase Dδ, a protein associated with the plasma membrane that hydrolyses membrane phospholipids to create phosphatidic acid, is overexpressed (Margutti et al. 2019). Improved salinity and osmotic stress tolerance were demonstrated by transgenic tobacco plants with constitutively overexpressed SbSRP genes (Udawat et al. 2017). Rising levels of lysine and threonine-encoded

Table 17.1 Effect of abiotic stresses on trees and their responses

Stress	Effects	Response
Salinity stress	Reduced chlorophyll a and chlorophyll b Reduced photosynthetic activity Reduced transpiration rate Reduced rate of surface expansion of leaves Decreased osmotic potential Nutrient deficiency Disturb potassium (K^+) nutrition Alkanoid increases Decrease in photochemical activities Decreased CO_2 fixation Accumulation of osmolytes and Osmo protectants	Salt avoidance through salt exclusion Phenological avoidance Dilution of high salt concentration in plant tissues by succulence and growth Nutritive tolerance Biochemical tolerance through adaptations of cell organelles
Heat stress	Male sterility Abnormalities in spikelet production Modification in callose walls of microspores Exine formation Impact on carbohydrate metabolism Reduced cell size Closure of stomata Cut down water loss Increased stomatal and trichome densities	Improved hormonal homeostasis Production of heat shock protein Retrograde signal pathways that protect cellular integrity and the normal growth of plants. Changes in water relation Antioxidant adjustment Changes in gene expression Osmotic adjustment
Cold stress	Cell dehydration Formation of a nucleation of ice Formation of ice crystals Water leakage	Decrease in secondary metabolism Increase in transmembrane activity Increase in chaperon and cytoskeleton protein Increase ROS scavenging Increase primary metabolism
Heavy metal stress	Alternation in spindle activity Cell injury/death Damage of lipids and DNA Reduction in mineral uptake Reduction in respiration	Activation of defence mechanism Alteration in nutrient uptake Change in root morphology Reduced growth
Drought stress	Reduced root growth Reduced leaf production Decreased gas exchange Decreased water and nutrient uptake Closure of stomata Reduced photosynthesis, chlorophyll and transpiration	Reduced root growth Increased root-shoot ratio Decreased gas exchange Closure of stomata Increase ROS scavenging Increase primary metabolism

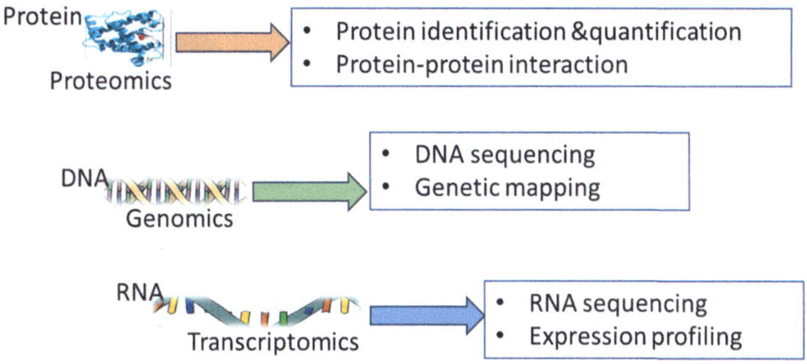

Fig. 17.1 Multi-omics approaches

genes were added to the rice genome, and transgenic types were created to boost the protein content in rice (Jiang et al. 2016). Additionally, high-quality protein maize was created, with seeds having twice lysine content (Babu et al. 2015). Similar experiments were also conducted in tobacco, canola, soybean, and Arabidopsis to improve the amount of lysine, threonine, and methionine. Transgenic variety development is still being researched to increase tolerance within the plant system (Fig. 17.2).

17.4.2 Epigenetic Regulation of Abiotic Stress

Environmental stressors, genomic pressures, or both may be responsible for epigenetic alterations (Bonasio et al. 2010). Mechanisms underlying epigenetic regulations includes acetylation, methylation, ubiquitylation, phosphorylation, sumoylation, reversible DNA methylation, and sRNAs. Only a limited aspect of the epigenetic regulation of response to abiotic stress has been studied so far. However, new findings are assisting in identifying the epigenetic mediators of plant responses to abiotic stress (Chinnusamy and Zhu 2009). Epigenetic changes most likely influence a plant's capacity to respond to stressors and, similar to environmental selection, act upon genetic diversity (Gutzat and Scheid 2012).

It has been reported that Norway spruce may possess epigenetic memory (Yakovlev et al. 2016), a long-lasting phenomenon, which is resolved by time. The seed reaches complete development, influences the progression of trees and enables quick adjustment to shift the environmental conditions. Because of this, epigenetics has applied ramifications for tree breeding. In response to climate change, the potential of epigenetics has been reported as one of the basis of adaptable characteristics for biotechnology, breeding trees, and ecosystem conservation (Bräutigam et al. 2013). Epigenetic modifications may play an important role in forest trees' potential for adaptation and phenotypic plasticity, as well as their resilience in changing climatic conditions.

Studies emphasize the significance of epigenetic mechanisms in relation to long-lived organisms' environmental adaption. In hybrid poplar (*P. deltoides* × *P. nigra*),

Fig. 17.2 Quality improvement under stress conditions using a proteomic approach

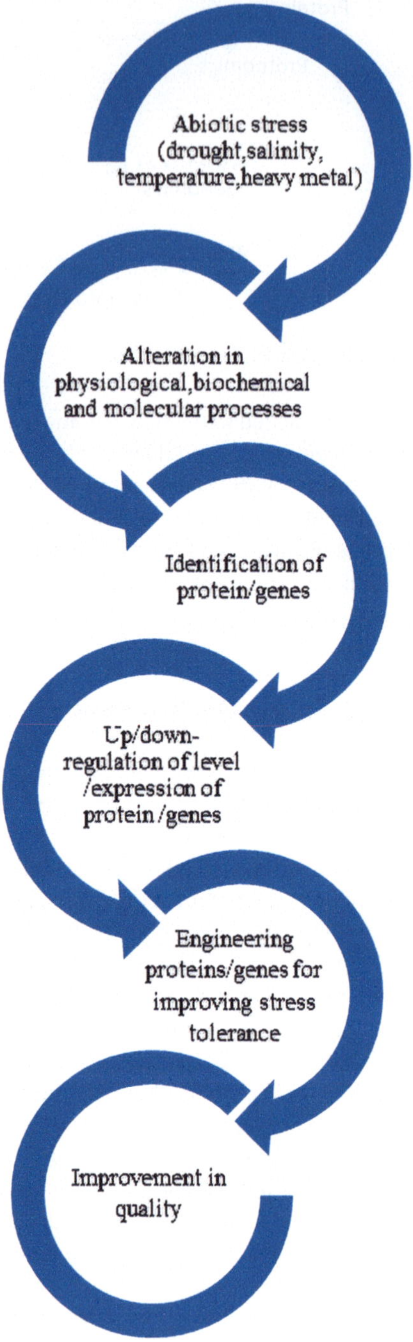

DNA methylation and features linked to productivity of biomass were analyzed (Hamanishi and Campbell 2011). Under well-watered conditions, these variables showed a positive connection. On the other hand, genotypes with decreased development in water-deficient environments displayed the most variance in DNA methylation, indicating that DNA methylation may cause the poplar plant's altered gene expression during water stress. These discoveries unveil various molecular defence mechanisms against abiotic stress, utilizing epigenetics to manage the trees.

17.4.3 Genetic Engineering

Genetic engineering and breeding techniques are employed to enhance tolerance to environmental stresses. Utilizing the abundant genetic diversity inherent in nature, researchers acquire extensive multi-omics data, which is instrumental in advancing breeding, genetic engineering, and genome editing (Fig. 17.3). Hence, it is imperative to delve into the molecular basis and identify genes related to stress tolerance to develop resilient trees. Trees can identify two distinct gene groups concerning stress management: protective genes and genes that control gene expression during stress conditions (Estravis-Barcala et al. 2020).

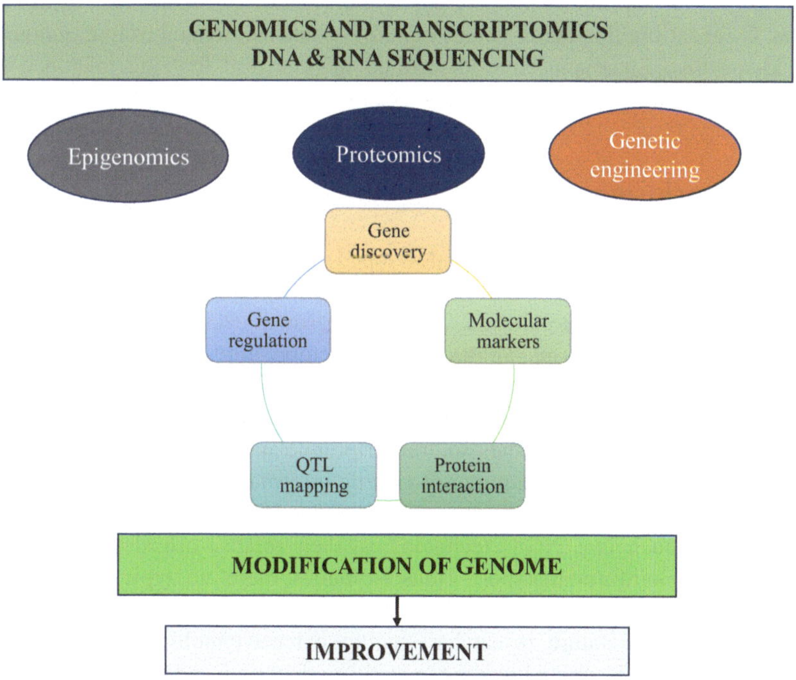

Fig. 17.3 Omics approaches to develop stress tolerance

Transgenic trees developed to withstand drought and saline stress exhibit greater resilience to these pressures than their non-transgenic counterparts. The studies revealed that the presence of CpFATB enhances drought tolerance in winter sweet trees. These carrier proteins are responsible for maintaining and regulating fatty acid composition, which has been confirmed as a key factor in drought tolerance. As a result, transgenic trees carrying CpFATB displayed elevated photosynthetic rates (Zhang et al. 2013). The overexpression of resistance genes, VGB, SacB, and JERF36, has been shown to confer tolerance to drought and salinity stress, as evidenced by improved growth in these trees (Su et al. 2011). Han et al. (2013) revealed that glycogen synthase kinase GSK3 in Thale cress deliberated drought and salt tolerance. In another study, Wang et al. (2010) identified the manganese superoxide dismutase gene from *Tamarix androssowii* as a contributor to salt tolerance in these species. Similarly, numerous studies have illustrated the use of genetic engineering to enhance cold stress tolerance in trees. One notable example is enhancing the AtCBF1 gene from *A. thaliana* in hybrid aspen (*P. tremula* × *P. alba*), which significantly increased the acceptance of its un-adapted stems and leaves to freezing (Benedict et al. 2006). By introducing the ThCAP gene from *Tamarix hispida*, which codes for a cold-acclimation protein, into transgenic poplar *(Populus davidiana* × *Populus bolleana)*, researchers observed an increased resistance to lower temperatures, surpassing that of non-transgenic plants (Guo et al. 2009). Enhanced cold-stress tolerance is achieved through the over-expression of two CBL genes obtained from *P. euphratica* in white poplar. Furthermore, the consistent expression of two C-repeat-binding factor genes, EguCBF1a and -b, isolated from *E. gunnii*, in a cold-sensitive Eucalyptus hybrid resulted in improved freezing tolerance and developmental outcomes (Navarro et al. 2011).

17.5 Conclusion

Over their complex lives, trees have developed complex mechanisms to deal with abiotic challenges. Abiotic stresses are expected to rise in the near future due to the growing impact of climate change. The longer growth cycle of trees makes them more prone to water stress, salinity, and low temperatures. Essential knowledge about some genes has been made available by genome sequencing, providing thoughts on potential research directions in the abiotic stress reduction in trees. The majority of studies on tree species are transcriptomic and omics in nature. This is to be expected since a massive number of genetic information can be generated even in the non-appearance of reference genomes thanks to advancements in Next Generation Sequencing (NGS) technologies, their relative affordability, and the availability of bioinformatics tools. Transcriptomics, therefore, typically represents Woody's initial strategy for molecular research. Utilizing the abundant genetic diversity inherent in nature, researchers acquire extensive multi-omics data, which is instrumental in advancing breeding, genetic engineering, and genome editing.

Acknowledgement The authors are thankful to the National Institute of Food Technology Entrepreneurship and Management, Kundli, India, for providing contingencies, facilities, and support.

References

Aukema JE, Leung B, Kovacs K, Chivers C, Britton KO, Englin J, Frankel SJ, Haight RG, Holmes TP, Liebhold AM, Von Holle B (2011) Economic impacts of non-native forest insects in the continental United States. PLoS One 6(9):e24587

Babu BK, Agrawal PK, Saha S, Gupta HS (2015) Mapping QTLs for opaque2 modifiers influencing the tryptophan content in quality protein maize using genomic and candidate gene-based SSRs of lysine and tryptophan metabolic pathway. Plant Cell Rep 34:37–45

Benedict C, Skinner JS, Meng R, Chang Y, Bhalerao R, Huner NP, Finn C, Chen THH, Hurry V (2006) The CBF1-dependent low temperature signalling pathway, regulon and increase in freeze tolerance are conserved in Populus spp. Plant Cell Environ 29(7):1259–1272

Berner LT, Law BE, Meddens AJ, Hicke JA (2017) Tree mortality from fires, bark beetles, and timber harvest during a hot and dry decade in the western United States (2003–2012). Environ Res Lett 12(6):065005

Bhattacharyya S (2022) Mechanism of temperature stress acclimation and the role of transporters in plants. In: Plant perspectives to global climate changes. Academic Press, pp 413–457

Bonasio R, Tu S, Reinberg D (2010) Molecular signals of epigenetic states. Science 330(6004):612–616

Bräutigam K, Vining KJ, Lafon-Placette C, Fossdal CG, Mirouze M, Marcos JG, Fluch S, Fraga MF, Guevara MÁ, Abarca D, Cervera MT (2013) Epigenetic regulation of adaptive responses of forest tree species to the environment. Ecol Evol 3(2):399–415

Chinnusamy V, Zhu JK (2009) Epigenetic regulation of stress responses in plants. Curr Opin Plant Biol 12(2):133–139

Cook BI, Mankin JS, Anchukaitis KJ (2018) Climate change and drought: from past to future. Curr Clim Chang Rep 4:164–179

Corwin DL (2021) Climate change impacts on soil salinity in agricultural areas. Eur J Soil Sci 72(2):842–862

De Palma M, Grillo S, Massarelli I, Costa A, Balogh G, Vigh L, Leone A (2008) Regulation of desaturase gene expression, changes in membrane lipid composition and freezing tolerance in potato plants. Mol Breed 21:15–26

Deska J, Jankowski K, Bombik A, Jankowska J (2011) Effect of growing medium pH on germination and initial development of some grassland plants. Acta Sci Pol 10:45–56

Esseen PA, Rytterstam J, Atrena A, Jonsson BG (2023) Long-term dynamics of the iconic old-forest lichen *Usnea longissima* in a protected landscape. For Ecol Manag 546:121369

Estravis-Barcala M, Mattera MG, Soliani C, Bellora N, Opgenoorth L, Heer K, Arana MV (2020) Molecular bases of responses to abiotic stress in trees. J Exp Bot 71(13):3765–3779

Faiola C, Taipale D (2020) Impact of insect herbivory on plant stress volatile emissions from trees: a synthesis of quantitative measurements and recommendations for future research. Atmos Environ 5:100060

Fensham RJ, Laffineur B, Collingwood TD, Beech E, Bell S, Hopper SD et al (2020) Rarity or decline: key concepts for the Red List of Australian eucalypts. Biol Conserv 243:108455

Guo XH, Jiang J, Lin SJ, Wang BC, Wang YC, Liu GF, Yang CP (2009) A ThCAP gene from *Tamarix hispida* confers cold tolerance in transgenic Populus (*P. davidiana*× *P. bolleana*). Biotechnol Lett 31:1079–1087

Gutzat R, Scheid OM (2012) Epigenetic responses to stress: triple defense? Curr Opin Plant Biol 15(5):568–573

Hamanishi ET, Campbell MM (2011) Genome-wide responses to drought in forest trees. Forestry 84(3):273–283

Han MS, Noh EW, Han SH (2013) Enhanced drought and salt tolerance by expression of AtGSK1 gene in poplar. Plant Biotechnol Rep 7:39–47

Hasanuzzaman M, Hossain MA, Fujita M (2012) Exogenous selenium pretreatment protects rapeseed seedlings from cadmium-induced oxidative stress by upregulating antioxidant defense and methylglyoxal detoxification systems. Biol Trace Elem Res 149:248–261

Hoover K, Riddle AA (2020) Forest carbon primer. Congressional Research Service, Washington, DC

Jiang SY, Ma A, Xie L, Ramachandran S (2016) Improving protein content and quality by over-expressing artificially synthetic fusion proteins with high lysine and threonine constituent in rice plants. Sci Rep 6(1):34427

Kidd PS, Proctor J (2001) Why plants grow poorly on very acid soils: are ecologists missing the obvious? J Exp Bot 52:791–799

Kolodziejek J, Patykowski J (2015) Effect of environmental factors on germination and emergence of invasive *Rumex confertus* in Central Europe. Sci World J 2015:170176

Kranner I, Minibayeva FV, Beckett RP, Seal CE (2010) What is stress? Concepts, definitions and applications in seed science. New Phytol 188(3):655–673

Kumar K, Mosa KA, Meselhy AG, Dhankher OP (2018) Molecular insights into the plasma membrane intrinsic proteins roles for abiotic stress and metalloids tolerance and transport in plants. Indian J Plant Physiol 23:721–730

Kumar P, Sharma PK (2020) Soil salinity and food security in India. Front Sustain Food Syst 4:533781

Liu C, Luan P, Li Q, Cheng Z, Xiang P, Liu D, Hou Y, Yang Y, Zhu H (2021) Biopolymers derived from trees as sustainable multifunctional materials: a review. Adv Mater 33(28):2001654

Majeed A, Muhammad Z (2019) Salinity: a major agricultural problem-causes, impacts on crop productivity and management strategies. In: Plant abiotic stress tolerance: agronomic, molecular and biotechnological approaches, pp 83–99

Margutti MP, Reyna M, Vilchez AC, Villasuso AL (2019) Lipid profiling shows tissue-specific differences in barley for glycerolipid composition in response to chilling. Environ Exp Bot 158:150–160

McDowell N, Allen CD, Anderson-Teixeira K, Brando P, Brienen R, Chambers J et al (2018) Drivers and mechanisms of tree mortality in moist tropical forests. New Phytol 219(3):851–869

Munns R (2011) Plant adaptations to salt and water stress: differences and commonalities. Adv Bot Res 57:1–32

Navarro M, Ayax C, Martinez Y, Laur J, El Kayal W, Marque C, Teulieres C (2011) Two EguCBF1 genes overexpressed in eucalyptus display a different impact on stress tolerance and plant development. Plant Biotechnol J 9(1):50–63

Nouri MZ, Komatsu S (2013) Subcellular protein overexpression to develop abiotic stress tolerant plants. Front Plant Sci 4:2

Paliy AE, Paliy IN (2018) Some biochemical features of *Olea europaea* L. frost resistance. In: VIII International Scientific and Practical Conference on Biotechnology as an Instrument for Plant Biodiversity Conservation 1324, pp 395–400

Palmer L (2021) How trees and forests reduce risks from climate change. Nat Clim Chang 11(5):374–377

Pandey P, Dubey RS (2019) Metal toxicity in rice and strategies for improving stress tolerance. In: Advances in rice research for abiotic stress tolerance. Woodhead Publishing, pp 313–339

Patra M, Bhowmik N, Bandopadhyay B, Sharma A (2004) Comparison of mercury, lead and arsenic with respect to genotoxic effects on plant systems and the development of genetic tolerance. Environ Exp Bot 52(3):199–223

Percival GC (2023) Heat tolerance of urban tree species – a review. Urban For Urban Green 86:128021. https://doi.org/10.1016/j.ufug.2023.128021

Petruccelli R, Bartolini G, Ganino T, Zelasco S, Lombardo L, Perri E, Durante M, Bernardi R (2022) Cold stress, freezing adaptation, varietal susceptibility of *Olea europaea* L.: a review. Plan Theory 11(10):1367

Sage RF (2020) Global change biology: a primer. Glob Chang Biol 26(1):3–30

Salehi-Lisar SY, Bakhshayeshan-Agdam H (2016) Drought stress in plants: causes, consequences, and tolerance. In: Drought stress tolerance in plants, Physiology and biochemistry, vol 1, pp 1–16

Sanzani SM, Schena L, Nigro F, Sergeeva V, Ippolito A, Salerno MG (2012) Abiotic diseases of olive. J Plant Pathol 94(3):469–491

Sapre S, Gontia-Mishra I, Tiwari S (2018) *Klebsiella sp.* confers enhanced tolerance to salinity and plant growth promotion in oat seedlings (*Avena sativa*). Microbiol Res 206:25–32

Su X, Chu Y, Li H, Hou Y, Zhang B, Huang Q et al (2011) Expression of multiple resistance genes enhances tolerance to environmental stressors in transgenic poplar (Populus × euramericana 'Guariento'). PLoS One 6(9):e24614

Sustainable Development Report (2023). https://www.sustainabledevelopment.report/reports/sustainable-development-report-2023/

Syvertsen JP, Garcia-Sanchez F (2014) Multiple abiotic stresses occurring with salinity stress in citrus. Environ Exp Bot 103:128–137

Tabari H (2020) Climate change impact on flood and extreme precipitation increases with water availability. Sci Rep 10(1):13768

Teskey R, Wertin T, Bauweraerts I, Ameye M, McGuire MA, Steppe K (2015) Responses of tree species to heat waves and extreme heat events. Plant Cell Environ 38(9):1699–1712

Tian F, Hu XL, Yao T, Yang X, Chen JG, Lu MZ, Zhang J (2021) Recent advances in the roles of HSFs and HSPs in heat stress response in woody plants. Front Plant Sci 12:704905

Tnay G (2019) Too much salt: the growing threat that salinity poses to global food production. Future Directions International Pty Ltd, Dalkeith

Udawat P, Jha RK, Mishra A, Jha B (2017) Overexpression of a plasma membrane-localized Sb SRP-like protein enhances salinity and osmotic stress tolerance in transgenic tobacco. Front Plant Sci 8:582

Wang YC, Qu GZ, Li HY, Wu YJ, Wang C, Liu GF et al (2010) Enhanced salt tolerance of transgenic poplar plants expressing a manganese superoxide dismutase from *Tamarix androssowii*. Mol Biol Rep 37:1119–1124

Welling A, Moritz T, Palva ET, Junttila O (2002) Independent activation of cold acclimation by low temperature and short photoperiod in hybrid aspen. Plant Physiol 129(4):1633–1641

Wisniewski M, Nassuth A, Arora R (2018) Cold hardiness in trees: a mini-review. Front Plant Sci 9:1394

Yakovlev IA, Carneros E, Lee Y, Olsen JE, Fossdal CG (2016) Transcriptional profiling of epigenetic regulators in somatic embryos during temperature induced formation of an epigenetic memory in Norway spruce. Planta 243:1237–1249

Yang Y, Saatchi SS, Xu L, Yu Y, Choi S, Phillips N, Kennedy R, Keller M, Knyazikhin Y, Myneni RB (2018) Post-drought decline of the Amazon carbon sink. Nat Commun 9(1):3172

Yi L, Li B, Korpelainen H, Yu F, Wu L, Tong L, Liu M (2020) Mechanisms of drought response in Populus. South For J For Sci 82(4):359–366

Zengin FK, Munzuroglu O (2005) Effects of some heavy metals on content of chlorophyll, proline and some antioxidant chemicals in bean (*Phaseolus vulgaris* L.) seedlings. Acta Biol Cracov Ser Bot 47(2):157–164

Zhang L, Liu M, Qiao G, Jiang J, Jiang Y, Zhuo R (2013) Transgenic poplar "NL895" expressing CpFATB gene shows enhanced tolerance to drought stress. Acta Physiol Plant 35:603–613

Ziogas V, Tanou G, Morianou G, Kourgialas N (2021) Drought and salinity in citriculture: optimal practices to alleviate salinity and water stress. Agronomy 11(7):1283

Abiotic Stress Tolerance: Role of Nitric Oxide Signalling in Developing Climate-Resilient Plants

18

Lata Sharma, Suman Parihar, Chandra Shekhar Seth, and G. S. Shekhawat

Abstract

Trees experience continuous exposure to climate change, which has become a critical issue worldwide. It accelerates the severity of abiotic stresses, including water deficit, salinity, UV radiation and temperature fluctuations. These changes exert devastating impacts on the progression and development of trees. Long-living trees have evolved different strategies to survive in changing climate. Nitric oxide (NO) is a more focused bioactive compound, and its endogenous synthesis and exogenous application in plants result in several physiological and molecular alterations that help ameliorate adverse impacts of abiotic stresses. NO regulates various stress-responsive gene expressions through comprehensive interconnected signalling pathways such as MAP-kinase-dependent, Ca^{2+}-dependent and phytohormones-dependent signalling. Moreover, NO can modulate antioxidative enzyme activities through post-translational modifications that are positively involved in the tolerance mechanisms. This is crucial for trees to be climate-smart, resilient and more tolerant in a fast-changing environment. NO could be a vital molecule to strengthen plant tolerance and resilience. This chapter addresses different abiotic stresses along with NO-mediated responses to various abiotic stresses.

Keywords

Climate change · Water deficit · Nitric oxide · Signalling · MAP-kinase · Post-translational modifications · Antioxidative

L. Sharma · S. Parihar · G. S. Shekhawat (✉)
Plant Biotechnology and Molecular Biology Laboratory, Department of Botany, Jai Narain Vyas University, Jodhpur, Rajasthan, India

C. S. Seth
Departments of Botany, University of Delhi, New Delhi, India

R. K. Kalia, R. Pathak (eds.), *Tree Biology and Biotechnology*,
https://doi.org/10.1007/978-981-96-0002-1_18

18.1 Introduction

Globally, the frequency and severity of prolonged droughts, salinity, heat waves, and other severe weather conditions are increasing due to climate change, which has adverse impacts on forest trees (Brichta et al. 2024) and agricultural crops. The primary drivers of these changes in climatic conditions are the increasing levels of carbon dioxide (CO_2) and rising temperatures (Eggleton 2012). The accumulation of CO_2 in the atmosphere is increasing steadily by 2–3 ppm per year and is predicted to increase to 550 ppm by 2050 (Semba et al. 2022). According to twenty-first-century climate simulations, the average annual temperature will rise by an additional 1.5 °C during the next 10 years. Also, there is a possibility that the highest temperature will rise by as much as 2.5 °C (Hassan and Nile 2021).

Plants are sessile organisms, rooted in the precise site at which their seed germinates and or they are planted. Throughout their lifecycle, from germination to seed production, they are subjected to a heterogeneous and fluctuating environment (Simontacchi et al. 2015). To survive under stress conditions, plants adapt on a cellular, molecular and physiological level, involving a sophisticated network that facilitates the sensitivity and diffusion of stress signals that sets off a cascade of reactions (Nakashima et al. 2009). Stress adaptation reactions are triggered by various types of stress, but they always result either in stress acclimation or programmed cell death (PCD) (Fancy et al. 2017). Developing trees with improved resistance to environmental changes could result in higher yields in the long run.

The common reaction of plants to these abiotic stresses is the generation of redox-active components, namely, reactive oxygen species (ROS) (Mahawar and Shekhawat 2019) and reactive nitrogen species (RNS) (Khator et al. 2024). However, the generation of these redox-active elements leads to oxidative damage to many components of the cell (Fancy et al. 2017). Hence, maintaining homeostasis of redox-active molecules is essential for abiotic stress tolerance (Khator and Shekhawat 2020). To maintain homeostasis of redox-active molecules and mitigate induced oxidative stress, plants have developed a regulated antioxidant system (Khator and Shekhawat 2018; Mahawar and Shekhawat 2018; Bulchandani and Shekhawat 2020). Nitric oxide (NO) is one of the major RNS that has dual functions as a downstream and upstream regulator in response to various abiotic stresses (Khator et al. 2024). Recent studies revealed that nitric oxide is a crucial signalling molecule in several acclimation reactions to both abiotic and biotic stresses (Simontacchi et al. 2015). Within this framework, this review will focus on the plants under various abiotic stress conditions influenced by climatic change and the function of NO in improving plant resilience.

18.2 Abiotic Stresses

Abiotic stresses are the major threats affecting the growth and development of plants by disturbing their morphological, physiological, metabolic and molecular functions (Mahawar and Shekhawat 2016; Mahawar et al. 2021) (Fig. 18.1). Several

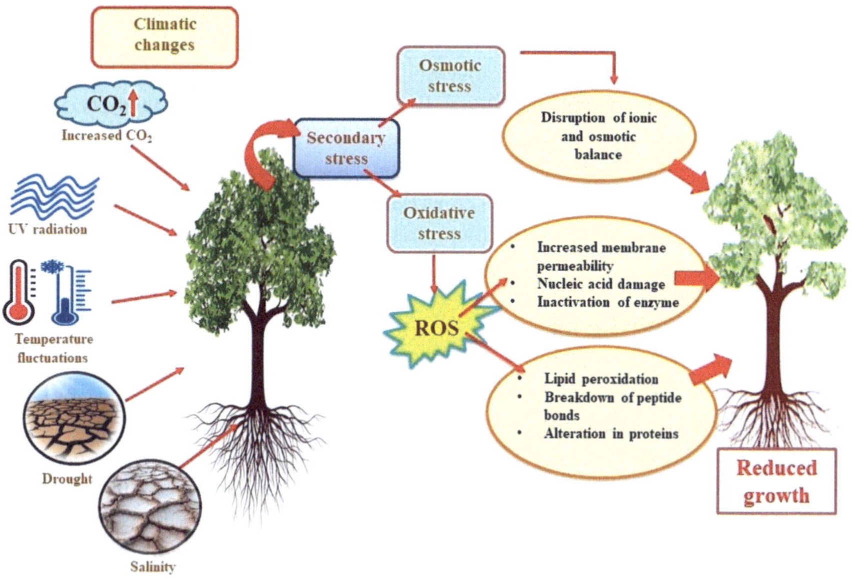

Fig. 18.1 Impact of different abiotic stress on plant

abiotic stresses, including salinity, drought, temperature fluctuations, high light intensity and UV radiation that affect the forest biome, are intensifying due to climate change (Wilmer et al. 2022; Abasi et al. 2024). These abiotic stresses lead to the generation of secondary stresses, such as osmotic stress and oxidative burst (Rao and Shekhawat 2016; Khator et al. 2021; Mahawar and Shekhawat 2023). Water deficit in plants occurs due to decreased water quantity in the soil and continued loss of water through transpiration (Abasi et al. 2024). Moreover, it occurs at increased salt levels in soil, which lowers the water potential and makes it challenging for roots to absorb water from the soil. Thus, this leads to the efflux of cytoplasmic water and cells get plasmolysed; it also inhibits photosynthesis by affecting the thylakoid membrane. On the other hand, the increased salt concentration in the soil causes salinity stress that has two significant impacts on plants, namely, ionic toxicity and osmotic stress, leading to the production of ROS (Mahawar et al. 2018a), which adversely affects membrane function, cytosolic metabolism and cell growth (Zhu 2001; Harfouche et al. 2014; Verma et al. 2015; Mahawar et al. 2018b). Many physiological functions, such as senescence, ion homeostasis, water relations and photosynthesis, are influenced by salinity (Sarwar et al. 2021). Seed germination, photomorphogenesis and growth of roots and leaves are also hindered by excessive salt (Hu et al. 2024).

The frequency and severity of high temperature have increased due to climatic changes (Amnuaylojaroen et al. 2022). High temperature also disturbs basic physiological, biochemical and metabolic processes of the plant life cycle (Sadiq et al. 2020) through the generation of ROS, which has a destructive impact on nucleic acids, proteins and lipids (Hasanuzzaman et al. 2020). High-temperature stress adversely affects membranes by causing the disintegration of lipids, which increases membrane permeability and leads to electrolyte loss (Zhao et al. 2020; Pandey et al. 2023). Stomatal conductance, photosynthesis, respiration and transpiration are impacted by heat stress. Water loss due to heat stress exerts negative impacts on cellular growth, fluidity of thylakoid membrane, and activity of photosystem II and rubisco (Sharma et al. 2020). Conversely, plants also experience adverse effects of low temperatures, including changes in enzyme activity cytoplasmic viscosity and induction of necrosis, chlorosis and membrane damage (Malhi et al. 2021).

18.3 NO-Biosynthesis

The mechanism of NO biosynthesis in plants remains unclear, despite the identification of significant oxidative processes by researchers. There are various oxidative and reductive methods involved in NO production. L-arginine, hydroxylamine, and polyamines are aminated molecules that are involved and oxidized in oxidative pathways to produce NO. NO is produced in mammals through three isoforms of NOS (nitric oxide synthase) such as endothelial, neuronal and inducible NOS in which NOS oxidizes arginine to produce NO (Wendehenne et al. 2001). Various studies indicate the presence of different forms in evolved plants, despite the lack of proof for homology between nitric oxide synthase (NOS) and plant genomic sequences. While reduction pathways involve nitrite being reduced to NO through the involvement of NR (nitrate reductase), cytochrome-c oxidase, membrane-bound NiR (nitrite reductase) and XOR (xanthine oxidoreductase). On the other hand, nitrite is reduced non-enzymatically in an acidic environment (Prochazkova et al. 2014; Pandey et al. 2023).

18.4 NO-Mediated Singalling in Abiotic Stress Response

The identification of the crucial role of NO in plant signalling accompanied the finding of sources of NO synthesis. NO-induced transcriptional changes or activation of secondary messengers regulate various physiological activities (Falak et al. 2021), such as organelle mobility, photosynthesis, PCD, photosynthesis, seed germination, pollen tube growth, hypersensitive response, cell wall lignification, and biotic-abiotic stress (Sami et al. 2018). NO relays its signals via some proteins which are triggered after post-translational modifications (PTMs), including tyrosine nitration and S-nitrosylation through NO or its derivative (Verma et al. 2013; Begara-Morales et al. 2016). NO regulates numerous signalling pathways, including protein kinase, calcium ions and cyclic nucleotides involved in abiotic stress tolerance mechanisms (Fig. 18.2). Heme oxygenase (HO) and carbon monooxide (CO) are also crosstalks with NO in tolerance mechanisms (Mahawar and Shekhawat 2018; Khator et al. 2020; Popek et al. 2022).

As biological messengers, NO and ROS are significantly involved in abiotic stressor responses (Anita et al. 2023; Khan et al. 2023). NO production is enhanced

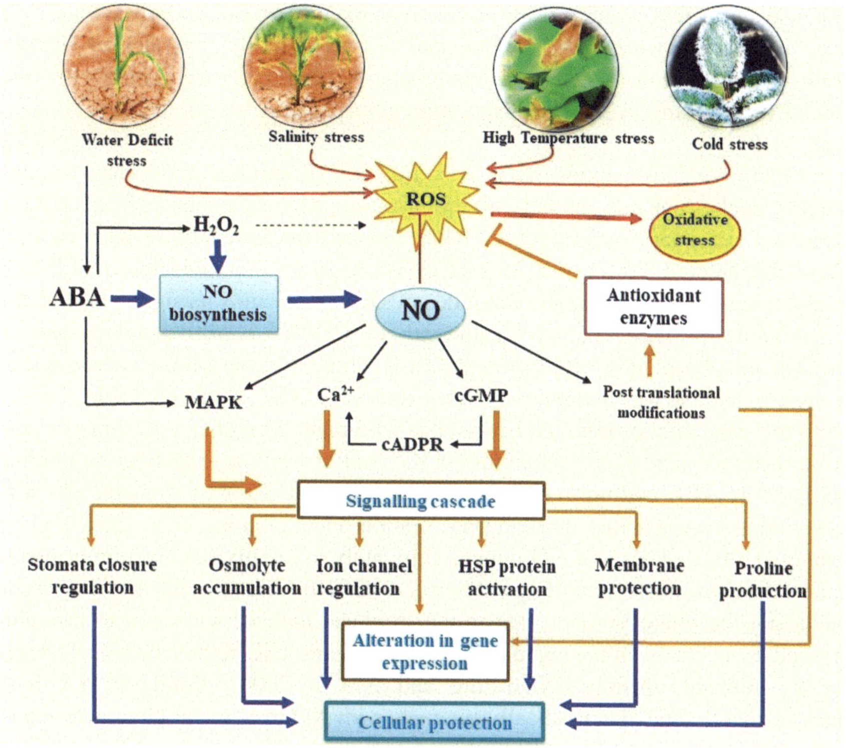

Fig. 18.2 Outline figure explaining NO mediated signalling in cellular protection under water deficit, salinity, high temperature and cold stress

because of the increased ROS production under abiotic stress, mainly through NR activity, which shows the relationship between ROS and RNS (Hasanuzzaman et al. 2018). The study also reported that the utilisation of H_2O_2 triggers NO synthesis, which involves the induction of antioxidative activity, which causes reduced ROS levels and the retention of redox balance in *Arabidopsis*. Furthermore, in the stress response, peroxynitrite ($ONOO^-$) is produced due to a reaction between NO and superoxide anion (O_2^-), which is involved in the activation of enzymes and proteins through nitration of Tyr-residue for further signalling (Saxena and Shekhawat 2013). Another derivative of NO is GSNO, produced by S-nitrosylation of NO with GSH, which regulates catalase and glycolate oxidase activities to maintain the H_2O_2 level in cells (Ortega-Galisteo et al. 2012).

Besides ROS and NO, other intricate molecular networks are also involved in signalling pathways, including calcium (Ca^{2+}), cyclic nucleotides, mitogen-activated protein kinases (MAPKs) and plant hormones (Farnese et al. 2016). The MAPK signalling pathway is activated through NO as well as by ROS and hormones; this involves a cascade of phosphorylation (Jonak et al. 2002). The activated MAPK phosphorylates other target molecules present in both cytosol and nucleus, including transcriptional factors and enzymes (Rodriguez et al. 2010). MAPK-mediated responses include activation of pro-oxidative enzymes or antioxidative, which can either intensify or weaken the initial signal that was initially generated by NO and ROS (Opdenakker et al. 2012). In mammals, NO indirectly triggers MAPK cascades by enhancing cyclic guanosine monophosphate synthesis (cGMP) (Francis et al. 2010).

Furthermore, nitric oxide (NO) may function in animals and plants through cADPR (cyclic ADP-ribose) and Ca^{2+} mobilization, which are induced by cGMP. An additional secondary messenger, cADPR, induced the intracellular level of Ca^{2+} (Wendehenne et al. 2001) (Fig. 18.2). Ca^{2+} is one of the key molecules that have a crucial role in response to environmental stresses. It transmits its signal through the Ca^{2+}-binding proteins, namely, calmoduline, CDPKs (Ca^{2+}-dependent protein kinases) and phosphatases (Gilroy et al. 2014). Thus, NO controls the appearance of genes accountable for sustaining the intrinsic level of Ca^{2+} via its channels.

To mitigate abiotic stress, NO crosstalks with plant hormones viz., salicylic acid (SA), jasmonic acid (JA), cytokinins (CKs) and abscisic acid (ABA). Signalling mediated by ABA and nitric oxide (NO) triggers the closing of stomata, which in turn minimizes water loss through transpiration under conditions of water scarcity (Lau et al. 2021). CKs and NO interact intricately to modify the reaction of plants against drought stress. According to Shao et al. (2010), CKs raise NO levels, thereby enhancing the photosynthetic rate in maize plants that are subjected to drought-stressed. Conversely, in the guard cells of broad beans, CKs reduce NO levels, leading to stomatal opening (Xiao-Ping and Xi-Gui 2006). Similarly, SA both antagonistically and synergistically interacts with NO to regulate plant reaction to water deficit stress (Asgher et al. 2017). Additionally, JA encourages the action of NO in guard cells and collaborates with NO to control the closing of stomata during water deficits (Liu et al. 2005).

18.5 Role of NO in Tolerance Mechanism Under Different Abiotic Stresses

Abiotic stresses affect various aspects of plants, including morphology, physiology and metabolism that affect growth, development and distribution (Harfouche et al. 2014). Mechanisms under stress response and tolerance include the inhibition of damage at a cellular level, homeostasis re-establishment and recovery of growth. Adaptation to abiotic stresses is regulated by various molecular networks or signalling, such as nitric oxide-mediated signalling. According to several studies, NO has been well reported as an antioxidant because of its role in developing tolerance against various stresses (Khator and Shekhawat 2019). The role of NO in the tolerance mechanism for different abiotic stresses is given below:

18.5.1 Water Deficit

Water deficit is the most detrimental abiotic stress for plants because it adversely affects physiological functions like photosynthesis and nutrient uptake (Siddiqui et al. 2011), as well as ROS homeostasis (Miller et al. 2010; Petrov et al. 2015). NO has a mitigation effect on drought stress in different plant species (Khator et al. 2024). Evidence from pharmacological studies on NO has shown that it is necessary for ABA-regulated closing of stomata, which improves plant resilience against dry conditions (Fancy et al. 2017). Furthermore, it has been proposed that the production of NO stimulated by ABA in guard cells of *Arabidopsis* is a prerequisite for stomatal closure by a combination of genetic and pharmacological methods (Fig. 18.2).

An increased level of endogenous PTM, specifically S-nitrosylationis, is found in plants with defective GSNOR activity that is not responsive to stomatal closure influenced by ABA, causing excessive water loss compared to wild-type (Fancy et al. 2017). Drought resistance is improved in cucumber plants through cytokinin- and polyamine-induced nitric oxide synthesis (Arasimowicz-Jelonek et al. 2009). Lu et al. (2009) reported enhancement in the drought resistance of Bermuda grass via elevated activities of CAT and SOD antioxidants with an increased level of ABA. Furthermore, in *Antiaris toxicaria*, seed germination is blocked by the over-accumulation of ROS in drought conditions that is recovered by applying NO, which reduces ROS through enhancement of the antioxidant defense system (Khator et al. 2024). NO is involved in the S-nitrosylation of dehydroascorbate reductase (DHAR), GR and APX (Khator et al. 2024). Thus, NO has an important function in the tolerance mechanism of drought response.

18.5.2 Salinity Stress

Salinization of soil is an inevitable environmental challenge that negatively affects plant growth (Khator et al. 2020). It disturbs ionic imbalance and causes osmotic stress and an oxidative burst (Bose et al. 2017). Salt stress can be mitigated through NO donor application as it prevents oxidative damage, maintains ion homeostasis and enhances plant growth (Zhang et al. 2006). A study on the AtNOA1 line of *Arabidopsis thaliana* with reduced activity of NOS showed more susceptibility to salinity and oxidative stress (Zhao et al. 2007). Moreover, sodium nitroprusside (SNP) treatment of mutant plants reduced the Na^+ to K^+ ratio that was caused by NaCl (Zhao et al. 2007). NO induced the activity of H^+-ATPase associated with the plasma membrane, which helped maintain a steady K^+:Na^+ ratio, thereby protecting plants against stress imposed by salinity (Zhao et al. 2004).

Tanou et al. (2009) identified a range of S-nitrosylated proteins in the stress of NaCl (Fancy et al. 2017). Notably, the degree of the amendment was decreased in a few of the proteins after NaCl stress (Ziogas et al. 2013). However, prior exposure to NO or H_2O_2 enhanced the S-nitrosylation of proteins caused by NaCl, which may have a protective effect in these stressful situations. In addition, it was observed that

S-nitrosylation of given protein decreased with both long- and short-term salt exposure (Ziogas et al. 2013). Camejo et al. (2013) suggested that the influence of prolonged salt stress mediates a high reduction in S-nitrosylation. Furthermore, the study has also documented that the activity of GSNOR also increased under prolonged salt exposure, particularly in the mitochondria of peas, which could potentially decrease S-nitrosylation.

A detoxification mechanism of H_2O_2 involves the ascorbate-glutathione cycle, which may be crucial in salt stress. According to the evidence, S-nitrosylation might potentially regulate this cycle. Furthermore, two key enzymes of this cycle, monodehydroascorbate reductase (MDAR) and ascorbate peroxidase (APX), are transcriptionally and translationally regulated due to salt stress (Begara-Morales et al. 2015). Thus, NO is essential for reducing the stress caused by salt. Various underlying mechanisms may be modulated by NO, including photorespiratory pathways, the redox alteration of ROS quenchers and the antioxidant defense system.

18.5.3 High-Temperature Stress

As global temprature continue to increase due to climate change, it's imporatnt to understand adaptaion mechanism towards stress in the plants. Nitrosothiol content increased in Brassica and peas under heat stress, which activated the metabolism of the main constituents of RNS (Abat and Deswal 2009). A significant adaption that plants exhibit under high-temperature stress is an enhanced production of NO (Gould et al. 2003). According to several research studies, heat stress tolerance may be correlated with the amount of NO or SNO (Hong et al. 2003). Antioxidant enzymes, namely, catalase, superoxide dismutase and ascorbate peroxidase, are activated by NO during high temperatures that lower the ROS level and confirm the NO function in the mitigation of heat stress (Song et al. 2006). Likewise, enhanced expression of heat shock protein (HSP)-related genes has been observed to reduce ROS levels and sustain cellular homeostasis under increased temperature (Konigshofer et al. 2008).

It is observed that the application of NO donor can decrease the cellular damage caused by high temperatures (Hasanuzzaman et al. 2012). Moreover, during high-temperature stress, SNP stimulates antioxidant activity (Yang et al. 2011) such as its application increases dehydro ascorbate reductase, monodehydroascorbate reductase and glyoxalase I and II activities and ascorbate and glutathione content in *T. aestivum* seedling to mitigate higher temperature stress (Hasanuzzaman et al. 2012). Several studies on different plants revealed the role of NO application in the recovery of electrolyte leakage, relative water content and chlorophyll content due to higher temperature stress (Pandey et al. 2023). Applying SNP to maize seedlings grown under high temperature enhances survival with reduced malondialdehyde content and electrolyte leakage (Bavita et al. 2012; Pandey et al. 2023).

18.5.4 Low-Temperature Stress

Trees have developed mechanisms to lessen the impact of stress caused by low temperature. NO regulates the mechanisms involved in cold response by ROS scavenging and reducing concentrations of flavonoids, sugars, anthocyanins and PAs. Additionally, phytohormones including jasmonates and ABA levels are also lowered by NO (Sanchez-Vicente and Lorenzo 2021). According to available data, the stress response involves the generation of NO through nitrate reductase (NR) and NOS-like enzymes (Cantrel et al. 2011). There is substantial evidence that osmolytes, viz., glycine betaine and proline, positively influence the stability of membrane and the function of enzymes at low temperature (Ashraf and Foolad 2007). According to Zhao et al. (2009), NO has the ability to cause cold acclimatization through the synthesizing proline.

Applying SNP exogenously lowers the impact of cold stress on seedlings of cucumber by decreasing the amount of malondialdehyde and raising the soluble-sugar content, chlorophyll and antioxidant enzymes, i.e., CAT, GR, SOD and POX (Liu et al. 2011). Moreover, S-nitrosylation, induced by cold stress, increases the enzymatic activity of GST (glutathione S-transferase) and DHAR. By producing ascorbate during the low-temperature response, this PTM facilitated ROS detoxification (Sehrawat and Deswal 2014). Collectively, research indicates that S-nitrosylation plays an important role in regulating signals related to cold stress by enhancing the remodelling of cell walls as well as the detoxification of reactive oxygen species.

18.6 Conclusion and Future Prospects

NO, declared the ‘Molecule of the Year’ in 1992, was initially observed as being released by plants in the 1970s (Pandey et al. 2023). However, research into its functions in plants gained momentum in the late 1990s (Neill 2002). Since then, numerous experiments have been performed and various aspects of NO have been observed, such as its synthesis, mode of action, signalling mechanisms, interactions with other hormones and functions due to abiotic stress. Thus, NO has key physiological, biochemical and molecular functions to ameliorate the negative impact of abiotic stress. Interaction between NO and PGRs, along with secondary signalling compounds, triggers the reprogramming of stress-associated gene expression, modulation of transcriptional processes, regulation of redox machinery, oxidative metabolism and involvement in multiple regulatory pathways during plant abiotic stress.

Despite the huge amount of data available, it is significant to highlight that studies on the NO role under abiotic stress in trees have been relatively less available in comparison to agricultural crops and Arabidopsis. For instance, tracing a sequential signalling cascade from receptors to effectors is challenging, and it is still unclear which specific sensors cause stress responses. There is a need to identify NO functions in tolerance mechanisms in trees under abiotic stresses. Transcriptomic investigations are the most common type of omics in tree species because they can provide genetic information even in the absence of reference genomes. This approach will boost our understanding of intricate networks of signalling pathways mediated by NO, triggered by abiotic stress and enhance our ability to design strategies to improve tree fitness under changing climatic conditions. The development of climate-resilient plants is crucial for ensuring a stable ecosystem in the face of climate change. Different biotechnological approaches such as genome or epigenome editing and alteration of key responsive genes like transcriptional factors involved in NO-mediated signalling could be attained to improve stress tolerance.

Acknowledgement Prof. G.S Shekhawat acknowledges RHTQIS-2023 for providing financial assistance.

References

Abasi F, Raza MU, Raja NI, Ehsan M, Ulfat A, Shahbaz M, Mehmood A (2024) Physiological mechanism and adaptation of plants to abiotic stresses. In: Improving stress resilience in plants. Academic Press, pp 447–458

Abat JK, Deswal R (2009) Differential modulation of S-nitrosoproteome of *Brassica juncea* by low temperature: change in S-nitrosylation of Rubisco is responsible for the inactivation of its carboxylase activity. Proteomics 9(18):4368–4380

Amnuaylojaroen T, Limsakul A, Kirtsaeng S, Parasin N, Surapipith V (2022) Effect of the near-future climate change under RCP8. 5 on the heat stress and associated work performance in Thailand. Atmos 13(2):325

Anita, Parihar S, Shekhawat GS (2023) Cadmium toxicity in plants: uptake, translocation and phytoremediation strategy. In: Heavy metal toxicity: environmental concerns, remediation and opportunities. Springer, Singapore, pp 1–18

Arasimowicz-Jelonek M, Floryszak-Wieczorek J, Kubis J (2009) Interaction between polyamine and nitric oxide signaling in adaptive responses to drought in cucumber. J Plant Growth Regul 28:177–186

Asgher M, Per TS, Masood A, Fatma M, Freschi L, Corpas FJ, Khan NA (2017) Nitric oxide signaling and its crosstalk with other plant growth regulators in plant responses to abiotic stress. Environ Sci Pollut Res 24:2273–2285

Ashraf MFMR, Foolad MR (2007) Roles of glycine betaine and proline in improving plant abiotic stress resistance. Environ Exp Bot 59(2):206–216

Bavita A, Shashi B, Navtej SB (2012) Nitric oxide alleviates oxidative damage induced by high temperature stress in wheat. Exp Dermatol 50:372–378

Begara-Morales JC, Sanchez-Calvo B, Chaki M, Mata-Perez C, Valderrama R, Padilla MN, Barroso JB (2015) Differential molecular response of monodehydroascorbate reductase and glutathione reductase by nitration and S-nitrosylation. J Exp Bot 66(19):5983–5996

Begara-Morales JC, Sanchez-Calvo B, Chaki M, Valderrama R, Mata-Perez C, Padilla MN, Barroso JB (2016) Antioxidant systems are regulated by nitric oxide-mediated post-translational modifications (NO-PTMs). Front Plant Sci 7:152

Bose J, Munns R, Shabala S, Gilliham M, Pogson B, Tyerman SD (2017) Chloroplast function and ion regulation in plants growing on saline soils: lessons from halophytes. J Exp Bot 68(12):3129–3143

Brichta J, Simunek V, Bílek L, Vacek Z, Gallo J, Drozdowski S, Fuchs Z (2024) Effects of climate change on scots pine (*Pinus sylvestris* L.) growth across Europe: decrease of tree-ring fluctuation and amplification of climate stress. Forests 15(1):91

Bulchandani N, Shekhawat GS (2020) Salicylic acid mediated up regulation of carvone biosynthesis during growth phase in cell suspension cultures of *Anethum graveolens*. 3 Biotech 10(11):482

Camejo D, del Carmen R-PM, Rodríguez-Serrano M, Sandalio LM, Lazaro JJ, Jimenez A, Sevilla F (2013) Salinity-induced changes in S-nitrosylation of pea mitochondrial proteins. J Proteome 79:87–99

Cantrel C, Vazquez T, Puyaubert J, Reze N, Lesch M, Kaiser WM, Baudouin E (2011) Nitric oxide participates in cold-responsive phosphosphingolipid formation and gene expression in *Arabidopsis thaliana*. New Phytol 189(2):415–427

Eggleton T (2012) A short introduction to climate change. Cambridge University Press, United Kingdom

Falak N, Imran QM, Hussain A, Yun BW (2021) Transcription factors as the "blitzkrieg" of plant defense: a pragmatic view of nitric oxide's role in gene regulation. Int J Mol Sci 22(2):522

Fancy NN, Bahlmann AK, Loake GJ (2017) Nitric oxide function in plant abiotic stress. Plant Cell Environ 40(4):462–472

Farnese FS, Menezes-Silva PE, Gusman GS, Oliveira JA (2016) When bad guys become good ones: the key role of reactive oxygen species and nitric oxide in the plant responses to abiotic stress. Front Plant Sci 7:181733

Francis SH, Busch JL, Corbin JD (2010) cGMP-dependent protein kinases and cGMP phosphodiesterases in nitric oxide and cGMP action. Pharmacol Rev 62(3):525–563

Gilroy S, Suzuki N, Miller G, Choi WG, Toyota M, Devireddy AR, Mittler R (2014) A tidal wave of signals: calcium and ROS at the forefront of rapid systemic signaling. Trends Plant Sci 19(10):623–630

Gould KS, Lamotte O, Klinguer A, Pugin A, Wendehenne D (2003) Nitric oxide production in tobacco leaf cells: a generalized stress response? Plant Cell Environ 26(11):1851–1862

Harfouche A, Meilan R, Altman A (2014) Molecular and physiological responses to abiotic stress in forest trees and their relevance to tree improvement. Tree Physiol 34(11):1181–1198

Hasanuzzaman M, Bhuyan MB, Zulfiqar F, Raza A, Mohsin SM, Mahmud JA, Fotopoulos V (2020) Reactive oxygen' species and antioxidant defense in plants under abiotic stress: revisiting the crucial role of a universal defense regulator. Antioxidants 9(8):681

Hasanuzzaman M, Nahar K, Alam MM, Fujita M (2012) Exogenous nitric oxide alleviates high temperature induced oxidative stress in wheat (*Triticum aestivum* L.) seedlings by modulating the antioxidant defense and glyoxalase system. Aust J Crop Sci 6(8):1314–1323

Hasanuzzaman M, Nahar K, Rahman A, Inafuku M, Oku H, Fujita M (2018) Exogenous nitric oxide donor and arginine provide protection against short-term drought stress in wheat seedlings. Physiol Mol Biol Plants 24:993–1004

Hassan WH, Nile BK (2021) Climate change and predicting future temperature in Iraq using CanESM2 and HadCM3 modeling. Model Earth Syst Environ 7:737–748

Hong SW, Lee U, Vierling E (2003) *Arabidopsis* hot mutants define multiple functions required for acclimation to high temperatures. Plant Physiol 132(2):757–767

Hu X, Zhang M, Liu J, Cheng X, Huang X (2024) Metabolomics and physiological changes underlying increased tolerance to salt stress induced by applied Nitric Oxide in *Fatsia japonica* seedlings. Forests 15(1):159

Jonak C, Okresz L, Bogre L, Hirt H (2002) Complexity, cross talk and integration of plant MAP kinase signaling. Curr Opin Plant Biol 5(5):415–424

Khan M, Ali S, Al Azzawi TNI, Yun BW (2023) Nitric oxide acts as a key signaling molecule in plant development under stressful conditions. Int J Mol Sci 24(5):4782

Khator K, Mahawar L, Shekhawat GS (2020) NaCl induced oxidative stress in legume crops of Indian Thar Desert: an insight in the cytoprotective role of HO1, NO and antioxidants. Physiol Mol Biol Plants 26:51–62

Khator K, Parihar S, Jasik J, Shekhawat GS (2024) Nitric oxide in plants: an insight on redox activity and responses toward abiotic stress signaling. Plant Signal Behav 19(1):2298053

Khator K, Saxena I, Shekhawat GS (2021) Nitric oxide induced cd tolerance and phytoremediation potential of *B. Juncea* by the modulation of antioxidant defense system and ROS detoxification. Biometals 34(1):15–32

Khator K, Shekhawat GS (2018) Regulatory role of thiols and proline in mitigation of Cu induced phytotoxicity in seven day's old hydroponically acclimatized seedling of *Cyamopsistetragonoloba*. Biotech Today: Int J Biol Sci 8(1):48–57

Khator K, Shekhawat GS (2019) Nitric oxide improved salt stress tolerance by osmolyte accumulation and activation of antioxidant defense system in seedling of *B. juncea* (L.) Czern. Vegetos 32(4):583–592

Khator K, Shekhawat GS (2020) Cd-and Cu-induced phytotoxicity on 2–3 leaf stage of *Cyamopsis tetragonoloba* and its regulation by nitrate reductase and ROS quenching enzyme. Acta Physiol Plant 42(7):120

Konigshofer H, Tromballa HW, Löppert HG (2008) Early events in signaling high-temperature stress in tobacco BY2 cells involve alterations in membrane fluidity and enhanced hydrogen peroxide production. Plant Cell Environ 31(12):1771–1780

Lau SE, Hamdan MF, Pua TL, Saidi NB, Tan BC (2021) Plant nitric oxide signaling under drought stress. Plan Theory 10(2):360

Liu X, Shi W, Zhang S, Lou C (2005) Nitric oxide involved in signal transduction of jasmonic acid-induced stomatal closure of *Vicia faba* L. Chin Sci Bull 50:520–525

Liu X, Wang L, Liu L, Guo Y, Ren H (2011) Alleviating effect of exogenous nitric oxide in cucumber seedling against chilling stress. Afr J Biotechnol 10(21):4380–4386

Lu S, Su W, Li H, Guo Z (2009) Abscisic acid improves drought tolerance of triploid Bermuda grass and involves H2O2-and NO-induced antioxidant enzyme activities. Plant Physiol Biochem 47(2):132–138

Mahawar L, Khator K, Shekhawat GS (2018a) Role of proline in mitigating NaCl induced oxidative stress in *Eruca sativa* Miller: an important oil yielding crop of Indian Thar Desert. Vegetos 31:55–63

Mahawar L, Kumar R, Shekhawat GS (2018b) Evaluation of heme oxygenase 1 (HO 1) in Cd and Ni induced cytotoxicity and crosstalk with ROS quenching enzymes in two to four leaf stage seedlings of *Vigna radiata*. Protoplasma 255:527–545

Mahawar L, Popek R, Shekhawat GS, Alyemeni MN, Ahmad P (2021) Exogenous hemin improves Cd2+ tolerance and remediation potential in *Vigna radiata* by intensifying the HO-1 mediated antioxidant defence system. Sci Rep 11(1):2811

Mahawar L, Shekhawat GS (2016) Salt induce oxidative stress and its tolerance mechanism in plant: morphological, biochemical and molecular perspective. Biotech Today: Int J Biol Sci 6(2):81–88

Mahawar L, Shekhawat GS (2018) Haem oxygenase: a functionally diverse enzyme of photosynthetic organisms and its role in phytochrome chromophore biosynthesis, cellular signaling and defense mechanisms. Plant Cell Environ 41(3):483–500

Mahawar L, Shekhawat GS (2019) EsHO 1 mediated mitigation of NaCl induced oxidative stress and correlation between ROS, antioxidants and HO 1 in seedlings of *Eruca sativa*: underutilized oil yielding crop of arid region. Physiol Mol Biol Plants 25:895–904

Mahawar L, Shekhawat GS (2023) Understanding the physiological mechanism of heme oxygenase for enhanced tolerance and phytoremediation of cd2+ in *Eruca sativa*: co-ordinated function of antioxidant defense system. J Plant Growth Regul 42(10):6307–6318

Malhi GS, Kaur M, Kaushik P, Alyemeni MN, Alsahli AA, Ahmad P (2021) Arbuscular mycorrhiza in combating abiotic stresses in vegetables: an eco-friendly approach. Saudi J Biol Sci 28(2):1465–1476

Miller GAD, Suzuki N, Ciftci-Yilmaz S, Mittler RON (2010) Reactive oxygen species homeostasis and signaling during drought and salinity stresses. Plant Cell Environ 33(4):453–467

Nakashima K, Ito Y, Yamaguchi-Shinozaki K (2009) Transcriptional regulatory networks in response to abiotic stresses in *Arabidopsis* and grasses. Plant Physiol 149(1):88–95

Neill SJ (2002) Hydrogen peroxide and nitric oxide as signaling molecules in plants. J Exp Bot 53:388–395

Opdenakker K, Remans T, Vangronsveld J, Cuypers A (2012) Mitogen-activated protein (MAP) kinases in plant metal stress: regulation and responses in comparison to other biotic and abiotic stresses. Int J Mol Sci 13(6):7828–7853

Ortega-Galisteo AP, Rodríguez-Serrano M, Pazmino DM, Gupta DK, Sandalio LM, Romero-Puertas MC (2012) S-Nitrosylated proteins in pea (*Pisum sativum* L.) leaf peroxisomes: changes under abiotic stress. J Exp Bot 63(5):2089–2103

Pandey A, Khan MK, Hamurcu M, Athar T, Yerlikaya BA, Yerlikaya S, Gezgin S (2023) Role of exogenous nitric oxide in protecting plants against abiotic stresses. Agronomy 13(5):1201

Petrov V, Hille J, Mueller-Roeber B, Gechev TS (2015) ROS-mediated abiotic stress-induced programmed cell death in plants. Front Plant Sci 6:125003

Popek R, Mahawar L, Shekhawat GS, Przybysz A (2022) Phyto-cleaning of particulate matter from polluted air by woody plant species in the near-desert city of Jodhpur (India) and the role of heme oxygenase in their response to PM stress conditions. Environ Sci Pollut Res 29(46):70228–70241

Prochazkova D, Haisel D, Pavlikova D (2014) Nitric oxide biosynthesis in plants-the short overview. Plant Soil Environ 60(3):129–134

Rao S, Shekhawat GS (2016) Phytotoxicity and oxidative stress perspective of two selected nanoparticles in Brassica juncea. 3 Biotech 6(2):244

Rodriguez MCS, Petersen M, Mundy J (2010) Mitogen-activated protein kinase signaling in plants. Annu Rev Plant Biol 61:621–649

Sadiq Y, Zaid A, Khan MMA (2020) Adaptive physiological responses of plants under abiotic stresses: role of phytohormones. In: Plant ecophysiology and adaptation under climate change: mechanisms and perspectives I: general consequences and plant responses. Springer, pp 797–824

Sami F, Faizan M, Faraz A, Siddiqui H, Yusuf M, Hayat S (2018) Nitric oxide-mediated integrative alterations in plant metabolism to confer abiotic stress tolerance, NO crosstalk with phytohormones and NO-mediated post translational modifications in modulating diverse plant stress. Nitric Oxide 73:22–38

Sanchez-Vicente I, Lorenzo O (2021) Nitric oxide regulation of temperature acclimation: a molecular genetic perspective. J Exp Bot 72(16):5789–5794

Sarwar M, Anjum S, Ali Q, Alam MW, Haider MS, Mehboob W (2021) Triacontanol modulates salt stress tolerance in cucumber by altering the physiological and biochemical status of plant cells. Sci Rep 11(1):24504

Saxena I, Shekhawat GS (2013) Nitric oxide (NO) in alleviation of heavy metal induced phytotoxicity and its role in protein nitration. Nitric Oxide 32:13–20

Sehrawat A, Deswal R (2014) S-nitrosylation analysis in *Brassica juncea* apoplast highlights the importance of nitric oxide in cold-stress signaling. J Proteome Res 13(5):2599–2619

Semba RD, Askari S, Gibson S, Bloem MW, Kraemer K (2022) The potential impact of climate change on the micronutrient-rich food supply. Adv Nutr 13(1):80–100

Shao R, Wang K, Shangguan Z (2010) Cytokinin-induced photosynthetic adaptability of *Zea mays* L. to drought stress associated with nitric oxide signal: probed by ESR spectroscopy and fast OJIP fluorescence rise. J Plant Physiol 167(6):472–479

Sharma A, Kumar V, Shahzad B, Ramakrishnan M, Singh Sidhu GP, Bali AS, Zheng B (2020) Photosynthetic response of plants under different abiotic stresses: a review. J Plant Growth Regul 39:509–531

Siddiqui MH, Al-Whaibi MH, Basalah MO (2011) Role of nitric oxide in tolerance of plants to abiotic stress. Protoplasma 248:447–455

Simontacchi M, Galatro A, Ramos-Artuso F, Santa-María GE (2015) Plant survival in a changing environment: the role of nitric oxide in plant responses to abiotic stress. Front Plant Sci 6:155160

Song L, Ding W, Zhao M, Sun B, Zhang L (2006) Nitric oxide protects against oxidative stress under heat stress in the calluses from two ecotypes of reed. Plant Sci 171(4):449–458

Tanou G, Job C, Rajjou L, Arc E, Belghazi M, Diamantidis G, Job D (2009) Proteomics reveals the overlapping roles of hydrogen peroxide and nitric oxide in the acclimation of citrus plants to salinity. Plant J 60(5):795–804

Verma K, Dixit S, Shekhawat GS, Alam A (2015) Antioxidant activity of heme oxygenase 1 in *Brassica juncea* (L.) Czern. (Indian mustard) under salt stress. Turk J Biol 39(4):540–549

Verma K, Mehta SK, Shekhawat GS (2013) Nitric oxide (NO) counteracts cadmium induced cytotoxic processes mediated by reactive oxygen species (ROS) in *Brassica juncea*: crosstalk between ROS, NO and antioxidant responses. Biometals 26:255–269

Wendehenne D, Pugin A, Klessig DF, Durner J (2001) Nitric oxide: comparative synthesis and signaling in animal and plant cells. Trends Plant Sci 6(4):177–183

Wilmer L, Trankner M, Pawelzik E, Naumann M (2022) Sufficient potassium supply enhances tolerance of potato plants to PEG-induced osmotic stress. Plant Stress 5:100–102

Xiao-Ping S, Xi-Gui S (2006) Cytokinin-and auxin-induced stomatal opening is related to the change of nitric oxide levels in guard cells in broad bean. Physiol Plant 128(3):569–579

Yang W, Sun Y, Chen S, Jiang J, Chen F, Fang W, Liu Z (2011) The effect of exogenously applied nitric oxide on photosynthesis and antioxidant activity in heat stressed *Chrysanthemum*. Biol Plant 55:737–740

Zhang Y, Wang L, Liu Y, Zhang Q, Wei Q, Zhang W (2006) Nitric oxide enhances salt tolerance in maize seedlings through increasing activities of proton-pump and Na+/H+ antiport in the tonoplast. Planta 224:545–555

Zhao J, Lu Z, Wang L, Jin B (2020) Plant responses to heat stress: physiology, transcription, non-coding RNAs, and epigenetics. Int J Mol Sci 22(1):117

Zhao L, Zhang F, Guo J, Yang Y, Li B, Zhang L (2004) Nitric oxide functions as a signal in salt resistance in the calluses from two ecotypes of reed. Plant Physiol 134(2):849–857

Zhao MG, Chen L, Zhang LL, Zhang WH (2009) Nitric reductase-dependent nitric oxide production is involved in cold acclimation and freezing tolerance in *Arabidopsis*. Plant Physiol 151(2):755–767

Zhao MG, Tian QY, Zhang WH (2007) Nitric oxide synthase-dependent nitric oxide production is associated with salt tolerance in *Arabidopsis*. Plant Physiol 144(1):206–217

Zhu JK (2001) Plant salt tolerance. Trends Plant Sci 6(2):66–71

Ziogas V, Tanou G, Filippou P, Diamantidis G, Vasilakakis M, Fotopoulos V, Molassiotis A (2013) Nitrosative responses in citrus plants exposed to six abiotic stress conditions. Plant Physiol Biochem 68:118–126

The manufacturer's authorised representative in the EU is Springer Nature Customer Service Centre GmbH, Europaplatz 3, 69115 Heidelberg, Germany. If you have any concerns regarding our products, please contact ProductSafety@springernature.com

Printed and bound by CPI Group (UK) Ltd, Croydon, CR0 4YY
01/10/2026
02225936-0011

The manufacturer's authorised representative in the EU is Springer Nature Customer Service Centre GmbH, Europaplatz 3, 69115 Heidelberg, Germany. If you have any concerns regarding our products, please contact ProductSafety@springernature.com

Printed and bound by CPI Group (UK) Ltd, Croydon, CR0 4YY
01/10/2026
02225937-0009